The Waste-to-Energy Nexus: Technologies for the Industries of the Future

Edited by

Harish Chandra Joshi
Department of Chemistry
Graphic Era (Deemed to be) University
Dehradun
Uttarakhand, India

Anand Chauhan
Department of Mathematics
Graphic Era (Deemed to be) University
Dehradun
Uttarakhand, India

Mikhail Vlaskin
Joint Institute for High Temperature of the Russian Academy
of Science Moscow, Russia
& Peoples' Friendship University of Russia named after
Patrice Lumumba,
Moscow, Russia

&

Maulin P. Shah
Department of Research Impact and Outcome Research
Research and Development Cell
Lovely Professional University
Phagwara, Punjab, India

The Waste-to-Energy Nexus: Technologies for the Industries of the Future

Editors: Harish Chandra Joshi, Anand Chauhan, Mikhail Vlaskin and Maulin P. Shah

ISBN (Online): 979-8-89881-189-1

ISBN (Print): 979-8-89881-190-7

ISBN (Paperback): 979-8-89881-191-4

Published by Bentham Science Publishers Pte. Ltd. Singapore, in collaboration with Eureka Conferences, USA. All Rights Reserved.
First published in 2026.

need for a court order if at any point you breach any terms of this License Agreement. In no event will any delay or failure by Bentham Science Publishers in enforcing your compliance with this License Agreement constitute a waiver of any of its rights.

3. You acknowledge that you have read this License Agreement, and agree to be bound by its terms and conditions. To the extent that any other terms and conditions presented on any website of Bentham Science Publishers conflict with, or are inconsistent with, the terms and conditions set out in this License Agreement, you acknowledge that the terms and conditions set out in this License Agreement shall prevail.

Bentham Science Publishers Pte. Ltd.
No. 9 Raffles Place
Office No. 26-01
Singapore 048619
Singapore
Email: subscriptions@benthamscience.net

CONTENTS

Kamini Pandey, Shubham Kumar, Pushpanjali SinghShuchi Verma and *Barkha, Singhal*

Sanskriti Thapliyal, Deepshikha Kothari, Alka N. Choudhary, Neetu Sharma, Bhawana, Harish Chandra Joshi and *Promila Sharma*

FOREWORD

As we stride deeper into the 21st century, humanity faces a dual-edged challenge: managing ever-increasing waste and securing sustainable energy sources. With a burgeoning global population and rapid industrialization, the mounting pressure on our planet's finite resources is undeniable. At the same time, our relentless pursuit of energy has driven innovation toward solutions that are both sustainable and environmentally conscious. The relationship between waste management and energy generation holds transformative potential in shaping a sustainable future. By extracting energy from waste materials, we not only reduce environmental harm but also preserve existing resources while creating clean, renewable energy. This integrated vision aligns seamlessly with the global commitment to the United Nations' 17 Sustainable Development Goals (SDGs), particularly the goal of achieving affordable and clean energy by 2030.

This book explores cutting-edge technologies and methodologies at the intersection of waste management and energy valorization. It presents an in-depth investigation into innovative processes and practical solutions designed to inspire a new generation of thinkers, researchers, and engineers. Each chapter meticulously unpacks the intricacies of various approaches, offering insights for both specialized readers and those new to the subject.

The journey begins with an exploration of waste-to-energy technologies, including incineration, gasification, pyrolysis, and anaerobic digestion, alongside renewable energy sources such as solar, wind, and biomass. Subsequent chapters explore the role of biorefineries in advancing circular economy principles, the immense potential of biomass and biochar in energy storage applications, and the revolutionary possibilities of hydrogen technology.

The book also sheds light on emerging fields, such as composite material recycling and CO_2 capture, offering practical insights into processes including chemical absorption, adsorption, and microbiological methods. The potential of alternative fuels, exemplified by India's Ethanol 100 initiative, is addressed, alongside strategies to embed circular economy principles into the lifecycle management of solar panels, wind turbines, and battery systems.

In addition to these innovative themes, this work emphasizes the importance of regional perspectives, particularly India's role in managing Municipal Solid Waste (MSW) and advancing renewable energy solutions, such as wind power and bioethanol derived from algal biomass. Together, these chapters provide a holistic framework for tackling global sustainability challenges.

It is my privilege to present this remarkable work to readers from diverse backgrounds, including graduate and postgraduate students, researchers, engineers, and policymakers. The comprehensive insights offered in this book ensure that even non-specialist readers can grasp the complexities of each topic and stay abreast of the latest advancements.

This book is a testament to the power of human ingenuity and collaboration in solving some of the most pressing challenges of our time. By bridging the gap between waste management and energy generation, it offers hope for a cleaner, greener, and more sustainable future.

Mika Sillanpaa
Department of Chemical Engineering
School of Mining Metallurgy and Chemical Engineering
University of Johannesburg
P. O. Box 17011
Doornfontein 2028, South Africa

PREFACE

As the world enters the 21st century, waste management, coupled with energy security, remains a serious challenge. With a growing global population and fast-paced industrialization, the rate of waste generation has skyrocketed, putting tremendous pressure on planetary resources. At the same time, an unbridled quest for energy has propelled innovation towards solutions that are not only effective but also sustainable and environmentally benign. In 2015, all 193 countries finally resolved to work together towards making the world healthier by committing to 17 Sustainable Development Goals for 2030. One of the goals addresses sustainability in terms of affordable and clean energy. The relationship between waste and energy remains a promising avenue for achieving a sustainable future. This encompasses all aims within a single framework: realizing the energy potential of wasted materials while reducing environmental harm, conserving resources, and generating clean energy. This book examines cutting-edge technologies that significantly transform approaches to waste management and energy valorization processes.

This book will be beneficial for graduate and postgraduate students, research scholars, and engineers. Even non-specialist readers have the opportunity to grasp the intricacies of each topic and access the latest content in this book.

Chapter 1 explores various waste-to-energy technologies, including incineration, gasification, pyrolysis, and anaerobic digestion, as well as renewable energy sources such as solar, wind, and biomass. This chapter also explores the optimization of these systems, including process improvements, lifecycle assessments, smart grid integration, and supportive policies.

Chapter 2 explains the role of biorefineries in waste valorization within the circular economy, offering significant potential for advancing sustainability.

Chapter 3 provides information on biomass and its derivatives, which offer great potential as renewable energy sources, making them excellent alternatives for sustainable energy.

Chapter 4 explores the Biomass-derived biochar emerges as a sustainable and efficient material for energy storage applications in supercapacitors.

Chapter 5 introduces the research on hydrogen production, which is crucial for the development of sustainable energy alternatives. Hydrogen technology, which has the potential to revolutionize various industrial applications and significantly contribute to a sustainable, low-carbon economy, is gradually gaining recognition.

Chapter 6 introduces new technologies for recycling composite materials, as well as improvements to existing procedures and their adaptation for alternative applications.

Chapter 7 presents an overview of the primary methods for MSW utilization, aiming to elucidate the current state of MSW management and to assess and compare the environmental impacts of different MSW utilization pathways.

Chapter 8 explores methods for capturing CO_2, including chemical absorption, adsorption using solid-phase porous materials, membrane separation, cryogenic separation, hydrate-based methods, and microbiological methods.

Chapter 9 explores "Ethanol 100: A New Approach for the Transportation Industry in India." This chapter outlines the policy initiatives undertaken, particularly in India, to develop and promote various alternative fuels for road transportation, highlighting the progress made in comparison to global advancements.

Chapter 10 presents the integration of circular economy ethics into the lifecycle management of solar panels, wind turbines, and battery storage systems, thereby ensuring a sustainable energy future.

Chapter 11 investigates the potential of algal biomass for bioethanol production and outlines a model for the complete conversion process.

Chapter 12 provides a comprehensive review of the characteristics, production, collection, disposal, and effective treatment technologies of MSW practiced in India. Incineration, pyrolysis, bio-refining, biogas facilities, recycling, and composting are among the waste management and treatment processes currently used.

Harish Chandra Joshi
Department of Chemistry
Graphic Era (Deemed to be) University
Dehradun, Uttarakhand, India

Anand Chauhan
Department of Mathematics
Graphic Era (Deemed to be) University
Dehradun
Uttarakhand, India

Mikhail Vlaskin
Joint Institute for High Temperature of the Russian Academy
of Science Moscow
Russia
&
Peoples' Friendship University of Russia named after
Patrice Lumumba
Moscow Russia

&

Maulin P. Shah
Department of Research Impact and Outcome Research
Research and Development Cell
Lovely Professional University
Phagwara
Punjab, India

List of Contributors

A. Harini Priya	School of Chemistry, Madurai Kamaraj University, Madurai-625021, India
Aash Mohammad	Department of Chemical and Biochemical Engineering, Rajiv Gandhi Institute of Petroleum Technology, Jais, Amethi, India
Abhinav Goel	Department of Mathematics, Graphic Era (Deemed to be) University, Dehradun, Uttarakhand, India
Alka N. Choudhary	Department of Pharmaceutical Science, ICFAI University, Dehradun, Uttarakhand, India
Anurag Tyagi	Department of Physics, Noida Institute of Engineering & Technology, Greater Noida, Affiliated to Dr. A.P.J Abdul Kalam Technical University, Lucknow, U.P, India
Barkha Singhal	School of Biotechnology, Gautam Buddha University, Greater Noida, Uttar Pradesh-201312, India
Bhawana	Department of Chemistry, Graphic Era (Deemed to be) University, Dehradun, Uttarakhand, India
Biplop Jyoti Hazarika	Department of Chemical Sciences, IISER-Kolkata, Nadia-741246, West Bengal, India
Biswajyoti Hazarika	Department of Chemistry, Arunachal University of Studies, Namsai-792103, Arunachal Pradesh, India
Deepak Kumar	Department of Physics, Graphic Era (Deemed to be) University, Dehradun, India
Deepshikha Kothari	Department of Biotechnology, Graphic Era (Deemed to be) University, Dehradun, Uttarakhand, India
Jeffy Jerusha	School of Chemistry, Madurai Kamaraj University, Madurai-625021, India
JV. Jaya Surya	School of Chemistry, Bharathidasan University, Trichy-620024, India
Kamini Pandey	School of Biotechnology, Gautam Buddha University, Greater Noida, Uttar Pradesh-201312, India
Kirill G. Ryndin	Joint Institute for High Temperatures of the Russian Academy of Sciences, Moscow, Russia
M.P. Laavanyaa Shri	School of Chemistry, Madurai Kamaraj University, Madurai-625021, India
Mansi Yadav	K.R. Mangalam University, Sohna Road, Gurugram, Haryana, India
Md. Juned K. Ahmed	Department of Chemistry, Arunachal University of Studies, Namsai-792103, Arunachal Pradesh, India
Mikhail S. Vlaskin	Joint Institute for High Temperatures of the Russian Academy of Sciences, Moscow, Russia
Neeraj Kumar	Nanoscience Laboratory, Department of Physics, University of Trento, Via Sommarive 14, Povo (TN) 38123, Italy Department of Humanities and Applied Science, RIT Roorkee, Roorkee-247668, India

Neetu Sharma — Department of Chemistry, Graphic Era (Deemed to be) University, Dehradun, Uttarakhand, India

Olesya A. Buryakovskaya — Joint Institute for High Temperatures of the Russian Academy of Sciences, Moscow, Russia

Pankaj K. Chauhan — Faculty of Applied Sciences and Biotechnology, Shoolini University, Solan, Himachal Pradesh-173229, India

Priyanka Banerji — The NorthCap University, Gurugram, Haryana, India

Promila Sharma — Department of Microbiology, Graphic Era (Deemed to be) University, Dehradun, Uttarakhand, India

Raj Kumar Goel — Department of Computer Application, North-Eastern Hill University, Tura Campus, 794002 Meghalaya, India

S. Ambika — PG and Research Department of Chemistry, Bishop Heber College, Trichy-620017, India

S. Allen Bimal Chander — Department of Chemistry, Madras Christian College, Tambaram, Chennai-600059, India

Sanskriti Thapliyal — Department of Biotechnology, Graphic Era (Deemed to be) University, Dehradun, Uttarakhand, India

Shubham Kumar — School of Biotechnology, Gautam Buddha University, Greater Noida, Uttar Pradesh-201312, India

Shuchi Verma — Department of Chemistry, Ramjas College, Delhi University-110007, India

Shweta Vishnoi — Department of Physics, Noida Institute of Engineering & Technology, Greater Noida, Affiliated, to Dr. A.P.J Abdul Kalam Technical University, Lucknow, U.P, India

Sujata Rathi — Department of Botany, Multanimal Modi College, Modinagar-201204, India

Sujeet Kumar Pandey — Department of Chemical and Biochemical Engineering, Rajiv Gandhi Institute of Petroleum Technology, Jais, Amethi, India

Vinod Kumar — Algal Research and Bioenergy Lab, Department of Food Science and Technology, Graphic Era (Deemed to be) University, Dehradun, Uttarakhand, India

Y. Manojkumar — PG and Research Department of Chemistry, Bishop Heber College, Trichy-620017, India

The Waste-to-Energy Nexus, 2026, 1-27

CHAPTER 1

Environmental Waste and Renewable Energy Optimization

Y. Manojkumar[1,*], S. Allen Bimal Chander[2], M.P. Laavanyaa Shri[3], A. Harini Priya[3], Jeffy Jerusha[3], JV. Jaya Surya[4] and **S. Ambika[1]**

[1] *PG and Research Department of Chemistry, Bishop Heber College, Trichy-620017, India*

[2] *Department of Chemistry, Madras Christian College, Tambaram, Chennai-600059, India*

[3] *School of Chemistry, Madurai Kamaraj University, Madurai-625021, India*

[4] *School of Chemistry, Bharathidasan University, Trichy-620024, India*

Abstract: The growing challenge of industrial waste management and the increasing global demand for sustainable energy solutions have driven the evolution of Waste-to-Energy (WtE) technologies. This chapter explores integrating environmental waste management with renewable energy generation, focusing on optimizing WtE systems. It covers various renewable energy sources, including solar energy, biomass energy, and wind energy, and examines their potential synergy with waste management practices. Promising WtE technologies are discussed, and successful implementations across different sectors are presented through case studies, showcasing both the benefits and challenges. Additionally, this chapter identifies barriers to the widespread adoption of WtE, suggests future research directions to improve efficiency, and addresses the economic and environmental sustainability of these technologies.

Keywords: Environmental remediation, Energy optimization, Renewable energy.

INTRODUCTION

Global discussions are increasingly dominated by two critical issues: Environmental concerns from industrial waste and the urgent need for renewable and sustainable energy systems. The escalating global waste generation, projected to reach nearly 4 billion tons within the next three decades, underscores the pressing need to transition to circular economy principles [1]. The increasing waste problem, caused by industrialization, consumerism, and population growth, is severely impacting our ecosystems and public health [2]. It offers a favorable

---

* **Corresponding author Y. Manojkumar:** PG and Research Department of Chemistry, Bishop Heber College, Trichy-620017, India; Tel: +91-9894601253; E-mail: yrmanojkumar@gmail.com

Harish Chandra Joshi, Anand Chauhan, Mikhail Vlaskin & Maulin P. Shah (Eds.)

solution to address these issues by converting waste into usable energy while minimizing environmental impact. Different types of waste are produced by industries, including gaseous, solid, and liquid types, and each poses a unique challenge for disposal and management. For example, solid waste contributes significantly to landfills, and through methods such as open burning and unregulated dumping, approximately 70% of municipal solid waste globally is being inadequately managed. This unhealthy mismanagement contributes to the emission of greenhouse gases and environmental hazards, making it necessary to adopt sustainable waste treatment strategies [3, 4].

Sustainable waste management emphasizes the role of WtE conversion. Space scarcity due to landfilling, substantial energy demands, and environmentally harmful CH_4 emissions represent an increasingly unsustainable approach [1]. Various WtE techniques, namely incineration, pyrolysis, gasification, and anaerobic digestion, offer a dual benefit comprising waste reduction and producing electricity (clean energy), contributing to the global shift towards renewable energy sources. The steel industry, renowned for its substantial waste generation, presents a unique opportunity for transformation. Slag, a residual material from steel manufacturing processes, presents an opportunity for industrial waste to contribute to a circular economy through its extensive utilization in construction and road infrastructure projects [5]. By decreasing dependence on traditional fossil fuels, blast furnace gas contains gases such as CO, H_2, and CH_4, offering opportunities for energy recovery through advanced combustion systems [6]. Toxic heavy metals pose a significant challenge to the effective management of industrial wastewater. Industries such as electroplating, battery manufacturing, and textile production contribute substantially to water contamination by releasing hazardous metals, including lead, cadmium, and mercury, which contaminate aquatic ecosystems and threaten environmental health [5]. The effective reduction of this pollution is achieved through innovative approaches, such as using industrial by-product ash as a low-cost adsorbent [2, 7]. However, the widespread implementation of WtE technologies faces challenges. The widespread adoption of WtE technologies faces several challenges, including substantial initial investment costs, regulatory obstacles, and the imperative for technological improvements to enhance efficiency and environmental compatibility. Furthermore, the integration of WtE systems with sustainable and renewable energy sources, such as solar, wind, and biomass, necessitates sophisticated infrastructure and a supportive policy framework to optimize energy output while minimizing ecological harm. This chapter explores the relationship between waste management and renewable energy optimization, focussing on the role of WtE technologies and their practical applications.

This chapter discusses successful WtE implementations across various industries and outlines promising directions for future research and development based on case studies. By identifying shortcomings in existing practices and proposing groundbreaking approaches, this chapter seeks to drive progress toward sustainable industrial and environmental solutions. To understand the potential of integrating waste management with renewable energy, it is essential to examine key renewable energy sources and how they align with waste-to-energy objectives.

Integration of Renewable Energy with Waste-to-Energy Technologies

To develop effective Waste-to-Energy (WtE) strategies, it is crucial to consider how renewable energy sources, such as solar, wind, and biomass, can complement WtE technologies. Integrating these renewables can help optimize energy output, reduce reliance on fossil fuels, and create more sustainable hybrid systems. The following sections explore each of these energy sources in this context. Currently, the most effective sustainable energy resources are solar, wind, and biomass, which offer potential replacements for hydrocarbon-based fuels and provide sustainable solutions to the energy crisis and climate change. These energy sources not only help to decrease the environmental effects of energy production but also align with waste management practices by transforming waste into a useful resource. The subsequent sections provide an in-depth exploration of each energy source, highlighting their advantages, technologies, and applications in the renewable energy landscape.

Solar Energy

Notably, the Earth receives 173,000 terawatts (TW) of solar radiation daily, 17,000 times more than the current global energy consumption [8, 9]. This indicates that solar energy is an abundant and environmentally friendly energy source. This makes solar power a key player in meeting the escalating global energy needs. Solar energy can be effectively harnessed to generate electricity by employing photovoltaic cells or concentrated solar power systems. Recent advancements in technology, such as the development of multi-junction photovoltaic cells, have significantly increased their efficiency, with some reaching approximately 34.1% [9]. Solar energy also benefits from its minimal environmental impact, as it uses little to no water, making it a sustainable energy source [8]. Around the globe, many countries can generate electricity using solar energy during the daytime. However, a key limitation is that the energy generated during peak periods must be stored for use during times of low solar energy production. Hence, we need efficient batteries that can store the excess solar energy produced during the day for use at night. Efficient batteries can provide a stable power supply to households and industries, reducing the use of fossil fuels [10]. According to reports, currently, nearly 13% of the world's power consumption is contributed by renewable energies, including solar, which will promise a complete reliance on green energies within the next twenty years [11]. While solar power offers an effective renewable solution, wind energy complements it by providing consistent energy generation in suitable environments.

Wind Energy

Next to solar power, wind energy is another promising source of green and renewable energy. Wind is generated by the unequal heating of the Earth's surface by the Sun and can be harnessed to produce electricity. It is eco-friendly, sustainable, and reduces greenhouse gas emissions and fossil fuel dependence. Denmark aims to produce 50% of its power consumption using windmills within the next 20 years. One of the setbacks in wind energy is that irregular wind flow causes fluctuation in power generation. Excessive electricity produced during high wind flow can be mitigated by utilizing hydro storage and batteries [12]. High wind flows around the clock are a favourable condition for effective power production. Further wind energy-based industries give job opportunities in the turbine manufacturing and maintenance sectors. In addition to solar and wind, biomass offers another vital renewable resource, especially significant in agricultural and rural contexts.

Biomass Energy

Biomass, primarily obtained from living and dead plant materials, can be converted into combustion material and biofuels. It is one of the sustainable energy resources. In recent years, it has been used as an effective alternative to fossil fuels. This can help reduce greenhouse gas emissions. Different types of biomass energies are as follows.

Types of Biomass Energy

Thermal Energy

Biomass is an excellent source to produce thermal energy. Compared to direct combustion, processes such as pyrolysis and gasification are promising techniques for obtaining the maximum calorific value from this biomass. During this process, due to the limited supply of air, a synthetic gas is produced, which is used for electricity generation [13].

Electrical Energy

In place of coal, many countries—particularly in Africa and Asia—use biomass directly for combustion. The heat energy produced converts water into steam, which then drives a turbine to generate electricity. However, one major drawback is the emission of carbon and other gases, which must be controlled or purified before being released into the environment [14].

Biofuels

In many countries, biodiesel and bioethanol are produced from various biomass sources. For transportation, it can be used as a standalone fuel or blended with conventional fossil fuels like petrol and diesel [15]. Biomass is a promising renewable energy resource that can be derived from living or waste plant materials, algae, and dead organisms, which can help reduce carbon emissions in a notable amount. Ethanol is one of the main products derived from biomass. It is mixed with gasoline, which increases the octane number and reduces carbon emissions.

Biogas

Biogas is a form of biomass energy produced through the decomposition of organic matter in the absence of oxygen. It is a renewable gaseous fuel that can replace natural gas for power and thermal energy [16]. Biogas production can be optimized by pre-treating materials such as switchgrass, a high-yielding crop that is well-suited for biogas production. To enhance the sustainability of biomass and other renewable energy systems, the circular economy framework offers a strategic model for long-term environmental and economic resilience. Solar, wind, and biomass each contribute uniquely to energy optimization goals. By integrating these renewable sources into waste management frameworks, nations can reduce dependency on fossil fuels and promote a circular energy economy.

The Circular Economy

Several models have been developed to understand sustainability, including the three-tiered model, the prism model, concentric circles, the two-tiered equilibrium model, and the bio-economic model. Among these, the three-tiered, or triple bottom line, model is the most widely recognized. It emphasizes the ESG (Environmental, Social, and Governance) factors, often referred to as "pillars" or "legs of the stool," and forms the basis for creating sustainability-focused business models [17].

Closed-Loop Systems and Sustainability

The circular economy aims to minimize waste through the principles of Reduce, Reuse, and Recycle (3Rs). By designing closed-loop systems, materials can remain in use for longer, reducing pressure on raw resources and ecosystems. These strategies are central to the regenerative economy model, which focuses on sustainability through the continuous recovery of resources.

Implementation of the Circular Economy

The circular or regenerative economy functions at multiple levels: micro (specific materials, businesses, and consumers), meso (Sustainable industrial parks), and macro (cities, regions, and countries). At each level, the objective is to achieve economic and social benefits while preserving the stability of the global ecosystem. This model operates within the environmental boundaries of our planet, ensuring that economic growth occurs without depleting natural resources [18].

It is possible to facilitate system innovation by shifting from a traditional linear model of production and consumption to a circular model. This transition supports the elimination of waste and the improvement of resource efficacy. Further, it maintains things in circulation and decouples economic growth from the utilization of finite resources. This transition is crucial for fostering a more sustainable equilibrium between the economy, the ecosystem, and society. It has thus garnered widespread recognition for its potential to redefine industrial practices and propel the advancement of sustainability. A key enabler of the circular economy is the reintegration of materials through recycling, which also supports the broader WtE objectives.

Recycling and Resource Reintegration

Recycling, as discussed earlier, supports the regenerative and circular economy models. Recycling has evolved beyond its traditional role as an end-of-life process and has become a vital component of the supply chain, significantly contributing to the sustainability of industries [19]. Incorporating recycled materials into manufacturing processes offers significant advantages, including reduced production costs and a smaller carbon footprint, thereby enhancing overall sustainability. This aligns with the global movement towards greener economies, which underscores the critical importance of embracing circular principles across the entire lifecycle of products and materials. Beyond recycling, regenerative development enhances environmental restoration and aligns with WtE systems aiming for carbon neutrality.

Regenerative Development and CO_2 Sequestration

Regenerative development cultivates a holistic understanding and evolving stakeholder collaboration. These strategies place a significant emphasis on environmental restoration, such as through carbon sequestration, which plays a crucial role in mitigating climate change. The inherent self-correcting mechanisms of our environment can be triggered by the emission of CO_2 by human activities, leading to the natural sequestration of carbon dioxide through carbon sinks such as oceans, forests, and soils. The magnitude and rapidity of human-induced climate change frequently exceed the capacity of these natural processes to mitigate its effects effectively [20]. These regenerative systems constitute a vital component of the circular economy, as they not only contribute to mitigating climate change but also support the long-term viability of the environment. By incorporating regenerative principles into the framework of the circular economy, industries and communities can enhance their resilience and provide significant support to global climate action [21]. Building on these sustainability principles, Waste-to-Energy (WtE) technologies represent a practical means to convert waste into usable forms of energy. The circular economy fosters resource efficiency, resilience, and sustainability by minimizing waste and promoting regenerative practices. This model supports WtE integration through structured reuse, recycling, and closed-loop systems.

Waste to Energy (WtE) Technologies

The rapid growth of Municipal Solid Waste (MSW) is damaging the environment worldwide. Waste-to-Energy technologies (WtE) offer an excellent solution by converting waste materials into energy sources, thereby reducing waste volume and supporting green development goals. Combustion, pyrolysis, gasification, landfill gas recovery, and anaerobic digestion are some of the important WtE technologies. Accordingly, these technologies are mainly classified into thermal and biochemical processes. In the thermochemical process, waste is converted into energy through the use of heat, for example, *via* gasification and pyrolysis. In the biochemical process, the gas and liquid products are yielded from biological waste through the biological activities of microorganisms. One example is anaerobic digestion. Presently, incineration and Landfill Gas Recovery (LFG), in which the methane is recovered, are the two most broadly integrated WtE technologies. Besides, mixed MSW incineration is the most economically workable option for future energy systems. Anaerobic digestion and landfill gas recovery are significant strategies that align with the energy recovery hierarchy in MSW management [22].

Incineration

According to studies, Incineration is the most widely used WtE method around the world. Approximately 1,400 incineration plants operate globally, producing heat, water, CO_2, and ash by maximizing the oxidation of carbon compounds through the high-temperature burning of waste. The MSW incineration ash produced from this method has numerous applications in various industries, such as construction, serving as an effective replacement for conventional raw materials. The ability to recover valuable materials, such as metals, from the ash for further recycling supports a circular economy. In addition to that, captured Flue gases will have a use in certain industries or for safety measures, where CO_2 and NOx are present. However, harmful substances, such as sulfur and nitrogen compounds, along with heavy metals like cadmium and mercury, were also produced in this method. Although it causes environmental damage, this method is considered the most efficient for large-scale waste management, as it effectively reduces waste volume by up to 90% [23]. Combustion, energy recovery, and Air Pollution Control (APC) are the three main stages of this process. This method not only recovers valuable resources but also significantly reduces the environmental burden of waste management [24]. The effective elimination of hazardous chemicals and bacteria present in harmful and medical waste is considered one of the primary advantages of this method. However, the incineration of these materials does not always generate energy directly; it plays an important role in waste management by effectively reducing the hazardous nature of these materials. Another key benefit is the substantial reduction in waste mass and volume, with some studies indicating that incineration can reduce waste mass by up to 70% and volume by up to 90% [25].

Moreover, this method offers a valuable opportunity to recover metals, which can be extracted and recycled. The ash produced can also be used in the construction sector, supporting sustainable building practices. Even though MSW ash may contain harmful pollutants, it has also been utilized in wastewater treatment, demonstrating its versatile applications [26]. The Municipal Solid Waste (MSW) incineration process is illustrated in Fig. (1), outlining three key stages: Combustion, energy recovery, and air pollution control. This schematic emphasizes how waste volume is significantly reduced while valuable byproducts, such as ash and heat, are harnessed for energy recovery, thus contributing to circular economy goals.

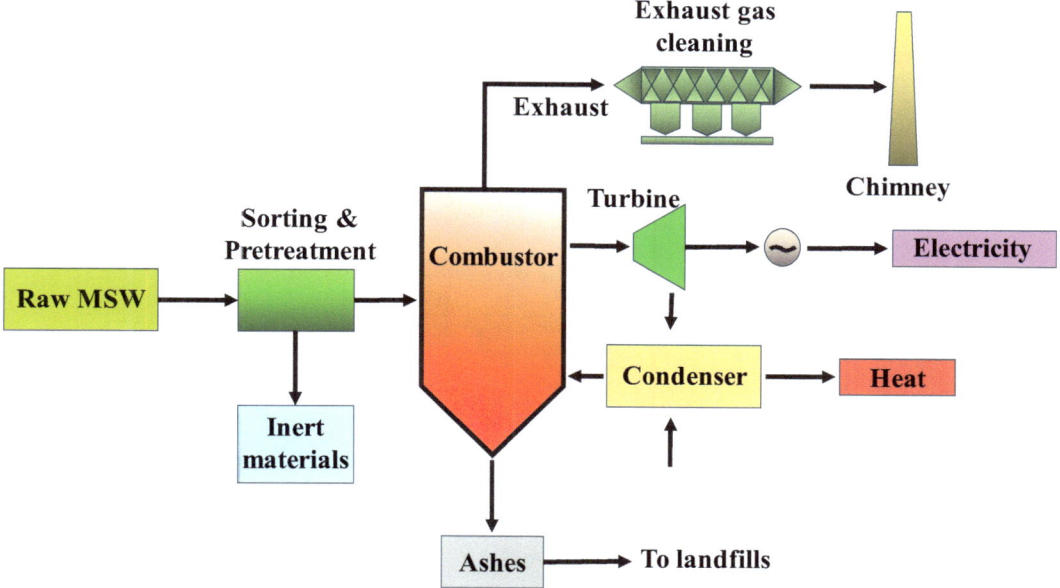

Fig. (1). Schematic diagram of the MSW incineration process [27].

Pyrolysis

Among various waste-to-energy methods, thermochemical waste valorization stands out for its ability to convert solid waste into useful energy products using high-temperature processes. Pyrolysis and gasification are the two principal thermochemical technologies widely applied for energy recovery and waste reduction. The following sections provide a detailed examination of these methods. The thermal degradation of organic materials in an anaerobic condition is known as pyrolysis. Fig. (**2**) depicts a schematic of the pyrolysis method, where organic waste is thermally decomposed in the absence of oxygen. The figure highlights the formation of syngas, bio-oil, and biochar, showcasing how this thermochemical technique supports waste-to-energy goals by converting heterogeneous waste into versatile energy carriers. In the modern WtE system, due to its potential to handle mixed waste and transform it into beneficial energy resources, pyrolysis is a desirable process. Here, waste materials are thermally decomposed at high temperatures (between 300°C and 800°C), producing valuable outputs such as syngas (synthetic gas), liquid bio-oil, and solid biochar. Pyrolysis produces energy-rich outputs, including bio-oil (a liquid fuel), biochar (a soil amendment and carbon sequestration agent), and syngas (for power generation). These products have a high calorific value, supporting a diverse range of energy applications. Among them, solid and liquid residues can be further processed or utilized in industries, while the syngas formed will have applications in power production. Pyrolysis can be enhanced by modifying factors such as temperature and residence time. Catalysts such as nickel, zeolites, and dolomite enhance pyrolysis efficiency and product yield [28]. For optimal functionality, the removal of glass, metals, and inert materials from the waste is necessary during the pre-treatment process.

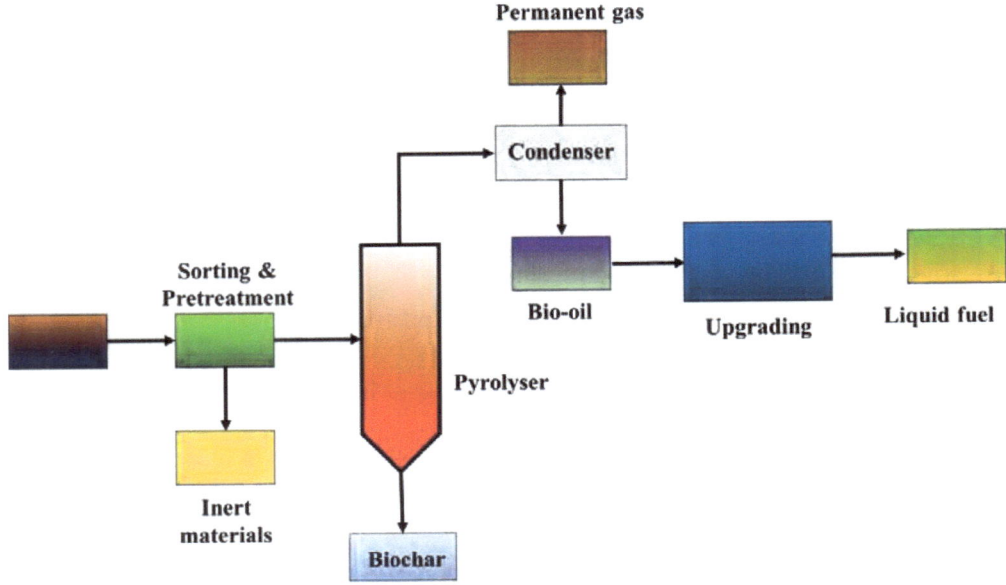

Fig. (2). Schematic diagram of the pyrolysis method [27].

Types of Pyrolysis

Pyrolysis of municipal solid waste liberates gases such as CH_4, CO_2, H_2, and CO, which can be used as fuel. The process, which occurs in a restricted, oxygen-free environment, enables the extraction of significant materials, such as bio-oil, which has potential as an alternative fuel source.

The versatility of pyrolysis enables the utilization of various types of waste, including plastics and rubber, thereby reducing landfill waste and enhancing resource recovery [29]. Fig. (3) illustrates the various types of pyrolysis, categorized by temperature and heating rate, including slow, fast, and flash pyrolysis. This classification helps in understanding which type is most suitable, depending on the desired output, such as maximizing bio-oil or biochar yield, thereby aiding in process optimization in WtE applications. The feedstocks and their major products are enumerated in Table 1. It summarizes the major products derived from various feedstocks used in pyrolysis and gasification. It provides a comparative view of input-output efficiency, aiding in the selection of appropriate materials for targeted energy recovery.

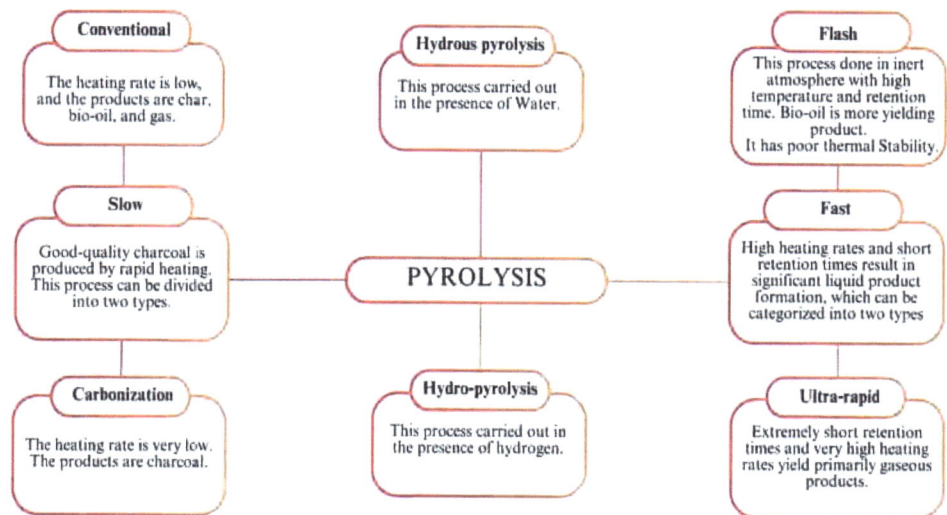

Fig. (3). Various types of pyrolysis.

Table 1. Various types of feedstock and their major products.

Feedstock	Materials	Maximum Yield	References
Organic and plastic waste	Herbaceous and Tree	Synthetic gas and Liquid biofuel	[36]
	Hazelnut shells and walnut shells.	Biochar	[37]
	Algal biomass	bio-oil	[38]
	Woody	Charcoal (bio-char)	[30]
	Polyethylene terephthalate (PET)	Liquid oil and Solid	[39, 40]
	Polypropylene (PP)	Gas	[39]
	polyvinyl chloride (PVC)	Solid and Gas	[39, 41]
Medical wastes	Face masks and gloves	Oil	[37]
	Syringe	Oil	[31, 42]
	Latex	Oil	[43]
	Saline bottle	Char	[44]
Others	Tires	Tire pyrolysis oil	[37]
	Paper	Bio-oil	[45]

The pyrolysis process offers numerous benefits, enabling waste to be utilized beneficially for both energy and chemical purposes. Furthermore, the char developed during pyrolysis can be a better replacement for carbon black, which is commonly used in various industries. Fast pyrolysis offers high efficiency and sustainability, producing hydrogen-rich gas suitable for clean energy applications. In addition to that, it uses the 2nd generation of bio-oil feedstocks, including industrial and MSW. These facts provide effective storage and transport options for the liquid fuels generated by this process, further developing its feasibility [30]. Moreover, this process has been utilized in various industries, making it a flexible option for waste management [31]. Additionally, this pyrolysis addresses the critical waste management challenges in areas with insufficient disposal facilities, providing a more effective solution to handle waste materials in an eco-friendly manner [32]. The fuel properties of the Pyrolysis Oil blend produced through pyrolysis suggest its potential as a feasible replacement for fossil fuels. After desulfurization and distillation, the properties of this blend closely resemble those of diesel fuel, making it suitable for use in automobiles [33].

Gasification

Municipal Solid Waste (MSW) is a complex and heterogeneous mixture that includes plastics, metals, and hazardous materials. Due to its high sulfur content, gasification requires careful selection of operating conditions for effective conversion [34]. Thermochemical technologies encompass a range of processes, including liquefaction, carbonization, combustion, pyrolysis, and gasification. Other methods, such as incineration, torrefaction, and hydrogenation, also fall within this category. While gasification has been used for over 100 years, it has recently witnessed a revival due to its scalability and flexibility in feedstock usage [35]. As shown in Fig. (**4**), the gasification process involves partial oxidation of waste at high temperatures (550–1,000°C), producing syngas rich in CO and H_2. This figure underscores the importance of controlled oxygen input and pre-treatment for efficient conversion, reinforcing gasification as a cleaner alternative to incineration.

This heat treatment breaks down molecules into their elemental components with the aid of an oxidizing agent, producing a gas mixture known as syngas. Syngas consists of a mixture of flammable gases, along with a small quantity of char and ash. It can then be further utilized for power generation, chemical production, hydrogen production, and the synthesis of liquid fuels [46, 47]. In addition, tars, heavy metals, halogens, alkaline, NOx and SOx, NH_3, H_2S, dioxins, furans, hydrocarbons, *etc.*, are released as by-products [47].

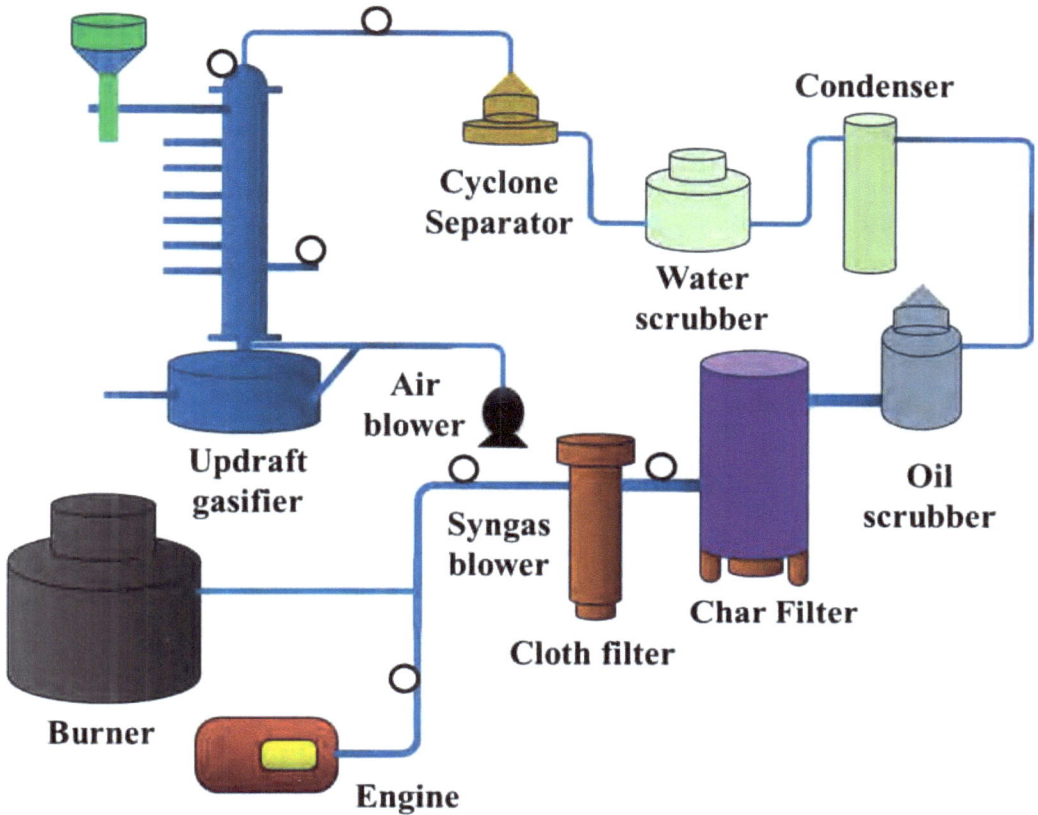

Fig. (4). Schematic diagram of gasification [52].

Gasification is a partial oxidation process that, after an initial heat input, generates its heat and prevents combustion. Because gasification occurs with minimal oxygen, the syngas formed is primarily composed of CO, H_2, CO_2, and N_2, along with small amounts of CH_4, tar, and ash [48]. Maximum efficiency in the conversion of feedstock to syngas is achieved when carbon is primarily oxidized to carbon monoxide [49]. Syngas can be utilized in combustion engines, for power generation, as a heat source in chemical production, and as a feedstock for the Fischer–Tropsch synthesis process to produce liquid fuels. Gasification is traditionally used with coal and biomass, which have relatively uniform compositions. However, utilizing waste materials from municipalities, various industries, solid and refuse-derived fuels, construction debris, or electronic waste can be more challenging due to the heterogeneous nature of these materials [50]. Downdraft gasifiers are the most commonly employed type of gasifier for biomass gasification [51].

In laboratory-scale applications, steam, CO_2, or a mixture of both is typically utilized as the gasification agent. In contrast, industrial-scale operations generally rely on air, oxygen, or steam [35]. Catalytic gasification reduces the required temperatures and helps minimize tar formation. The metal species used as catalysts can either be introduced *via* a precursor or be naturally present in the feedstock. To date, pyrolysis/gasification-based Waste-to-Energy (WtE) technologies have not been widely implemented on a commercial scale, with only around a hundred plants reported to process Municipal Solid Waste (MSW) [53]. Fig. (5) provides a flowchart outlining the different types of gasifiers used in WtE systems, including downdraft, fluidized bed, and plasma gasifiers. Each design offers specific advantages, depending on the feedstock type and energy application, which aids in technology selection for diverse waste profiles.

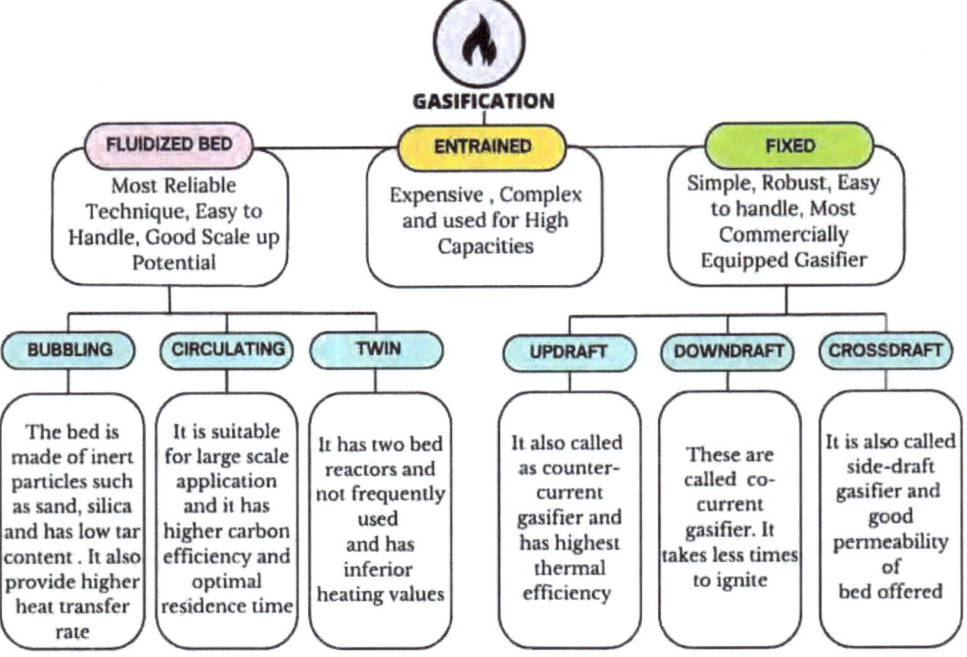

Fig. (5). Flow chart of various types of gasification.

Higher-quality producer gas can be obtained from raw materials such as coconut shells, wood blends, and rubber seed kernels. Among these, wood pellets and materials with high volatile solids content are generally considered better choices for gasification. At a specific moisture level, paper exhibited the highest hydrogen composition, while wood produced the highest levels of carbon monoxide. Additionally, the paper also showed the highest methane composition compared to other biomass materials tested. Other materials tested, including butter tree seeds, *Panicum virgatum* (switchgrass), pinewood shavings, and chicken litter, also produced oils with higher heating values, comparable to those of dairy manure oils [51]. Coal remains a primary feedstock for gasification systems [51]. Plastics, a primary type of waste with significant environmental impacts, have increasingly been considered as feedstocks for gasification [54]. Medical waste disposal has become particularly crucial following the COVID-19 pandemic, and plasma gasification presents a promising technique for effectively disposing of such waste materials [55]. The choice of a specific gasifier is mainly impacted by two factors: the type of feedstock and the desired properties of the producer gas. Fluidized bed technology is attractive due to its cost-effectiveness, ease of maintenance, suitable operating temperatures, high efficiency, and the ability to handle a wide range of particle sizes. Additionally, it allows for the implementation of effective cleaning processes and offers considerable potential for scale-up [48, 56].

Gasification offers several advantages over traditional combustion methods for treating Municipal Solid Waste (MSW). It requires minimal oxygen, which helps limit the formation of byproducts, such as dioxins, and high amounts of oxides of nitrogen and sulfur. Because only a small amount of gas is used in the process, the volume of process gas is less, which in turn requires smaller and more cost-effective gas cleaning equipment. Furthermore, gasification can be integrated with combined cycle turbines and reciprocating engines, which produce electricity more effectively from fuel energy than traditional steam boilers [50].

Types of Gasifier

Fluidized bed reactors offer varying levels of energy performance. Compared to fixed-bed reactors, bubbling fluidized bed reactors generally exhibit lower energy efficiency. However, they are typically less expensive and require lower maintenance costs due to their more straightforward operation. Circulating fluidized bed reactors, while offering higher energy efficiency, generally have higher capital, operational, and maintenance costs. Overall, fluidized bed gasification methods, particularly bubbling fluidized bed gasification, are often associated with lower capital costs compared to other gasification technologies. Together, pyrolysis and gasification demonstrate the potential of thermochemical waste valorization technologies to recover energy and materials from diverse waste streams. While pyrolysis relies on oxygen-free decomposition, gasification uses partial oxidation to produce syngas. Both play vital roles in sustainable and scalable WtE systems.

Anaerobic Digestion

Anaerobic digestion, also known as anaerobic fermentation or biomethanation, is a natural biochemical process where bacteria degrade organic matter under anaerobic conditions [57]. The various steps involved in this process are illustrated in Fig. (**6**). This process converts lignocellulosic (organic) materials into a flammable biogas comprising methane and carbon dioxide. It can be used for the synthesis of various chemical products or for producing electricity, heat, or fuel. The increasing interest in Anaerobic digestion stems from its potential as an unconventional energy material and its ability to support waste management and the production of renewable energy [58, 59]. Anaerobic digestion has demonstrated significant potential for producing high volumes of biogas. These systems, which were once primarily used for sanitation, are now employed for the management of agricultural, municipal, and industrial waste. Anaerobic digestion systems are notably more efficient than landfills, producing 2 to 4 times more methane per tonne of Municipal Solid Waste in just three days, compared to seven years in a landfill [60].

ANEROBIC DIGESTION

HYDROLYSIS

Organic Matter (Carbohydrates, Protein, and Fat)

Monomers (Sugar, Amino acid, a long-chain fatty acid)

ACETOGENESIS

The third phase, acetogenesis, involves the metabolism and transformation of organic acids and alcohols into acetate.

ACIDOGENESIS

These bacteria convert monomers into short-chain organic acids (C1-C5 carbon atoms) like butyric, propionic, acetate, acetic acid, and alcohols.

METHANOGENESIS

Methanogenesis is the final step of ad which is carried out by methanogenic archaea. Among others, H_2 oxidation, CO_2 reduction, and utilization are the unique characteristics of methanogens.

Fig. (6). Various steps involved in anaerobic digestion.

MSW plants apply anaerobic digestion to mitigate odors, reduce waste volume, and enhance treatment capabilities. Fig. (**6**) outlines the multi-step anaerobic digestion process, including hydrolysis, acidogenesis, acetogenesis, and methanogenesis. This visual representation helps clarify the sequential microbial breakdown of organic matter, highlighting how each phase contributes to optimal biogas production.

Feedstocks should be readily biodegradable and free from inhibitory compounds that may hinder microbial activity. Suitable feedstocks include sewage sludge, livestock manure, organic compounds from municipal solid waste, and slaughterhouse waste. Certain feedstocks, such as biomass with high lignin content or materials containing toxic components, may require pretreatment to enhance biodegradability. For example, wood-based biomass can be challenging to digest in anaerobic digesters and may need pretreatment for structural breakdown. Co-digestion, which involves combining different feedstocks, can enhance biogas production by optimizing the nutrient balance within the digester.

To address the limitations of traditional WtE methods, emerging technologies offer innovative solutions for complex and hazardous waste streams. WtE technologies, such as incineration, pyrolysis, gasification, and anaerobic digestion, serve as transformative tools to convert waste into energy. Their implementation enhances energy recovery and reduces environmental burdens across sectors.

EMERGING TECHNOLOGIES IN WASTE-TO-ENERGY

Plasma Gasification

Plasma gasification is a modern WtE technology that uses extreme heat to decompose complex waste into syngas. This method enables efficient energy conversion from difficult-to-recycle materials such as plastics, sewage sludge, and hazardous waste. This results in syngas with high energy content, suitable for power generation or as a raw material for chemical production [61]. One important advantage of plasma gasification is its ability to handle waste that other technologies, such as conventional incineration, cannot process effectively [62]. Despite its potential, plasma gasification faces challenges, including the formation of tar and high energy consumption. However, it is more energy-efficient than conventional fluidized bed gasifiers.

Hydrothermal Carbonization (HTC)

It offers a promising route for transforming organic waste into products like hydrochar. The process involves subjecting waste to the hydrothermal environment, resulting in carbonization, polymerization, and dehydration reactions [63]. HTC offers significant advantages for treating biomass and plastics that are difficult to process using traditional pyrolysis methods. The carbonized product, hydrochar, has applications as a solid fuel, supercapacitor, or sorbent [64]. While HTC technology offers the potential for sustainable waste management, it faces challenges such as increased construction and operational costs. Additionally, the quality of hydrochar may vary depending on feedstock composition and contamination levels. While emerging technologies address technical challenges, real-world implementations demonstrate how WtE can be adapted to diverse geographical and industrial contexts. Innovations such as plasma gasification and hydrothermal carbonization expand the capability of WtE to handle diverse and complex waste streams, reinforcing the need for continuous technological advancement.

Waste-to-Energy Systems in Urban, Agricultural, and Industrial Settings

Urban areas worldwide are increasingly turning to WtE technologies to handle growing waste volumes while generating clean energy. Agricultural waste-to-energy solutions are particularly effective in rural areas, where large quantities of biomass waste are generated. Gasification and other WtE technologies are being implemented in industrial settings to reduce waste volume, mitigate emissions, and enhance sustainability.

Incineration in Istanbul, Turkey

In Istanbul, a large-scale MSW incineration project processes waste from 37 municipalities, producing 53.72 MW of electrical power and nearly 100 MW of thermal energy for district heating. This process improves urban energy resilience by reducing waste volume and contributing to both electricity and thermal energy needs. The efficiency and optimization of this system have been analyzed in terms of energy and exergy, with a focus on minimizing environmental impact [65].

Anaerobic Digestion in Sweden

Sweden has been a pioneer in biogas production through the anaerobic digestion of food and agricultural waste. This process involves breaking down organic matter using microorganisms in the absence of oxygen. The resulting biogas is used as fuel for public transportation in several cities, including Stockholm, Västra Götaland, and Skåne. Despite its benefits, the expansion of biogas use in Sweden is hindered by economic instability and a lack of long-term policy support, highlighting the need for sustained investment to maintain and further grow this sector [66].

Gasification in Cambodia

In Cambodia, biomass gasification is being explored as a solution to increase rural electricity supply. By utilizing agricultural residues and woody biomass, this technology aims to address energy shortages in the country's rural areas, where nearly 76% of villages were without electricity as of 2010 [67].

Anaerobic Digestion in Germany

Germany has made significant strides in adopting anaerobic digestion technology for converting agricultural by-products, such as animal manure, plant residues, and food scraps, into biogas. This biogas is utilized to produce electricity and thermal energy, while the resulting digestate is repurposed as a valuable fertilizer, thus benefiting both energy production and agricultural practices [68].

Combustion of Agricultural Waste in Denmark

Denmark has successfully utilized agricultural residues, such as wheat straw, for combustion-based energy generation. An environmental impact assessment of using straw for energy production, compared to coal and natural gas, demonstrates reduced impacts on global warming and a decreased reliance on non-renewable energy resources. However, the assessment also highlights potential trade-offs, such as increased eutrophication and acidification, underscoring the need for a balanced approach that considers all environmental costs [69].

Gasification in the Netherlands

In the Netherlands, gasification is gaining traction as a sustainable method for converting waste into energy, particularly in industrial applications. The life cycle assessment indicates that gasification-to-methanol provides an important short-term reduction of greenhouse gas compared to various waste treatment technologies. Furthermore, gasification of municipal solid waste is increasingly being adopted as a viable solution for mitigating climate change impacts in the long run, exemplified by projects such as the development of a gasification plant for converting waste into methanol in the Port of Rotterdam [70]. Despite these successful implementations, multiple technical, economic, and environmental barriers remain to be addressed through targeted research and policy innovations. Real-world case studies demonstrate the adaptive and scalable nature of WtE across various geographies, underscoring its potential to meet both energy and waste management objectives.

CHALLENGES AND FUTURE DIRECTIONS

Technical Challenges

The diversity in feedstocks presents unique challenges to each WtE technology. Variations in feedstock composition (*e.g.*, moisture content, volatile matter, and fixed carbon) significantly affect process efficiency, particularly in gasification. Efficient characterization and management of feedstock, including addressing impurities and ensuring consistent quality, are critical to increasing the overall performance of pyrolysis, gasification, and other WtE processes.

Gasification Challenges

Gasification faces challenges related to feedstock diversity, including issues with plastic, biomass, and coal.

- Plastic waste: Impurities and variations in plastic types complicate syngas upgrading.
- Biomass: The low bulk density, along with issues related to soot formation and salt accumulation, reduces the energy efficiency of biomass gasification.
- Coal: Emissions reduction and the implementation of Clean Coal Technologies (CCT) are critical for reducing the environmental impact of coal gasification.

Anaerobic Digestion

Anaerobic digestion encounters challenges such as low biodegradability of feedstocks, process inhibition, digester toxicity, and the need for pre-treatment of lignocellulosic biomass. Addressing these barriers, particularly the chemical barriers posed by plant toxins, is a key area of future research.

Pyrolysis

Pyrolysis encounters feedstock-associated challenges that significantly influence product distribution and energy recovery. Variations in feedstock characteristics, including particle size, moisture content, and chemical composition, necessitate process optimization to achieve better control over product quality and yield.

Economic Barriers

Despite its growth, the biomass gasification market has faced significant economic challenges. Careful cost estimation, including capital expenses and revenue projections, requires greater consideration for the economic viability of gasification technologies. The cost structure for these technologies varies globally. Advanced systems, such as fluidized bed gasification, require more attention due to their high investment and operational costs. On the other hand, pyrolysis also presents challenges, including high initial investment costs, limited access to feedstock, economic hurdles, and market price uncertainty. This implements a complex pyrolysis technology for municipal solid waste and other feedstocks, despite having high potential for energy recovery.

Environmental Concerns

Incineration presents a range of ecological challenges, including the release of hazardous gases such as dioxins and CO_2, as well as the management of residual ash. Although emission control systems are effective in mitigating these emissions, they often entail significant costs and present complex technological challenges. Gasification processes likewise generate contaminants, including tars, heavy metals, and halogens. Without proper management, these byproducts can cause damage to equipment and pose substantial environmental risks. Addressing these environmental concerns necessitates the advancement of cleaner and more efficient gasification methods. Additionally, the pyrolysis of certain plastics, notably polyvinyl chloride (PVC) and polyethylene terephthalate (PET), can lead to the generation of harmful chlorinated hydrocarbons and a decline in oil quality. Research into improved feedstock management and catalytic pyrolysis methods is essential to mitigate these challenges.

FUTURE RESEARCH DIRECTIONS

Advancements in Technology

Future research work will focus on improving the efficiency and environmental sustainability of WtE technologies. Techniques such as plasma-assisted gasification and the integration of pyrolysis with gasification processes have the potential to enhance energy output while minimizing the formation of pollutants significantly. Moreover, the development of decentralized systems, such as small-scale rotary kilns for incineration, can facilitate the implementation of effective local waste management strategies.

Hybrid Systems

The incorporation of multiple WtE technologies possesses the potential to increase the efficiency of WtE conversion substantially. Hybrid systems that combine methodologies such as gasification, pyrolysis, and anaerobic digestion with renewable energy sources offer the potential to maximize energy recovery while simultaneously minimizing environmental impact.

Policy and Economic Models

The successful scaling up of WtE technologies will necessitate the establishment of robust regulatory frameworks and the implementation of effective policies. Research into financial models, with a particular focus on those that provide incentives for the generation of clean energy and the promotion of sustainable waste management practices, plays a crucial role in overcoming economic barriers and driving the adoption of WtE technologies in both developed and developing economies. Advancing WtE technologies requires a collaborative approach involving technology, policy, and community engagement to realize their full potential for sustainable development. While WtE systems present immense opportunities, technical, economic, and regulatory barriers must be systematically addressed to unlock their full global impact.

CONCLUSION

The advancement in WtE technologies is of paramount interest to the scientific community, as it provides an advantage over traditional waste management and the utilization of renewable energy resources for sustainable development. As discussed, WtE technologies contribute significantly to environmental sustainability by reducing pollution. In this chapter, various successful and implemented technologies, including incineration, gasification, pyrolysis, and anaerobic digestion, are described with examples and case studies. Various optimization methods, challenges, and ongoing development have also been discussed to get more efficient, affordable, and environmentally friendly WtE technologies in the near future.

REFERENCES

[1]   H.B. Sharma, K.R. Vanapalli, B. Samal, V.R.S. Cheela, B.K. Dubey, and J. Bhattacharya, "Circular economy approach in solid waste management system to achieve UN-SDGs: Solutions for post-COVID recovery", *Sci. Total Environ.,* vol. 800, p. 149605, 2021.
[http://dx.doi.org/10.1016/j.scitotenv.2021.149605] [PMID: 34426367]

[2]   A. Minelgaitė, and G. Liobikienė, "Waste problem in European Union and its influence on waste management behaviours", *Sci. Total Environ.,* vol. 667, pp. 86-93, 2019.
[http://dx.doi.org/10.1016/j.scitotenv.2019.02.313] [PMID: 30826684]

[3]   M. Takata, K. Fukushima, M. Kawai, N. Nagao, C. Niwa, T. Yoshida, and T. Toda, "The choice of biological waste treatment method for urban areas in Japan—An environmental perspective", *Renew. Sustain. Energy Rev.,* vol. 23, pp. 557-567, 2013.
[http://dx.doi.org/10.1016/j.rser.2013.02.043]

[4]   S. Mor, and K. Ravindra, "Municipal solid waste landfills in lower- and middle-income countries: Environmental impacts, challenges and sustainable management practices", *Process Saf. Environ. Prot.,* vol. 174, pp. 510-530, 2023.
[http://dx.doi.org/10.1016/j.psep.2023.04.014]

[5]   M. Ahmaruzzaman, "Industrial wastes as low-cost potential adsorbents for the treatment of wastewater laden with heavy metals", *Adv. Colloid Interface Sci.,* vol. 166, no. 1-2, pp. 36-59, 2011.

[http://dx.doi.org/10.1016/j.cis.2011.04.005] [PMID: 21669401]

[6]   K.H. Vardhan, P.S. Kumar, and R.C. Panda, "A review on heavy metal pollution, toxicity and remedial measures: Current trends and future perspectives", *J. Mol. Liq.,* vol. 290, p. 111197, 2019.
        [http://dx.doi.org/10.1016/j.molliq.2019.111197]

[7]   H.S. Rangappa, I. Herath, C. Lin, and S. Ch, "Industrial waste-based adsorbents as a new trend for removal of water-borne emerging contaminants", *Environ. Pollut.,* vol. 343, p. 123140, 2024.
        [http://dx.doi.org/10.1016/j.envpol.2023.123140] [PMID: 38103712]

[8]   M.B. Hayat, D. Ali, K.C. Monyake, L. Alagha, and N. Ahmed, "Solar energy-A look into power generation, challenges, and a solar-powered future", *Int. J. Energy Res.,* vol. 43, no. 3, pp. 1049-1067, 2019.
        [http://dx.doi.org/10.1002/er.4252]

[9]   D.Y. Goswami, and F. Kreith, *Energy conversion.* CRC press, 2007.
        [http://dx.doi.org/10.1201/9781420044324]

[10]   F. Mohamad, J. Teh, and C.M. Lai, "Optimum allocation of battery energy storage systems for power grid enhanced with solar energy", *Energy,* vol. 223, p. 120105, 2021.
        [http://dx.doi.org/10.1016/j.energy.2021.120105]

[11]   J.L. Holechek, H.M.E. Geli, M.N. Sawalhah, and R. Valdez, "A global assessment: can renewable energy replace fossil fuels by 2050?", *Sustainability (Basel),* vol. 14, no. 8, p. 4792, 2022.
        [http://dx.doi.org/10.3390/su14084792]

[12]   A. Boretti, and S. Castelletto, "Cost of wind energy generation should include energy storage allowance", *Sci. Rep.,* vol. 10, no. 1, p. 2978, 2020.
        [http://dx.doi.org/10.1038/s41598-020-59936-x] [PMID: 32076061]

[13]   M. Baratieri, P. Baggio, L. Fiori, and M. Grigiante, "Biomass as an energy source: Thermodynamic constraints on the performance of the conversion process", *Bioresour. Technol.,* vol. 99, no. 15, pp. 7063-7073, 2008.
        [http://dx.doi.org/10.1016/j.biortech.2008.01.006] [PMID: 18296047]

[14]   I. Malico, R. Nepomuceno Pereira, A.C. Gonçalves, and A.M.O. Sousa, "Current status and future perspectives for energy production from solid biomass in the European industry", *Renew. Sustain. Energy Rev.,* vol. 112, pp. 960-977, 2019.
        [http://dx.doi.org/10.1016/j.rser.2019.06.022]

[15]   M. Balat, and H. Balat, "Recent trends in global production and utilization of bio-ethanol fuel", *Appl. Energy,* vol. 86, no. 11, pp. 2273-2282, 2009.
        [http://dx.doi.org/10.1016/j.apenergy.2009.03.015]

[16]   M. Rehan, M. Amir Raza, A. Ghani Abro, M. M Aman, I. Mohammad Ibrahim Ismail, A. Sattar Nizami, M. Imtiaz Rashid, A. Summan, K. Shahzad, and N. Ali, "A sustainable use of biomass for electrical energy harvesting using distributed generation systems", *Energy,* vol. 278, p. 128036, 2023.
        [http://dx.doi.org/10.1016/j.energy.2023.128036]

[17]   S. Kara, M. Hauschild, J. Sutherland, and T. McAloone, "Closed-loop systems to circular economy: A pathway to environmental sustainability?", *CIRP Ann.,* vol. 71, no. 2, pp. 505-528, 2022.
        [http://dx.doi.org/10.1016/j.cirp.2022.05.008]

[18]   S. Podvalny, D. Logunov, and E. Vasiljev, "Modeling of systems with a closed loop of material resources circulation",
        [http://dx.doi.org/10.1088/1742-6596/2131/3/032115]

[19]   P. Rogetzer, L. Silbermayr, and W. Jammernegg, "Sustainable sourcing of strategic raw materials by integrating recycled materials", *Flex. Serv. Manuf. J.,* vol. 30, no. 3, pp. 421-451, 2018.
        [http://dx.doi.org/10.1007/s10696-017-9288-4]

[20]   S. Potrč, A. Nemet, L. Čuček, P.S. Varbanov, and Z. Kravanja, "Synthesis of a regenerative energy system – beyond carbon emissions neutrality", *Renew. Sustain. Energy Rev.,* vol. 169, p. 112924,

2022.
[http://dx.doi.org/10.1016/j.rser.2022.112924]

[21]   M.A. Camilleri, "The circular economy's closed loop and product service systems for sustainable development: A review and appraisal", *Sustain. Dev. (Bradford)*, vol. 27, no. 3, pp. 530-536, 2019.
[http://dx.doi.org/10.1002/sd.1909]

[22]   R. Hosseinalizadeh, H. Izadbakhsh, and H. Shakouri G, "A planning model for using municipal solid waste management technologies- considering Energy, Economic, and Environmental Impacts in Tehran-Iran", *Sustain Cities Soc.,* vol. 65, p. 102566, 2021.
[http://dx.doi.org/10.1016/j.scs.2020.102566]

[23]   C. Ram, A. Kumar, and P. Rani, "Municipal solid waste management: A review of waste to energy (WtE) approaches", *BioResources,* vol. 16, no. 2, pp. 4275-4320, 2021.
[http://dx.doi.org/10.15376/biores.16.2.Ram]

[24]   A.H. Kanhar, S. Chen, and F. Wang, "Incineration fly ash and its treatment to possible utilization: A review", *Energies,* vol. 13, no. 24, p. 6681, 2020.
[http://dx.doi.org/10.3390/en13246681]

[25]   V. Blahuskova, J. Vlcek, and D. Jancar, "Study connective capabilities of solid residues from the waste incineration", *J. Environ. Manage.,* vol. 231, pp. 1048-1055, 2019.
[http://dx.doi.org/10.1016/j.jenvman.2018.10.112] [PMID: 30602228]

[26]   K.A. Shukla, A.D.A.B.A. Sofian, A. Singh, W.H. Chen, P.L. Show, and Y.J. Chan, "Food waste management and sustainable waste to energy: Current efforts, anaerobic digestion, incinerator and hydrothermal carbonization with a focus in Malaysia", *J. Clean. Prod.,* vol. 448, p. 141457, 2024.
[http://dx.doi.org/10.1016/j.jclepro.2024.141457]

[27]   L. Chand Malav, K.K. Yadav, N. Gupta, S. Kumar, G.K. Sharma, S. Krishnan, S. Rezania, H. Kamyab, Q.B. Pham, S. Yadav, S. Bhattacharyya, V.K. Yadav, and Q-V. Bach, "A review on municipal solid waste as a renewable source for waste-to-energy project in India: Current practices, challenges, and future opportunities", *J. Clean. Prod.,* vol. 277, p. 123227, 2020.
[http://dx.doi.org/10.1016/j.jclepro.2020.123227]

[28]   C. Mukherjee, J. Denney, E.G. Mbonimpa, J. Slagley, and R. Bhowmik, "A review on municipal solid waste-to-energy trends in the USA", *Renew. Sustain. Energy Rev.,* vol. 119, p. 109512, 2020.
[http://dx.doi.org/10.1016/j.rser.2019.109512]

[29]   M.J.B. Kabeyi, and O.A. Olanrewaju, "Review and design overview of plastic waste-to-pyrolysis oil conversion with implications on the energy transition", *J. Energy,* vol. 2023, no. 1, pp. 1-25, 2023.
[http://dx.doi.org/10.1155/2023/1821129]

[30]   C.Z. Zaman, K. Pal, W.A. Yehye, S. Sagadevan, S.T. Shah, G.A. Adebisi, E. Marliana, R.F. Rafique, and R.B. Johan, "Pyrolysis: a sustainable way to generate energy from waste", *Pyrolysis,* vol. 1, pp. 3-36, 2017.
[http://dx.doi.org/10.5772/intechopen.69036]

[31]   G. Su, H.C. Ong, S. Ibrahim, I.M.R. Fattah, M. Mofijur, and C.T. Chong, "Valorisation of medical waste through pyrolysis for a cleaner environment: Progress and challenges", *Environ. Pollut.,* vol. 279, p. 116934, 2021.
[http://dx.doi.org/10.1016/j.envpol.2021.116934] [PMID: 33744627]

[32]   J.W. Jang, T.S. Yoo, J.H. Oh, and I. Iwasaki, "Discarded tire recycling practices in the United States, Japan and Korea", *Resour. Conserv. Recycling,* vol. 22, no. 1-2, pp. 1-14, 1998.
[http://dx.doi.org/10.1016/S0921-3449(97)00041-4]

[33]   H. Yaqoob, Y.H. Teoh, F. Sher, M.A. Jamil, D. Murtaza, M. Al Qubeissi, M. UI Hassan, and M.A. Mujtaba, "Current status and potential of tire pyrolysis oil production as an alternative fuel in developing countries", *Sustainability (Basel),* vol. 13, no. 6, p. 3214, 2021.
[http://dx.doi.org/10.3390/su13063214]

[34] Y. Zhang, Y. Cui, P. Chen, S. Liu, N. Zhou, K. Ding, L. Fan, P. Peng, M. Min, and Y. Cheng, "Gasification technologies and their energy potentials", In: *Sustainable resource recovery and zero waste approaches.* Elsevier, 2019, pp. 193-206.

[35] R.A. Arnold, and J.M. Hill, "Catalysts for gasification: a review", *Sustain. Energy Fuels,* vol. 3, no. 3, pp. 656-672, 2019.
[http://dx.doi.org/10.1039/C8SE00614H]

[36] A. Friedl, E. Padouvas, H. Rotter, and K. Varmuza, "Prediction of heating values of biomass fuel from elemental composition", *Anal. Chim. Acta,* vol. 544, no. 1-2, pp. 191-198, 2005.
[http://dx.doi.org/10.1016/j.aca.2005.01.041]

[37] K.W. Chew, S.R. Chia, W.Y. Chia, W.Y. Cheah, H.S.H. Munawaroh, and W.J. Ong, "Abatement of hazardous materials and biomass waste *via* pyrolysis and co-pyrolysis for environmental sustainability and circular economy", *Environ. Pollut.,* vol. 278, p. 116836, 2021.
[http://dx.doi.org/10.1016/j.envpol.2021.116836] [PMID: 33689952]

[38] S. Pourkarimi, A. Hallajisani, A. Alizadehdakhel, and A. Nouralishahi, "Biofuel production through micro- and macroalgae pyrolysis – A review of pyrolysis methods and process parameters", *J. Anal. Appl. Pyrolysis,* vol. 142, p. 104599, 2019.
[http://dx.doi.org/10.1016/j.jaap.2019.04.015]

[39] S. Papari, H. Bamdad, and F. Berruti, "Pyrolytic conversion of plastic waste to value-added products and fuels: A review", *Materials (Basel),* vol. 14, no. 10, p. 2586, 2021.
[http://dx.doi.org/10.3390/ma14102586] [PMID: 34065677]

[40] R. Prurapark, K. Owjaraen, B. Saengphrom, I. Limthongtip, and N. Tongam, "Effect of temperature on pyrolysis oil using high-density polyethylene and polyethylene terephthalate sources from mobile pyrolysis plant", *Front. Energy Res.,* vol. 8, p. 541535, 2020.
[http://dx.doi.org/10.3389/fenrg.2020.541535]

[41] E.C.R. Lopez, "Pyrolysis of polyvinyl chloride, polypropylene, and polystyrene: current research and future outlook", *Eng. Proc.,* vol. 56, no. 1, p. 44, 2023.
[http://dx.doi.org/10.3390/ASEC2023-15376]

[42] A. Dash, S. Kumar, and R.K. Singh, "Thermolysis of medical waste (waste syringe) to liquid fuel using semi batch reactor", *Waste Biomass Valoriz.,* vol. 6, no. 4, pp. 507-514, 2015.
[http://dx.doi.org/10.1007/s12649-015-9382-3]

[43] M. Paraschiv, R. Kuncser, M. Tazerout, and T. Prisecaru, "New energy value chain through pyrolysis of hospital plastic waste", *Appl. Therm. Eng.,* vol. 87, pp. 424-433, 2015.
[http://dx.doi.org/10.1016/j.applthermaleng.2015.04.070]

[44] A. Abedeen, "Recovery and characterization of fuel from pyrolysis of medical wastes: an alternative source of energy", In: *Khulna University of Engineering & Technology.* KUET: Khulna, Bangladesh, 2019.

[45] M.M. Hasan, M.G. Rasul, M.M.K. Khan, N. Ashwath, and M.I. Jahirul, "Energy recovery from municipal solid waste using pyrolysis technology: A review on current status and developments", *Renew. Sustain. Energy Rev.,* vol. 145, p. 111073, 2021.
[http://dx.doi.org/10.1016/j.rser.2021.111073]

[46] U. Arena, "Process and technological aspects of municipal solid waste gasification. A review", *Waste Manag.,* vol. 32, no. 4, pp. 625-639, 2012.
[http://dx.doi.org/10.1016/j.wasman.2011.09.025] [PMID: 22035903]

[47] H.H. Shah, M. Amin, A. Iqbal, I. Nadeem, M. Kalin, A.M. Soomar, and A.M. Galal, "A review on gasification and pyrolysis of waste plastics", *Front Chem.,* vol. 10, p. 960894, 2023.
[http://dx.doi.org/10.3389/fchem.2022.960894] [PMID: 36819712]

[48] A. Ramos, E. Monteiro, and A. Rouboa, "Numerical approaches and comprehensive models for gasification process: A review", *Renew. Sustain. Energy Rev.,* vol. 110, pp. 188-206, 2019.

[http://dx.doi.org/10.1016/j.rser.2019.04.048]

[49] S.M. Frolov, "Organic waste gasification: A selective review", *Fuels,* vol. 2, no. 4, pp. 556-650, 2021.
[http://dx.doi.org/10.3390/fuels2040033]

[50] S. M. Santos, A. C. Assis, L. Gomes, C. Nobre, and P. Brito, "Waste gasification technologies: a brief overview," in Waste, 2022, vol. 1, no. 1: MDPI, pp. 140-165.
[http://dx.doi.org/10.3390/waste1010011]

[51] S.J. Suryawanshi, V.C. Shewale, R.S. Thakare, and R.B. Yarasu, "Parametric study of different biomass feedstocks used for gasification process of gasifier—a literature review", *Biomass Convers. Biorefin.,* vol. 13, no. 9, pp. 7689-7700, 2023.
[http://dx.doi.org/10.1007/s13399-021-01805-2]

[52] L. Ding, M. Yang, K. Dong, D.V.N. Vo, D. Hungwe, J. Ye, A. Ryzhkov, and K. Yoshikawa, "Mobile power generation system based on biomass gasification", *Int. J. Coal Sci. Technol.,* vol. 9, no. 1, p. 34, 2022.
[http://dx.doi.org/10.1007/s40789-022-00505-0]

[53] D. Panepinto, V. Tedesco, E. Brizio, and G. Genon, "Environmental performances and energy efficiency for MSW gasification treatment", *Waste Biomass Valoriz.,* vol. 6, no. 1, pp. 123-135, 2015.
[http://dx.doi.org/10.1007/s12649-014-9322-7]

[54] P.N.T. Pilapitiya, and A.S. Ratnayake, *The world of plastic waste: a review.* Clean. Mater, 2024, p. 100220.

[55] K. Yin, R. Zhang, M. Yan, L. Sun, Y. Ma, P. Cui, Z. Zhu, and Y. Wang, "Thermodynamic and economic analysis of a hydrogen production process from medical waste by plasma gasification", *Process Saf. Environ. Prot.,* vol. 178, pp. 8-17, 2023.
[http://dx.doi.org/10.1016/j.psep.2023.08.007]

[56] T.K. Patra, and P.N. Sheth, "Biomass gasification models for downdraft gasifier: A state-of-the-art review", *Renew. Sustain. Energy Rev.,* vol. 50, pp. 583-593, 2015.
[http://dx.doi.org/10.1016/j.rser.2015.05.012]

[57] M. Laiq Ur Rehman, A. Iqbal, C.C. Chang, W. Li, and M. Ju, "Anaerobic digestion", *Water Environ. Res.,* vol. 91, no. 10, pp. 1253-1271, 2019.
[http://dx.doi.org/10.1002/wer.1219] [PMID: 31529649]

[58] V. Kamperidou, and P. Terzopoulou, "Anaerobic digestion of lignocellulosic waste materials", *Sustainability (Basel),* vol. 13, no. 22, p. 12810, 2021.
[http://dx.doi.org/10.3390/su132212810]

[59] M.M. Uddin, and M.M. Wright, "Anaerobic digestion fundamentals, challenges, and technological advances", *Physical Sciences Reviews,* vol. 8, no. 9, pp. 2819-2837, 2023.
[http://dx.doi.org/10.1515/psr-2021-0068]

[60] T.H. Tsui, and J.W.C. Wong, "A critical review: emerging bioeconomy and waste-to-energy technologies for sustainable municipal solid waste management", *Waste Disposal & Sustainable Energy,* vol. 1, no. 3, pp. 151-167, 2019.
[http://dx.doi.org/10.1007/s42768-019-00013-z]

[61] S. Achinas, "An overview of the technological applicability of plasma gasification process," Contemporary environmental issues and challenges in era of climate change, pp. 261-275, 2020.
[http://dx.doi.org/10.1007/978-981-32-9595-7_15]

[62] İ. Yayalık, A. Koyun, and M. Akgün, "Gasification of municipal solid wastes in plasma arc medium", *Plasma Chem. Plasma Process.,* vol. 40, no. 6, pp. 1401-1416, 2020.
[http://dx.doi.org/10.1007/s11090-020-10105-y]

[63] E. Bevan, J. Fu, and Y. Zheng, "Challenges and opportunities of hydrothermal carbonisation in the UK; case study in Chirnside", *RSC Advances,* vol. 10, no. 52, pp. 31586-31610, 2020.
[http://dx.doi.org/10.1039/D0RA04607H] [PMID: 35520654]

[64] Y. Shen, "A review on hydrothermal carbonization of biomass and plastic wastes to energy products", *Biomass Bioenergy,* vol. 134, p. 105479, 2020.
[http://dx.doi.org/10.1016/j.biombioe.2020.105479]

[65] M. Ozturk, and I. Dincer, "An efficient waste management system with municipal solid waste incineration plant", *Greenh. Gases Sci. Technol.,* vol. 10, no. 4, pp. 855-864, 2020.
[http://dx.doi.org/10.1002/ghg.1955]

[66] A. Hagstroem, "Prospects for continued use and production of Swedish biogas in relation to current market transformations in public transport," ed, 2019.

[67] H. Abe, A. Katayama, B.P. Sah, T. Toriu, S. Samy, P. Pheach, M.A. Adams, and P.F. Grierson, "Potential for rural electrification based on biomass gasification in Cambodia", *Biomass Bioenergy,* vol. 31, no. 9, pp. 656-664, 2007.
[http://dx.doi.org/10.1016/j.biombioe.2007.06.023]

[68] P. Weiland, "Anaerobic waste digestion in Germany – Status and recent developments", *Biodegradation,* vol. 11, no. 6, pp. 415-421, 2000.
[http://dx.doi.org/10.1023/A:1011621520390] [PMID: 11587446]

[69] T.L.T. Nguyen, J.E. Hermansen, and L. Mogensen, "Environmental performance of crop residues as an energy source for electricity production: The case of wheat straw in Denmark", *Appl. Energy,* vol. 104, pp. 633-641, 2013.
[http://dx.doi.org/10.1016/j.apenergy.2012.11.057]

[70] M. Hardy, "An Assessment of the Global Warming Potential of Municipal Solid Waste Treatment Scenarios in the Netherlands," 2018.

CHAPTER 2

Valorizing Waste: The Role of Biorefineries and Bioproducts in the Circular Economy

Kamini Pandey[1], Shubham Kumar[1], Pushpanjali Singh[1], Shuchi Verma[2] and Barkha Singhal[1,*]

[1] *School of Biotechnology, Gautam Buddha University, Greater Noida, Uttar Pradesh-201312, India*

[2] *Department of Chemistry, Ramjas College, Delhi University-110007, India*

Abstract: In the rising quest for a sustainable future, the circular economy has emerged as a transformative model that prioritizes the continuous use of resources, minimizing waste and environmental impact. Central to this model are biorefineries, which offer an innovative solution for valorizing waste streams by converting them into valuable bioproducts and bioenergy. This chapter explores the pivotal role of biorefineries in the circular economy, examining their ability to transform agricultural residues, industrial by-products, and municipal waste into a diverse array of bio-based products, including biofuels, biochemicals, and biomaterials. The chapter opens with an overview of the circular economy framework, emphasizing the shortcomings of traditional linear production models and the environmental burden posed by escalating waste accumulation. It then delves into the principles and technologies underlying biorefineries, with particular attention to the integrated processes that enable the efficient conversion of biomass into a variety of high-value products. Special attention is given to selecting feedstocks, pre-treatment methods, and bioconversion techniques that maximize resource efficiency and product yield. Through case studies and real-world examples, the chapter demonstrates how biorefineries are actively contributing to the circular economy by closing resource loops, decreasing dependence on fossil fuels, and generating new economic opportunities. The discussion also addresses the scalability of biorefinery technologies and the financial, regulatory, and technological barriers that must be overcome to realize their full potential. By valorizing waste, biorefineries help to decouple economic growth from resource depletion, paving the way for a more sustainable and resilient industrial future.

Keywords: Biorefineries, Bio-based products, Bioconversion, Circular economy, Sustainable production, Waste valorisation.

---

* **Corresponding author Barkha Singhal:** School of Biotechnology, Gautam Buddha University, Greater Noida, Uttar Pradesh-201312, India; E-mails: arkha@gbu.ac.in and gupta.barkha@gmail.com

Harish Chandra Joshi, Anand Chauhan, Mikhail Vlaskin & Maulin P. Shah (Eds.)

INTRODUCTION

The world is confronting the severe consequences of massive waste generation resulting from a wide range of human activities. Growing population demands have intensified environmental challenges, including resource depletion, climate change, and rising pollution levels. Therefore, there is a dire need for urgent and innovative solutions to overcome these challenges, and it also requires shifting from a 'take-make-use-dispose' economic model to a circular economic model [1]. In the current context, the circular economy framework emerges as a transformative paradigm for promoting resource efficiency, minimizing waste, and fostering sustainability, while recovering value-added products from various waste streams [2]. Within this framework, biorefineries play a crucial role in transforming waste into various value-added bioproducts that can be utilized for multiple applications. Biorefineries are integrated facilities that convert waste biomass derived from plants, animals, and microbial resources into various products, including biofuels, bioplastics, and biochemicals [3]. These facilities utilize renewable biological resources to produce valuable bioproducts, paving the way for reduced reliance on fossil fuels and generating wealth from waste. Adopting biorefineries facilitates sustainable production, reduces the emission of greenhouse gases, and promotes economic independence by managing various biological wastes. The application of advanced technologies, such as fermentation, anaerobic digestion, and gasification, for optimizing biomass into high-value products highlights the crucial role of technology and innovation in achieving sustainable development. Biorefineries play a significant role in job creation and economic growth by transforming waste into valuable bioproducts [4]. They stimulate the emergence of new industries and markets, thereby bolstering local economies while simultaneously tackling global sustainability challenges. Therefore, it is essential to evaluate the role of biorefineries in valorizing waste. This chapter will therefore explore the diverse array of bioproducts generated from waste, with a focus on their applications across various sectors, including agriculture, packaging, and pharmaceuticals. The case studies demonstrating successful implementations, cutting-edge technologies, and the socio-economic advantages of integrating biorefineries in local and global contexts have been included. Ultimately, this investigation seeks to underscore the pivotal role of biorefineries and their associated bioproducts in advancing the circular economy. It emphasizes that waste should not be regarded merely as a burden to be managed, but as a valuable resource capable of enhancing sustainability and resilience in modern societies.

The Concept of Circular Economy

The circular economy is a sustainable production and consumption model that emphasizes reducing waste, reusing materials, and recycling resources to extend the lifecycle of products. Thus, the circular economy aims to recover resources through a regenerative use cycle that traditional waste management techniques would otherwise discard. It aims to tackle global challenges such as climate change, biodiversity loss, and pollution, and paves the way for a future that supports ecological balance, social equity, and economic prosperity. It replaces the linear economy that relies on "take, make, dispose of" (Fig. (**1**) through conceptual design, new technology approaches, scale-up, and commercialization, emphasizing and proposing waste design from the outset and maintaining a balance between progress and sustainability [5]. According to the UNECE (2024), waste valorization has played a crucial role in promoting the circular economy. This approach encourages a more sustainable and efficient use of resources by converting waste into secondary raw materials. According to a survey by the United Nations Industrial Development Organization (UNIDO, 2021), global trade in various types of waste experienced steady growth between 2002 and 2019, indicating the existence of an active and expanding market for these materials. In recent years, worldwide, it has been observed that there is a tremendous acceleration toward a circular economy. The market is predicted to reach USD 517.79 billion in 2025 and USD 798.3 billion by 2029, with a remarkable compound annual growth rate of 11.4%. This surge highlights not only increasing awareness but also strong policy support, technological innovation, and private-sector investment aimed at reducing waste and maximizing the production of valuable products [6].

The circular economy is based on three core pillars:

Creative process design: It focuses on redesigning the production system to reduce waste and pollution.

Circular Experimental Data: This involves experimental data or design that keeps resources in use longer through reuse, repair, and remanufacturing, thereby reducing the need for new raw materials.

Regenerating Modelling and analysis: The circular model aims to restore and enhance natural ecosystems, ensuring long-term environmental sustainability and resilience. Utilizing methods such as LCA and system dynamics enables data-driven decision-making, optimizes resource utilization, and facilitates the design of an efficient circular system.

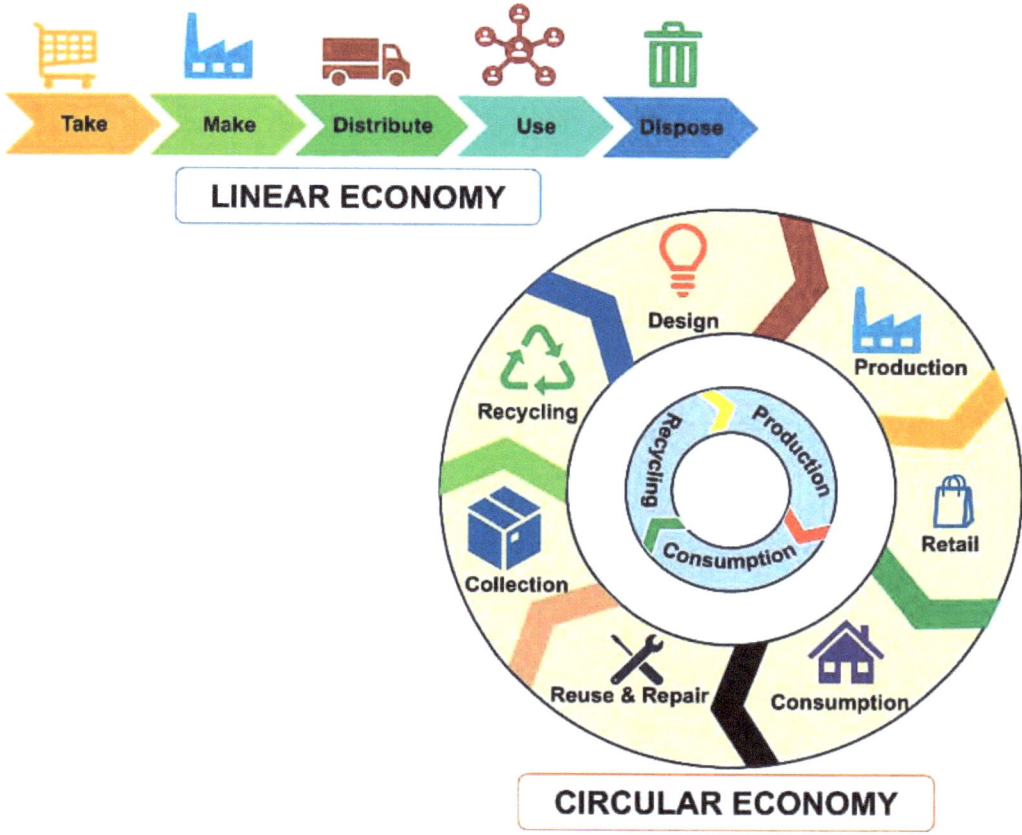

Fig. (1). Diagrammatic representation of linear and circular economy in waste valorization.

Benefits of the Circular Economy

The circular economy offers numerous advantages and potential benefits for various sectors of industry. According to the World Economic Forum (WEF, 2021), the circular economy is estimated to offer $4.5 trillion in economic potential through job creation, innovation, and waste reduction. It also has the potential to reduce greenhouse gas emissions by 79–99% and industrial waste in certain sectors by 80–99% (UNEP, 2021). Businesses can save money, increase process efficiency, and lessen their environmental impact by valuing industrial waste [7]. Furthermore, circular systems improve the quality of life in communities by reducing pollution and promoting more steady economic growth (European Commission, 2015) [8].

Challenges and Barriers to Implementing the Circular Economy

One of the main challenges is the lack of funding and resources for creative projects, such as product design, testing, and integration into technical and natural systems. The wide variety of waste types also poses a significant challenge for effective management. Each category of waste requires a distinct strategy for handling, treatment, transportation, and disposal. Addressing this complexity calls for expert technical knowledge and strict adherence to environmental laws. Efficient industrial waste management also requires substantial investments in infrastructure, specialized equipment, and training for the workforce. These expenses can be considerable and often present a financial hurdle for businesses [9].

Biorefineries: A Platform for Waste Valorization

Global urbanization has led to an increase in anthropogenic waste output and material and energy consumption. Due to the increasing global population and waste output, the world must adopt sustainable waste management practices to utilize its resources effectively. Few efforts have been made to address these issues, which are negatively impacting the environment and human well-being. International organizations have established several goals and guidelines to minimize the adverse impacts of the excessive consumption of non-renewable resources. Biorefineries play a vital role in converting biomass into valuable products, making them central to the implementation of a circular bioeconomy—a strategic approach to addressing global environmental and resource challenges. Achieving a green and low-carbon environment involves replacing petroleum with bioenergy and biomaterials as production feedstock while maintaining the energy-environment interaction. Despite a reduction in the use of fossil fuels for heat and electricity, the transportation sector remains heavily reliant on them. Another method for fostering a sustainable circular bioeconomy is by integrating waste into bioprocesses that produce valuable products and metabolites. Biorefineries are dynamic facilities that produce a range of beneficial products, including biofuels, bioplastics, biochemicals, and biomaterials, by transforming various biomass sources such as algae, forestry leftovers, and agricultural residues [10]. As a known renewable carbon source, biomass provides benefits such as carbon sequestration, bioenergy generation, and bioproduct manufacture [11]. In addition to providing sustainable alternatives to their fossil-based counterparts, the biorefining of the Organic Fraction of Municipal Solid Waste (OFMSW) into biofuels, bioenergy, and biomaterials also advances the circular economy, meets the Sustainable Development Goals (SDGs), and addresses waste management issues, especially in emerging and developing countries. Biowaste biorefineries are still in development and have not yet been widely implemented due to their technological constraints and low efficiency in reaction design. More precisely, microbial-based biorefineries face significant obstacles, including the need for substantial energy and a high degree of dependence on the efficiency and establishment of the microbes. Scientists and technologists categorize biorefineries based on their feedstock type. Energy crops, food crops, animal fats, edible oil seeds, and other materials are used in first-generation biorefineries; lignocellulosic biomass is used in second-generation biorefineries; and algae and other microorganisms are used in third- or fourth-generation biorefineries.

Types of Biorefineries

Biorefineries are categorized based on several factors, including the type of feedstock used, the conversion technologies employed, and the final products produced. Their diversity enables them to manage various waste streams and produce a range of bio-based products. Fig. (**2**) represents a diagrammatic overview of biorefineries.

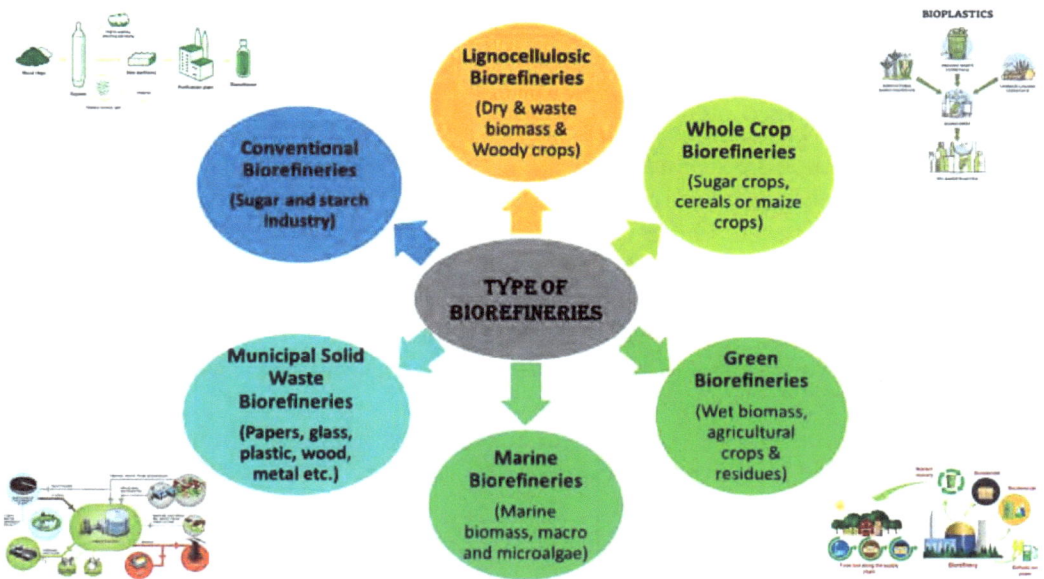

Fig. (2). Types of Biorefineries based on biowaste.

Lignocellulosic Biorefineries

As an extension of pulping refineries, lignocellulosic biorefineries are considered the most developed biorefineries in terms of value addition. A second-generation lignocellulosic biorefinery is built on the utilization of dry biomass from woody crops such as eucalyptus, eastern cottonwood, and maple, which are used as raw materials by most contemporary pulping industries and fiber-line, as well as various waste biomasses such as DDGS and bagasse, to produce bio-based fuels and chemicals through thermochemical conversion [12]. Several platforms for thermochemical conversion include rapid pyrolysis, gasification, torrefaction, and combustion. Lignocellulosic biorefineries produce bio-based products by combining process engineering with biotechnology instruments [13]. The biomass is heated to high temperatures (800-900°C) in the presence of little oxygen or air during the gasification process, which turns it into synthesis gas (a mixture of CO and H_2) [14]. Following cleaning, the gas is either used to generate electricity (as fuel for diesel generators) or used as a chemical platform to produce olefins, dimethyl ether (DME), and methanol (*via* Fischer-Tropsch synthesis). The standard methods for obtaining these bioproducts from biomass involve several valorization procedures that purify lignocellulosic biomass into more than 200 value-added compounds. Biorefineries convert the polysaccharides cellulose and hemicellulose into energy-based products, such as biofuels, bioethanol, biogas, and bio-oils, as well as non-energy-based food ingredients, including fibers and proteins, and animal feed.

Whole Crop Biorefineries

Whole crops are considered the first-generation feedstocks in the biorefinery industry. These include a variety of sugar crops, such as beets, sugarcane, and broomcorn, as well as a range of cereal grains, including maize, wheat, oats, barley, and potatoes. A whole-crop biorefinery is a technique that supports the production of food and bioenergy without competing for resources by utilizing whole crops as feedstocks to create enhanced market value, including industrial products, health benefits, and household appliances. It could address a shortage of fossil fuels in the future. Following their separation into fibers, sugars, and starches, the crops are either fermented, chemically treated, or subjected to thermochemical treatment. Oleochemistry applies oils and their derivatives to prepare valuable oleochemical products. The primary sources of oil with no feed applications are oils obtained from palm, algae, soy, coconut, and waste from food industries [15]. Oleochemicals are widely recognized for producing a range of bioproducts, including adhesives, household products, cosmetics, nutraceuticals, and industrial cleaning supplies. Compared to the harmful products of the petrochemical sector, the items created are both economically effective and environmentally beneficial. Green biorefineries are currently manufacturing these non-toxic products [16], with a primary focus on addressing the main challenge of valorizing waste oils and fats, which are hazardous to the environment. In biorefineries, extraction processes from the products of different crops have been employed as feedstocks to create various value-added goods from waste, including biobased items such as bioplastics, textiles, food, and pharmaceuticals.

Green Biorefineries

The sustainable conversion of green, renewable resources into a range of commercial goods is known as a "green biorefinery." Crops and their residues, known as wet biomass, such as grass, leaves, clover, *etc.*, are used in green biorefineries. These plants are abundant in essential nutrients, including proteins, carbohydrates, fats, and other substances such as vitamins, pigments, and minerals. In addition to leaf protein concentrate, the plant generates nutrient-rich residual pulp that can be utilized as fertilizer and biogas, as well as fiber pulp that can be used for livestock feed, biomaterials, or bioenergy [17]. Different grasses, in both fresh and waste forms, are the most commonly used feedstocks for Green Biorefining. These biorefineries are particularly beneficial in rural regions with abundant green biomass. They promote regional agriculture while assisting in the valorization of fresh plant matter into various bio-based goods. Comparing the green biorefinery to the lignocellulosic biorefinery, the analysis shows comparable mean performance and importance values. At the same time, we observe a trend toward lower performance ratings for economic and policy elements that are more significant. Green biorefineries produce low-cost feed bioproducts, increasing the competitiveness of biorefinery products against imported soybean feed. They can also define the economic dimensions of the demand as low-value products. Green Biorefinery fibers can be used as a raw material for end products such as animal feed, horticulture products, packaging, pulp, paper, and building materials [18].

Marine Biorefineries

Most of the fish waste produced by the seafood business is dumped into the ocean. Effective waste valorization enhances the industry's production, improves marine resource management, and reduces ocean contamination. Industrial facilities that utilize marine resources to produce bio-based products and biofuels are referred to as marine biorefineries. They can help mitigate the effects of climate change and address the shortages of food, energy, and water, as well as promote sustainable agriculture. Marine biorefineries use various marine resources, including seawater, marine biomass, marine microorganisms, macroalgae, and microalgae. Marine biorefineries can also support the fishing and aquaculture industries by managing the substantial volumes of waste they generate. Marine biorefineries employ various biochemical techniques, including fermentation, anaerobic digestion, and enzymatic hydrolysis, to transform marine biomass into biofuels, biochemicals, and nutraceuticals. Due to their high protein and amino acid profiles, microalgae are often used as a substitute for fishmeal [19]. In this study, algae biorefineries can be considered the most recent biorefinery technology. Additionally, the economic factors for commercialization reveal the most significant disparity between its recognized importance and experimental performance. The high costs of feedstock cultivation have been identified as the biggest problem in the economic sector; even higher investment in biorefineries is the actual issue. Concerns about the sustainability of cultivation technologies, particularly in relation to societal and environmental issues, have long been highlighted [20].

Municipal Solid Waste (MSW) Biorefineries

Municipal Solid Waste (MSW) is produced in vast quantities worldwide, making the sustainable processing and disposal of this waste difficult and expensive to handle. However, considering its energy and nutrient content opens up numerous possibilities for use as a cost-reducing, plentiful, and sustainable source for renewable energy production, such as biofuels (*e.g.*, syngas), compost, synthetic fuels, chemicals, and value-added products through waste biorefinery-based methodologies. The United States produced around 292.4 million tons of MSW in 2018, from which approximately 50% of the waste was valorized through recycling and composting. Still, approximately 146 million tons of waste were disposed of in landfills [21]. Environment, nature, human health, and safety are the primary concerns and are constantly under significant stress due to landfills and improper waste disposal methods. Numerous techniques, such as thermochemical conversion (gasification, pyrolysis) and biological conversion (anaerobic digestion), can rapidly transform MSW into valuable energy [22].

Technologies for Waste Valorization in Biorefineries

Anaerobic Digestion

Anaerobic digestion is a significant biological technology for waste valorization, offering a cost-effective and straightforward method for converting agricultural waste into biogas and digestate, thereby reducing greenhouse gas emissions and generating renewable energy without requiring external power. It involves the breakdown of organic materials into biogas, a process that occurs without oxygen, facilitated by a consortium of microorganisms [23]. Various organic wastes can be utilized as fuel for anaerobic digestion. Examples of carbon or protein-rich feedstocks include energy crops, agricultural residues, manure, home waste, food market waste, and sewage sludge. Anaerobic digestion produces biogas and digestate. Biogas can be utilized to generate heat and power or to manufacture biomethane. Digestate contains the nutrients in feedstock, particularly nitrogen and phosphorus, and can be used as crop fertilizer or for soil conditioning [24].

Composting

Composting is an age-old biological method that uses aerobic breakdown of organic waste to produce nutrient-rich compost. It supports beneficial microbes for plant growth and utilizes feedstocks such as kitchen waste, yard waste, farm waste, and industrial waste. Hyperthermophilic composting enhances this process by utilizing high temperatures and specific bacteria to accelerate the breakdown of lignocellulose, thereby improving compost quality and efficiency [25]. Modern polymer composites, such as wood-plastic and mycelium-based composites, utilize synthetic and organic waste elements to provide sustainable alternatives to conventional materials. For instance, mycelium-based composites develop environmentally beneficial materials for packaging and insulation purposes by combining organic waste, such as agricultural waste residue or coffee grounds, with fungal mycelium [26].

Fermentation

The anaerobic process of fermentation breaks down glucose into organic molecules. Sugar undergoes several chemical reactions to produce acid or alcohol. The biomass material is supplemented with bacteria or yeast, which consume the sugar to produce carbon dioxide and ethanol. Corn and sugarcane, the most common agricultural feedstocks, are purposefully converted into industrial ethanol; however, some recently advanced techniques have been developed that utilize lignocellulosic waste materials as feedstock, such as sugarcane bagasse, corn stover, straw, and paper mill wastes. The biomass can be fermented into valuable products such as lactic acid, ethanol, acetic acid, dihydroxyacetone, succinic acid, and 1,2-propanediol [27]. There are two primary fermentation pathways: homofermentation and heterofermentation. The most common products of homofermentation are lactic acid, acetic acid, and ethanol [28]. An overview of the contribution of technologies for biorefineries has been represented in Fig. (**3**).

Thermochemical Waste Valorization Technologies

Pyrolysis

Pyrolysis is a thermochemical treatment process that produces valuable compounds by decomposing organic waste, plastic waste, and biomass, such as syngas, biogas, and biochar, in the absence of oxygen at temperatures typically between 300°C and 700°C. There are two primary types of the process: slow pyrolysis and quick pyrolysis. Slow pyrolysis produces primarily biochar over several hours at approximately 300°C. On the other hand, fast pyrolysis produces a larger yield of bio-oil by heating the material quickly to approximately 500°C within a period of less than one second [29].

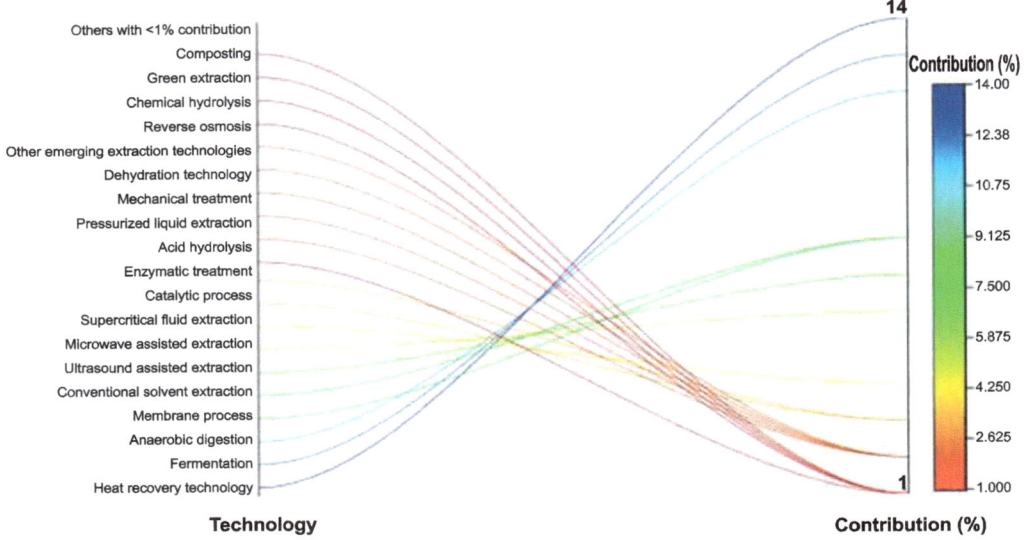

Fig. (3). An overview of the contribution of various technologies for the bioconversion of biowaste in biorefineries.

Moreover, pyrolysis is highly beneficial for converting waste plastic into smaller hydrocarbons, producing char and pyrolysis oil as byproducts. To recover monomers such as ethylene and propylene from plastic waste, pyrolysis has been refined and optimized. This allows for the creation of new polymers with a minimal amount of virgin material input. Recent developments have led to the widespread adoption of catalysts, such as zeolites (*e.g.*, H-ZSM-11), to enhance pyrolysis efficiency and improve the quality of the end products. Reaction temperatures are lowered, and yields of particular hydrocarbons, including olefins and aromatics, are increased by catalytic pyrolysis [30].

Gasification

In integrated biorefineries, gasification converts solid or liquid biomass into char and volatile gases at high temperatures. The volatiles, mainly CO and H_2, can be utilized for power generation and chemical production through fuel cells, engines, or boilers. Co-gasification is a significant technological advancement that reduces tar formation and improves process efficiency by combining biomass and fuels derived from waste. For example, the co-gasification of pine and refuse-derived fuel biomass increased the gas heating value from 5.8 to 6.4 MJ/Nm^3 and raised methane and ethylene yields by 78.2% [31]. Another key advancement is plasma gasification, a high-temperature waste-to-energy technology that converts waste into syngas with exceptional efficiency. It achieves one of the highest energy outputs at 816 kWh per tonne of waste, while producing minimal emissions and solid residue. Although it requires significant initial investment and energy input, its scalability of up to 1,000 MW makes it a promising solution for large-scale waste management [32].

Hydrothermal Liquefaction

Hydrothermal liquefaction (HTL) is a thermal depolymerization technology that converts wet biomass or solid waste into an energy-dense bio-crude. Current efforts are focused on enhancing its applicability for converting damp food waste and wet waste-activated sludge [33]. The HTL conversion efficiency is typically enhanced when using solvents and heterogeneous catalysts. Combining these conditions, solvent and catalyst loading further improve the process yield [34]. A recent addition to this paradigm is Fast Low-temperature Advanced Steam Hydropyrolysis Hydrothermal Liquefaction (FLASH-HTL), which reduces reactor fouling through the use of steam-based heat recovery and is a more effective method of converting waste into usable fuels, particularly when paired with advanced catalysts [35].

Microwave-assisted pyrolysis

Microwave-assisted pyrolysis is a novel approach that utilizes microwave energy in a traditional process. It differs significantly from classical thermal pyrolysis because microwave pyrolysis employs electromagnetic energy rather than conductive or convective heat transfer. Biomass conversion to syngas, biochar, and bio-oil [36] *via* microwave-assisted pyrolysis has garnered significant research interest. The function of MAP in turning plastic waste into valuable goods, including liquid oils, gases, and carbon compounds, has been highlighted by recent studies [37].

Combustion

Combustion is the process of burning biomass in the presence of air. The chemical energy of the biomass is converted into heat, mechanical power, or electricity. The direct burning of agro-industrial waste materials produces geomaterials as waste. Electric furnaces, boilers, or turbines are commonly utilized for combustion [38]. Burning biomass in boilers generates high-pressure steam that powers a turbine connected to a generator, enabling the production of electricity. It was discovered that, for biomass combustion, the conversion efficiency ranged from 20% to 40%, with the system exceeding 100 MWe.

Physiochemical Waste Valorization Technologies

Ultra-assisted Extraction

Ultrasound techniques using solvent waves ranging from 10 kHz to 20 MHz can disrupt the structure and constituent parts of lignocellulosic biomass. Recent studies have shown that ultrasound-assisted methods can effectively enhance lignin extraction by disrupting the dense lignocellulosic matrix, thereby allowing for easier separation and conversion of lignin into valuable products. The UAE has previously been shown to improve cellulose and hemicellulose hydrolysis rates by exposing a larger surface area, which helps convert them into fermentable sugars, thus facilitating bioethanol production and other bio-based materials [39]. Acoustic cavitation, used in this technology, creates micro-bubbles or cavities in fluids, potentially improving the efficiency of the biomass valorization process while having a chemical and physical impact. Furthermore, the duration of the ultrasound pretreatment, which ranges from 0 to 150 seconds, directly correlates with the improved accessibility of cellulose [40].

Supercritical Fluid Extraction

The extraction process in SFE technology occurs above the critical temperature and pressure points of the extractive solvents, where liquid and gaseous phases coexist. SFE technology is well-known for its effectiveness, minimal solvent use, and rapid extraction periods compared to conventional extraction methods. Higher yields are produced when essential or unconventional oils are extracted from fruit and vegetable industry waste using critical fluid extraction techniques, as opposed to traditional extraction methods. The SFE technique extracts lipids, biorepellents, and phenolic compounds from tropical biomass [41].

Hydrothermal Carbonization

HTC is a thermochemical process that occurs in subcritical water under mild parameters. The reaction temperature typically ranges from 180°C to 250°C, while the pressure ranges from 10 to 50 bar. In these cases, the biomass is broken down by various chemical reactions, including hydrolysis, polymerization, dehydration, and aromatization, which occur when liquid water acts as both a reactant and a solvent or reaction medium. The end product is a hydrochar that has applications in the agriculture sector (*e.g.*, as a solid amendment), energy generation (*e.g.*, co-combustion with conventional fossil fuels [42]), adsorption, and as a foundation for advanced carbon materials [43].

Chemical Processing Waste Valorization Technologies

Transesterification

Glycerol and esters are produced when oil or fat combines with an alcohol in a process known as transesterification. Biodiesel is made from biomass using a catalyst, such as methanol or ethanol, and sodium or potassium hydroxide [44]. Diesel engines and gasoline vehicles can be powered by biodiesel. It is produced by transesterifying or esterifying animal fats, oils, and other similar feedstocks. This technique yields glycerol as a byproduct, which is used in cosmetics, pharmaceuticals, and other sectors. The utilization of different types of catalysts made from waste biomass-including eggshells, banana peels, and other lignocellulosic materials-is a noteworthy application of transesterification. These catalysts are attractive for industrial-scale applications because they have been demonstrated to produce high yields of biodiesel [45].

Acid and Base treatment

Acid and base treatments have been utilized in biomass waste valorization research to convert waste into valuable goods. When lignocellulosic materials are broken down by acid hydrolysis, sugars can be extracted for the generation of biofuel or other bioproducts. For example, cardboard waste has been converted into fermentable sugars through a two-step acid hydrolysis process, which shows promise for producing bioethanol or other compounds [46]. Similarly, base treatment, often in conjunction with alkali hydroxides, enhances biomass digestibility and releases organic components for further processing, which facilitates the breakdown of complex organic wastes, such as food or agricultural residues [47].

Electrochemical Conversion

Electrochemical conversion of biomass utilizes electrochemical cells to transform biomass waste into valuable chemicals and energy-rich products. It offers a cleaner alternative to traditional thermal or chemical processes that often require high temperatures and harsh chemicals. In an electrochemical setup, renewable electricity (often from solar or wind sources) drives reactions that decompose the compound in biomass [48]. The main component of this technique is the application of specific electrocatalysts designed for biomass constituents. Advances in catalyst design focus on improving efficiency, selectivity, and stability, especially when processing complex biomass-derived compounds rich in functional groups [49]. In biomass valorization, the oxidation reaction can convert biomass derivatives into valuable chemicals, such as carboxylic acids, which can be used to produce pharmaceuticals, food additives, or polymers.

Emerging Integrated Technologies for Waste Valorization in Biorefinery

A novel approach in biorefineries involves integrating various technologies based on waste characteristics to maximize conversion efficiency. The integrated thermo-biochemical process combines methods such as Anaerobic Digestion (AD), pyrolysis, and hydrothermal liquefaction (HTL) to enhance energy recovery. Anaerobic digestion and thermal processes have been used in numerous procedures to enhance the energy recovery of biomass [50]. Studies have reported that bio-oil production from biomass pre-treatment before pyrolysis significantly improves the quality of pyrolytic oils and biochar [51]. The Anaerobic Digestion process, combined with pyrolysis of raw food waste, yields 60.3 percent bio-oil and 7.4 percent biogas [52]. Combining Anaerobic Digestion (AD) and hydrothermal liquefaction (HTL) techniques can enhance the recovery of liquid fuels from biomass waste. Combining these two approaches yielded high-quality diesel fuel with a heating value of 43.1 MJ kg^{-1} and an energy recovery of 62-84% by the combined plant [53]. Table **1** summarizes all types of technology and the products from their feedstocks.

Table 1. Technologies used for waste valorization in biorefineries.

S. No.	Technology	Feedstock	Products	References
1.	Anaerobic digestion	Food waste / Agricultural waste, Industrial waste, Vegetable, crops and fruit waste	Biogas; CH_4 called biogas and digested slurry as manure	[54]
2.	Extraction	Food waste	Food additives, therapeutics, cosmetics	[55]
3.		Electronic waste	Lead recovery	[56]
4.	Fermentation	Food waste	Biofuels, organic acids, enzymes, single-cell protein, Bioethanol and other valuable chemicals.	[57]
5.		Agricultural waste	Biofuels, biogas, organic acids	[58]
6.		Textile waste	Biofuels, cellulase, succinate	[59]
7.		Sugarcane bagasse	Gluconic acid, Xylooligosaccharides	[60]
9.	Gasification	Agricultural waste	Gas, char	[61]
10.	Gasification	Plastic waste	Biofuels	[62]
11.	Hydrothermal liquefaction	Agricultural waste	Biofuels	[63]

(Table 1) cont.....

S. No.	Technology	Feedstock	Products	References
12.	Pyrolysis	Food waste	Biofuels	[64]
13.		Agricultural waste	Biofuels	[65]
14.		Textile waste	Activated carbon fibre, bio-oil, char, syngas	[66]
16.	Transesterification	Waste oil	Biodiesel	[67]
17.	Torrefaction	Food waste / Agricultural waste	Biocoal, gas	[68]
18.	Combustion	Black Liquor	Oxidized products, often gaseous products, Smoked	[69]
19.	Enzymatic hydrolysis	Sawdust	Bioethanol	[70]
20.	Saccharification and Solid-state fermentation	Corn stover	Bioethanol	[71]
21.	Solid-state fermentation	Grape pomace	Bioethanol	[72]
22.	Fed batch fermentation	Corn cob molasses	2,3-Butanediol	[73]
23.	Microwave-assisted extraction	Food waste	Phenolic compounds	[74]
24.	Ultrasound assisted extraction	Agro-industrial waste	Proteins	[75]
25.	Supercritical fluid extraction	Dairy, fish and meat-products waste, Industrial waste	Oil and extracts	[76]
26.	Conventional solvent extraction	Dairy, fish and meat-products waste	Bio energy and Bio oil.	[77]

The next part examines the various bioproduct categories that link waste valorization technologies with their respective outputs, providing a deeper understanding of how this technology efficiently transforms valuable bioproducts from waste. Every technology section describes the intermediate or end products that are produced from the processing of specific waste types. All high-value bioproducts, including fertilizers, platform chemicals, biofuels, and bioplastics, are then covered in the upcoming part of the chapter, with a focus on how these goods support the objectives of the circular economy.

Types of Bioproducts from Waste Valorization

With a growing population, increasing consumption, and expanding industry, waste management is one of the significant problems in the modern world. The World Bank estimates worldwide waste production will be 2.24 billion tons in 2020 (0.79 kg per inhabitant per day), potentially increasing to 3.88 billion tons by 2050. Fig. (**4**) illustrates the types of biowaste generated worldwide. Much waste is buried in dump yards and water bodies, and then treated chemically or burned. These poor methods can lead to the release of hazardous chemicals into the environment, harming ecosystems and living organisms. Now, the scientific community and industries are realizing the profitability of waste, shifting their focus to a circular economy from a linear one. The circular economy is a concept that aims to produce valuable products from waste and reintroduce them into the economic cycle. Biowaste is produced when biological items are processed in industries. Industries such as agriculture, food processing, and animal husbandry generate significant amounts of biowaste, including husks, peels, pomace, feathers, skin, shells, and hair. This biowaste can serve as low-cost raw materials for producing bioproducts such as bioplastics, biofuels, organic acids, pigments, and biomaterials Fig. **5**).

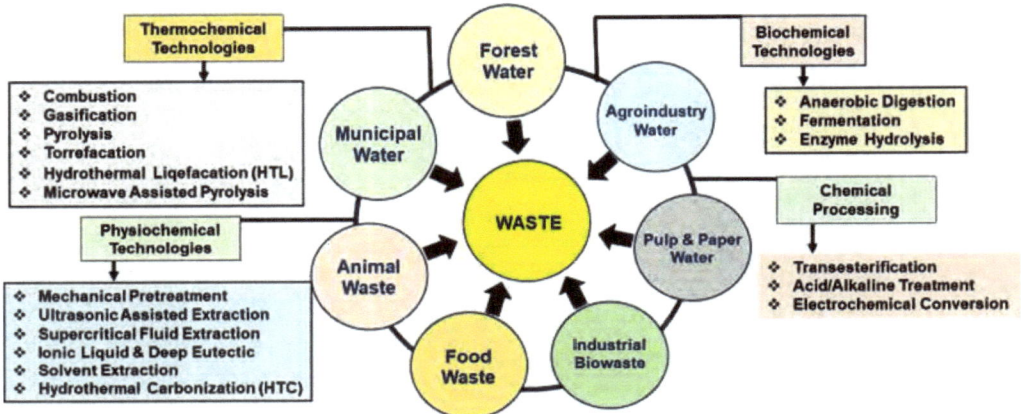

Fig. (4). Types of biowaste used in biorefineries.

Biofuel- Bioethanol and Biodiesel

Bioethanol

It is commonly known that biofuel may save energy by lowering fossil fuel requirements. The two largest producers of bioethanol fuel are the United States and Brazil, accounting for over 62% of the global output. Fermentation accounts for over 60% of the bioethanol produced globally. Liquid biofuels that can be created are the most common ones, such as bioethanol [78]. A significant biofuel made by valorizing sugars, starches, and lignocellulosic biomass is bioethanol. Since cellulose, a type of sugar, and starch are glucose polymers, producing bioethanol from lignocellulosic biomass is in demand [79]. Therefore, lignocellulosic bioethanol production can provide energy, particularly in nations that employ agricultural and forestry wastes as inputs. *Saccharomyces cerevisiae* was used to digest vegetable waste and make bioethanol. In this investigation, potato, carrot, and onion peels were employed.

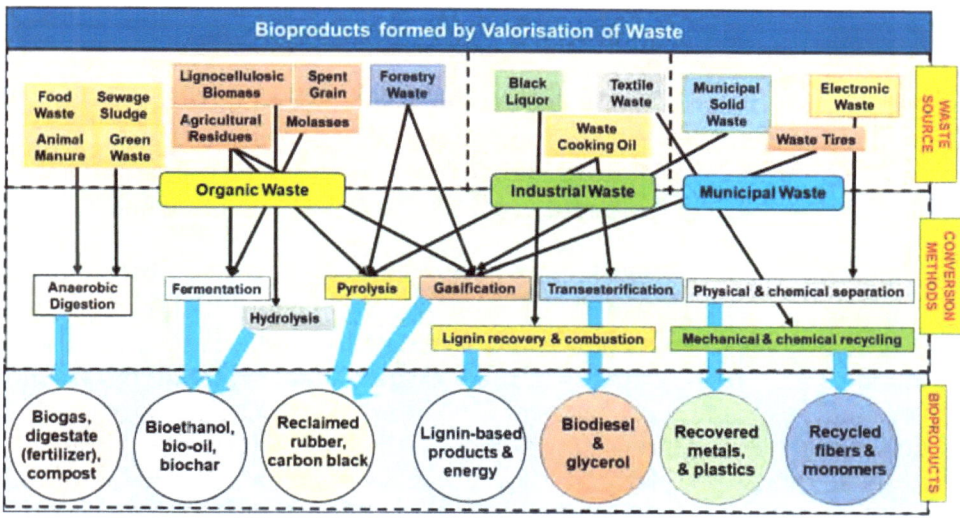

Fig. (5). An overview of types of bioproducts produced in biorefineries.

Fermentation is the most commonly used method for producing bioethanol. There are three steps in the process of producing bioethanol. First, a pre-treatment phase is necessary to make the hydrolytic enzymes readily available for conversion into sugars. Second, cellulose, hemicellulose, and starch undergo enzymatic hydrolysis. Thirdly, microbial fermentation transforms carbohydrates into ethanol. Food waste has the advantages of being readily accessible, inexpensive, and having a recyclable substrate when used as a raw material. Developments in enzymatic hydrolysis, microbial fermentation, and related immobilization resulted in the commercialization of "second-generation bioethanol." An alternate feedstock for the synthesis of biofuel is microalgal biomass. Algal biofuels may have a more significant energy potential than the current generation of fossil fuels due to their faster growth and higher biomass output [80]. The third-generation biofuels, which are emerging technologies, employ sustainable feedstocks such as algae, crop leftovers, or municipal solid waste to ferment sugars and produce biodiesel. The bioethanol production method aims to transform carbohydrate-based waste into a fuel that can be transported and stored, while recovering some of the energy used in the proposed process.

Biodiesel

Fatty acid alkyl esters comprise the chemical composition of biodiesel, a promising alternative to traditional petroleum-based fuels for the future. This biofuel may be contributing to the global energy issue due to its various benefits, including high efficiency in combustion engines, low greenhouse gas emissions, low toxicity, and biodegradability. Four traditional techniques are often used to produce biodiesel: transesterification, thermal cracking/pyrolysis, microemulsion, and direct mixing or dilution [81]. The choice of feedstock has a significant impact on several factors, including chemical composition, yield, and ultimate product quality. Edible plant oils, including palm, coconut, canola, rapeseed, peanut, olive, mustard, sunflower, and soybean, are widely utilized in biodiesel production. These oils are commonly used due to their low Free Fatty Acid (FFA) content, low water content, and ease of biodiesel conversion [82]. Biodiesel is gaining popularity as a diesel engine fuel due to its renewable nature and environmental benefits. However, the rapid development in biodiesel production has raised significant concerns about waste management due to the generation of large amounts of crude glycerol. Crude glycerol, an essential by-product of biodiesel synthesis, can be further processed to produce valuable compounds such as glycerin, propylene glycol, and polyols [83]. Catalytic transesterification is now one of the most popular and often utilized procedures due to its simplicity and the capacity to use a wide range of lipid sources and catalysts, resulting in high conversion and biodiesel yields at moderate reaction conditions. To produce biodiesel, various innovative synthesis techniques have been developed, including microwave transesterification, *in situ* transesterification, ultrasound-assisted transesterification, membrane transesterification, reactive distillation, and supercritical fluid technology (a non-catalytic process) [84]. Biodiesel is widely used in the industrial sector, particularly in power plants, agriculture, machinery, and water heating. Table **2** presents the biomass waste used in various methods for biodiesel production.

Table 2. Different types of biomass used for biodiesel production.

S. No.	Waste Biomass Type	Method of Biodiesel Production	Significance/Benefits	References
1.	Used Cooking Oil	Transesterification	Abundant waste material reduces environmental waste and greenhouse gas emissions.	[85]
2.	Animal Fat	Esterification and Transesterification	Provides a high-energy alternative to fossil fuels, utilizing waste from the meat industry.	[86]
3.	Waste Vegetable Oil	Enzymatic Transesterification	Low-cost feedstock reduces disposal issues and contributes to a circular economy.	[87]
4.	Sewage Sludge	Lipid Extraction and Transesterification	Utilizes municipal waste; prevents land and water pollution	[88]

(Table 2) cont.....

S. No.	Waste Biomass Type	Method of Biodiesel Production	Significance/Benefits	References
5.	Algae Biomass	Microbial Lipid Accumulation	Fast-growing and renewable; high lipid yield	[89]
6.	Waste Grease	Acid and Base Catalysis	Converts urban waste into energy; cost-effective and efficient	
7.	Agricultural Residues	Pyrolysis and Transesterification	Reduces agricultural waste; a source of renewable energy	[90]

Bioplastics

One type of bio-based material is bioplastic, which was developed primarily to address the need for society to reduce plastic pollution caused by the manufacture of plastics using a significant amount of non-renewable resources. The circular economy has become increasingly investigated as a potential solution to the shortage of raw materials for creating goods that might eventually replace traditional plastics. Bioplastics appear to be a good substitute because of the estimated properties of the finished goods. The process of creating bioplastics can be either biological or chemical. For example, biodegradable aliphatic polyesters are the most widely used substance for making bioplastics. Lactic acid is fermented from various renewable resources to develop biodegradable and renewable aliphatic polyesters [91]. As the biodegradable industries have grown, these materials have garnered much attention as one of the most promising alternatives to petrochemical-based plastics. Petroleum-based polymers are replaced by fermenting microorganisms to create polymer molecules [92]. Renewable resources, including food waste, lignocellulose residues, and vegetable oils, are utilized to create bioplastics such as polylactic acid and polyhydroxyalkanoates (Table **3**) [93].

Table 3. Different types of biowaste used for the production of bioplastics

S. No.	Type of Waste	Bioplastic Produced	Advantages	Disadvantages	References
1.	Agricultural Waste	Polylactic Acid (PLA), PHAs	Abundant and renewable source	Requires significant preprocessing and treatment	[94]
2.	Food Waste	PLA, Polyhydroxyalkanoates	Reduces landfill waste	High variability in composition	[95]
3.	Food Waste	(PHAs)	Lowers greenhouse gas emissions	Collection and preprocessing can be expensive	[95]

(Table 3) cont.....

S. No.	Type of Waste	Bioplastic Produced	Advantages	Disadvantages	References
4.	Forestry Waste	Cellulose-based plastics	Utilizes non-food waste	Processing is energy-intensive	[96]
5.	Municipal Solid Waste (MSW)	PLA, PHAs	Reduces the burden on landfills. Can integrate with existing waste systems	Complex separation and treatment processes are required High contamination risk	[97]
6.	Wastewater Sludge	Polyhydroxybutyrate (PHB)	Utilizes waste from wastewater treatment	Variable composition and treatment costs	[98]
7.	Sugarcane Bagasse	PLA, Bio-PET	Renewable and widely available byproduct of the sugar industry Reduces carbon footprint	Seasonal availability Competes with other bagasse applications (*e.g.*, bioenergy)	[99]

Biochemicals

These organic acids and chemicals play essential roles in various industries, including cosmetics, healthcare, the food industry, and pharmaceuticals. Biochemicals can be produced through fermentation, where biowaste can serve as a substrate. Biochemicals are widely distributed compounds that microorganisms can produce. Propionic, itaconic, and fumaric acids are bio-based compounds that represent emerging parts of bioprocessing and bioeconomy [100]. Fermentation of carbohydrates is crucial for producing these compounds; both wild and genetically modified microorganisms are used to synthesize a wide range of biochemicals. Numerous anaerobic microbes frequently use fermentation to produce biochemicals such as butyric acid, propionic acid, lactic acid, citric acid, acetic acid, and succinic acid. Many fungi can produce organic acids efficiently in high concentrations. It is commonly known that filamentous fungi and yeast may generate organic acids [101].

Succinic Acid

Succinic acid is a byproduct of anaerobic fermentation and an intermediate product of the Krebs cycle. The pharmaceutical and food industries are the primary domains where succinic acid is utilized. Still, it significantly contributes to the production of bio-based plasticizers, coatings, resins, lubricants, cosmetics, and personal care products. Additionally, succinic acid is crucial in the production of chemicals such as tetrahydrofuran, γ-butyrolactone, and 1,4-butanediol. The synthesis of succinic acid has been studied in numerous microorganisms, including various species of fungi (*Lentinus degener*, *Byssochlamys nivea*, *Penicillium vinifera*, and *Paecilomyces varius*), as well as yeast (*Saccharomyces cerevisiae*) [102] and many Aspergillus species. Additionally, some Gram-positive bacteria, such as *Enterococcus faecalis* and *Corynebacterium glutamicum*, have been investigated. *Actinobacillus succinogenes* [103] *and Anaerobiospirillum succiniciproducens* [104] are the most widely used microbes. A 2020 study suggests that using organic waste as feedstock can significantly reduce the production cost of succinic acid. This study utilized hydrolysates of the Organic Fraction of Municipal Solid Waste (OFMSW) for succinic acid production. Two bacteria were evaluated: *Actinobacillus succinogenes* and *Basfia succiniciproducens*. *Actinobacillus succinogenes* was found to be more efficient, with a yield of 0.56 g/g and productivity of 0.89 g/L/h, in batch fermentation. Additionally, it achieved a yield of 0.47 g/g, with a productivity of 1.27 g/L/h, in continuous fermentation [105].

Lactic acid

Lactic acid is known for its applications in food industries due to its acidic, emulsifying, and antibacterial properties. Still, lactic acid is also beneficial in other industries, such as paper, textiles, and detergents. Lactic acid is also being utilized to form acrylic acid and biodegradable plastic polylactic acid (PLA) [106]. Lactic acid is also used to produce medicines and beauty products. Lactic acid is an integral and essential part of living systems. It is a product of anaerobic respiration. Lactic acid can be produced by various microorganisms, including *E. coli*, Lactic Acid Bacteria (LAB), and *Bacillus* strains. LAB are widely used commercial microorganisms. These include species of *Lactobacillus*, such as *L. delbrueckii* spp., *L. casei*, and *L. helveticus*. These bacteria are considered safe for use in the workplace, as they do not pose a risk to human health [107]. Lactic acid production has been observed in a study using bagasse from banana peduncles, carob, and sugarcane as agricultural waste. Sugarcane and carob residues were found to be the most suitable feedstocks for lactic acid production [108]. Different biowastes from agro-food industries have been utilized to produce lactic acid, with productivity reaching up to 1.12 g/L/h [109].

Furfural

Furfural is an organic compound that mainly comes from hemicellulose and cellulose. Furfural works as a precursor for the sustainable production of several value-added products, including biofuels. The commercial importance of furfural was first noticed by researchers in 1921 [110]. This organic molecule has been recognized as an essential platform chemical reported for producing many commercial products, including petrochemicals. More than 80 chemicals derived from furfural, such as maleic acid, succinic acid, and 2-methyl furan, are useful industrial products [111]. Physical and chemical methods, such as pyrolysis and acid hydrolysis, can convert lignocellulose-based biowaste into furfural. It has been reported that the furfural yield can be increased by up to 74% in the presence of porous solid acid catalysts, such as polytriphenylamine-SO_3H [112].

Levulinic acid

Levulinic acid is a yellow brow crystalline compound. It is a 5-carbon-containing short-chain fatty acid known as gamma-ketovaleric acid or 4-oxopentanoic acid. It has crucial applications in various fields, including ink, textiles, antifreeze, animal feed, solvent, resin, cosmetics, synthetic rubber, food flavoring, pharmaceutical compounds, and coating materials. The production of levulinic acid has been reported from various agro-industrial waste sources, including starch industry waste, brewery liquid waste, pomace solid waste, and brewery spent grain [113]. Additionally, household waste, such as spent coffee grounds and cooked tea leaves, has also been investigated for levulinic acid production [114].

Nutraceuticals and Pharmaceuticals

With the increasing population, waste production is also increasing exponentially, while essential items for living, such as food, nutrients, and medicine, are becoming increasingly expensive. Renewable resources can be a boon in such scenarios, where valuable products like bioactive compounds can be produced using zero-cost substrates, such as biowaste.

Omega-3 fatty acids

Omega-3 fatty acids show a crucial impact in a wide range of conditions such as infant development, cancer treatment, neurological diseases, heart diseases, Attention Deficit Hyperactivity Disorder (ADHD), Alzheimer's disease, depression, cognitive functioning, muscular degeneration, rheumatoid arthritis, and eye diseases [115]. Docosahexaenoic acid (DHA) and eicosapentaenoic acid (EPA) are well-known and extensively studied omega-3 fatty acids. Valorization of biowastes, such as seed residues and fish waste, can be utilized for the production of omega-3 fatty acids. Olive pomace can also serve as an excellent feedstock for obtaining omega-3 fatty acids [116].

Vitamins and antioxidants

Vitamins and antioxidants are bioactive compounds that actively participate in our physiological reactions. Vitamins are the organic molecules that our body needs in small amounts to function and grow normally. Animal, food, agro-industrial, and food waste can be an efficient and sustainable source for producing vitamins and antioxidants to meet our nutritional requirements. Epicatechin, an antioxidant, has recently been isolated from biowaste, along with vitamins such as vitamin C, B12, and B9 [117].

Economic, Environmental, and Social Impacts of Waste Valorization

Economic Impacts

The concept of waste valorization offers an opportunity to gain economic benefits in several areas. Parties responsible for waste management can save costs associated with its disposal or pre-treatment, depending on the properties of the waste material. Additionally, municipalities and businesses that implement valorization can also save on these costs. In particular, selling products obtained from waste valorization holds enormous potential benefits because it can partly cover the costs. Specifically, within the scope of plastics valorization, waste valorization is viewed as a crucial tool for achieving the circular economy [118]. Plastic valorization can close the loop and contribute to end-product sustainability in a sector characterized by high polymer demand. By doing so, valorization can make economies more sustainable. Moreover, valorizing plastic waste can create favorable market opportunities for countries that are primarily dependent on imported raw materials and, therefore, vulnerable to the volatility of raw material prices [119]. The entire attention has shifted to the issue of food waste, with the predicted annual cost of food loss and waste in industrialized and developing nations being $680 billion and $310 billion, respectively. A suitable valorization strategy is essential for minimizing the cost of FW management and preserving resources, as the economic value of FW depends on the valorization processes. The researchers also claimed that high-value-added biogas applications will boost the economic benefits of the FW AD process.

Environmental Impacts

The conversion of waste valorization processes to a less polluting and energy-efficient mode of operation than normal operations can have several environmental benefits. The reduced ecological footprint of waste management operations, enhanced resource conservation, decreased greenhouse gas emissions, and decreased leachate toxicity and potential are some of the environmental advantages attributed to waste valorization processes [120]. Additionally, a life cycle assessment of valorization processes has been conducted to analyze the environmental impacts of the pollutants released by these processes within the boundaries of their product systems. Valorization procedures would, therefore, have little effect on the environment. The ecological consequences of some waste valorization procedures can be observed, as they are used as a pretreatment for subsequent valorization stages. It has been proposed that biochar can increase soil fertility by acting as a groundwater cleaner, soil enhancer, and protector while reducing environmental damage [121]. It would be feasible to accomplish sustainable development and further encourage the preservation of the planet's biodiversity by utilizing the most effective waste valuation systems. In addition to producing economic benefits, existing sustainable development models should not hurt the environment or public health. For example, the amount of heavy metal residues in bioplastics and bio-oils produced from the valorization of industrial waste should fall below permissible levels [122].

Social Impacts

Waste valorization offers substantial societal benefits. Turning waste into valuable resources promotes local job opportunities, particularly in green industries, which can improve health, working conditions, and overall well-being. Employment development fosters social cohesion in local communities and promotes economic equity. Waste valorization also promotes social inclusion by highlighting sustainable development from political, cultural, economic, and environmental viewpoints. Community ties can be strengthened, and sustainable attitudes can be fostered through holistic approaches that consider the requirements of all stakeholders. Waste valorization is still in its infancy, but it has the potential to provide significant, locally focused solutions on a global scale. Despite enough food being produced worldwide, around a billion people suffer from hunger and malnutrition. Large amounts of healthy, edible food that may benefit people in need, particularly socially disadvantaged populations inside these nations, are thrown away in many industrialized countries. Although 16 million people in the EU rely on food aid and 79 million live below the poverty line, up to 50% of edible food is wasted annually. Research indicates that rewarding employment, meaningful connections, and sufficient free time are more critical factors in achieving fulfillment than material possessions. Effective waste valorization, which enhances public health and reduces waste-related illnesses, relies on community cooperation. Involving residents reduces respiratory risks, promotes waste management ownership, and generates employment opportunities in recycling and waste collection [123].

Challenges in Waste Valorization and Biorefineries

Technical, legal, and financial obstacles hinder waste valorization, but continued research and development efforts aim to overcome them. Collaboration across disciplines is necessary for scaling up, as is an awareness that supply chains entail more than efficiency. Technology, legislation, and environmental trends are key areas for growth and expansion. Manufacturers may be encouraged to promote waste reduction through government initiatives, such as polluter-pays legislation. Businesses can profit from sustainable operations, gaining a competitive edge through public demand, insurance trends, and government support. For valorization to be widely accepted, old ways of thinking that disregard ecological effects must be abandoned, and science should be used to support sustainable lifestyle choices.

Case Studies and Success Stories

Second Generation (2G) Ethanol Plant, Panipat: Bioethanol from Agricultural Waste

While addressing the mismanagement of biowaste, we also face the issue of overconsumption of natural resources. A circular bioeconomy can solve this problem, where we can utilize biowaste to generate valuable products that serve as alternatives to natural resources like petroleum and coal. Globally, Brazil is one of the leading countries in the production of 2G bioethanol, with significant contributions from two commercial giants: Raízen and GranBio. Raízen started its first 2G bioethanol plant, named Costa Pinto Bioenergy Park, in 2014, producing up to 30 million liters of bioethanol per year. Raízen aims to have 20 bioethanol plants within the next 10 years, and an estimate suggests that the collective production of these plants would exceed 1.6 billion liters of ethanol per year. GranBio launched its first bioethanol production plant, named BioFlex, in 2017. This plant produces approximately 30 million liters of bioethanol annually. GranBio aims to achieve 60 million liters of annual productivity by 2026 [124]. While global efforts have demonstrated the potential of bioethanol production, countries like India have also started to recognize innovative solutions to solve problems through the circular bioeconomy.

Agricultural waste is a significant problem in northern India; disposable methods like burning exacerbate the issue. Burning stubble is one of the leading causes of pollution in North India, more precisely in the National Capital Region (NCR), between October and November. To address this issue and meet the requirement for ethanol, on World Biofuel Day 2022, the Indian government took the initiative in the direction of a circular bioeconomy. It established India's first 2G ethanol plant. Lignocellulose and cellulose-based biowaste, such as rice straw, were utilized as raw materials to produce bioethanol. This plant is operated by Indian Oil Corporation Limited (IOCL), a reputable government body in the energy sector, with its expansion into petrochemicals, gas, oil, and alternative energy sources. Using 2 lakh tons of agricultural biowaste as feedstock annually, this plant can produce three crore liters of bio-ethanol [125]. Produced ethanol is mainly used in ethanol blending. Ethanol blending is an exercise practiced by many countries, including India, in which ethanol is mixed with fuel to reduce CO_2 emissions. Thus, this project controls pollution in two ways: first, by preventing farmers from burning agricultural residues and utilizing them as feedstock for bioethanol production, and second, by using a significant portion of the produced ethanol in blending. It has been estimated that plants can absorb up to 3 lakh tons of CO_2 annually, equivalent to the pollution caused by 62,000 cars [126]. Currently, the ethanol blending percentage in India is 15% and aims to reach 20% in 2025. Biowaste valorization plays a vital role in achieving this goal.

Future Directions and Innovations in Waste Valorization

The future of waste valorization through biorefineries requires transformative advancements that will significantly change the perception of waste materials' utilization. Sustainability and resource efficiency are the prime focus in the global context; thus, innovations and emerging technologies will help biorefineries evolve and enhance their role in a circular economy. One of the most promising avenues for future development lies in advanced biomass processing technologies. Improving the pre-treatment process will facilitate the conversion of a broader range of waste materials, leading to more significant resource recovery and utilization. Advancements in enzymatic hydrolysis and integrated biorefining approaches are poised to revolutionize the conversion of complex biomass into high-value products.

Furthermore, the application of synthetic biology and metabolic engineering will enable the design of microbial strains with enhanced capabilities for bioproduct synthesis. These engineered strains can be tailored to produce high-value chemicals, pharmaceuticals, and biofuels more efficiently from waste-derived substrates. Moreover, circular supply chains should be established to facilitate the continuous flow of materials. Additionally, collocating biorefineries with renewable energy sources, such as solar and wind, can provide a sustainable energy source for biomass processing. Collaboration between biorefineries, agricultural producers, and industrial sectors will create synergies that optimize resource use and minimize waste generation. The integration of digital technologies, including big data analytics, AI, ML, and the Internet of Things (IoT), will enhance the monitoring and optimization of biorefinery operations. The expansion of bioproduct diversification is crucial for the continued success of biorefineries. Future research will likely explore the potential of valorizing waste to produce not just biofuels and bioplastics but also nutraceuticals, high-value biochemicals, and biodegradable materials tailored for specific applications. Realizing the potential of these innovations will only be possible through supportive policies and regulatory frameworks that the government and other public stakeholders prioritize. Lastly, the public will be more aware of waste valorization and bioproducts that can be initiated through various awareness programs and social media initiatives. By fostering a culture of sustainability and resource efficiency, we can cultivate a supportive environment for biorefineries and their role in waste valorization.

CONCLUSION

The role of biorefineries in waste valorization within the circular economy offers significant potential for advancing sustainability. Biorefineries help reduce environmental pollution, enhance resource efficiency, and promote energy recovery by transforming waste into valuable bioproducts. With advanced technologies such as fermentation, anaerobic digestion, and enzymatic hydrolysis, a range of valuable products can be generated, including biofuels, chemicals, bioplastics, and pharmaceuticals. This process minimizes landfill waste and opens new economic opportunities, supporting growth in emerging markets. The advantages of waste valorization extend beyond environmental and financial benefits to positive social impacts, including the creation of green jobs and the promotion of sustainable practices in local communities. Despite these benefits, challenges such as high initial costs, technological complexities, and the need for supportive policy frameworks remain. Successful real-world examples—such as converting agricultural residues into bioethanol or transforming organic waste into biogas—offer valuable insights into overcoming these challenges and demonstrate the viability of biorefineries. Looking to the future, innovations in waste valorization technologies, combined with advancements in biotechnology and system integration, will enable a broader range of waste to be converted into valuable resources. As research, investment, and collaboration between industries, academic institutions, and governments continue, biorefineries have the potential to play a crucial role in creating a more sustainable and waste-free future. Through these ongoing efforts, biorefineries will make a significant contribution to the transition to a circular economy.

ACKNOWLEDGEMENTS

The authors are grateful to Gautam Buddha University and Ramjas College, University of Delhi, for providing technical support for writing this chapter.

REFERENCES

[1] J.A. Mesa, L. Sierra-Fontalvo, K. Ortegon, and A. Gonzalez-Quiroga, "Advancing circular bioeconomy: A critical review and assessment of indicators", *Sustainable Production and Consumption,* vol. 46, pp. 324-342, 2024.
[http://dx.doi.org/10.1016/j.spc.2024.03.006]

[2] J. Rawat, M. Nanda, S. Kumar, N. Sharma, R. Sharma, H.C. Joshi, M.S. Vlaskin, A. Hussain, and V.

Kumar, "Integrating wastewater treatment to bio-stimulant & biochar generation for plant growth promotion using microalgae", *Process Biochem.*, vol. 145, pp. 187-194, 2024.
[http://dx.doi.org/10.1016/j.procbio.2024.06.031]

[3]   A. Barathi, P. P. Dagwar, D. Kundu, K. Rajendran, and S. Jacob, "Biorefineries and waste valorization in integrated biorefinery concepts and applications", *Biotechnol. Appl. Ind. Waste Valorization*, pp. 1-21, 2025.

[4]   A. Chauhan, Harish Chandra Joshi, "Sources for Biofuels Production from Biomass," Trends in Mathematics, Part F3197, pp. 1-64.
[http://dx.doi.org/10.1007/978-981-99-7250-0_1]

[5]   H. K. Lamba, N. S. Kumar, and S. Dhir, "Circular economy and sustainable development: a review and research agenda", *Int. J. Product. Perform. Manag.*, vol. 73, no. 2, pp. 497-522, 2024.

[6]   Filipović, N. Lior, and M. Radovanović, "The green deal – just transition and sustainable development goals Nexus," Renewable and Sustainable Energy Reviews, vol. 168, pp. 112759, 2022/10/01/, 2022.
[http://dx.doi.org/10.1016/j.rser.2022.112759]

[7]   W. C. Munonye, "Towards circular economy metrics: A systematic review", *Circ. Econ. Sustain.*, pp. 1-43, 2025.

[8]   Da Freiria Ferreira, N., Echevarria, N.E.P., Guerra, A. and Dias, F.T., The role of waste valorization in the transition to the circular economy.

[9]   K. H. Rashid, R. Al Aziz, C. L. Karmaker, A. M. Bari, and A. Raihan, "Evaluating the challenges to circular economy implementation in the apparel accessories industry: Implications for sustainable development", *Green Technol. Sustain.*, vol. 3, no. 2, p. 100140, 2025.

[10] V. Kumar, S. K. Malyan, W. Apollon, and P. Verma, "Valorization of pulp and paper industry waste streams into bioenergy and value-added products: An integrated biorefinery approach", *Renew. Energy*, vol. 228, p. 120566, 2024.

[11] A. Hayder, S. Mazhkoo, V. Patel, O. Norouzi, R. M. Santos, and A. Dutta, "Recent advances in hydrothermal carbonization of food waste derived bioproducts: Valorization approaches, applications, and the prospective assessment", *Waste Biomass Valori.*, pp. 1-39, 2025.

[12] A. Chauhan, and H.C. Joshi, "Recent developments and applications in bioconversion and biorefineries", *Trends in Mathematics,* no. Part F3197, pp. 247-307, 2024.
[http://dx.doi.org/10.1007/978-981-99-7250-0_6]

[13] D. B. Sulis, N. Lavoine, H. Sederoff, X. Jiang, B. M. Marques, K. Lan, C. Cofre-Vega, R. Barrangou, and J. P. Wang, "Advances in lignocellulosic feedstocks for bioenergy and bioproducts", *Nat. Commun.*, vol. 16, no. 1, p. 1244, 2025.

[14] D. D. Sapariya, U. J. Patdiwala, H. Panchal, P. V. Ramana, J. Makwana, and K. K. Sadasivuni, "A review on thermochemical biomass gasification techniques for bioenergy production", *Energy Sources A*, vol. 47, no. 2, p. 2000521, 2025.

[15] M. A. Abdullah, M. S. Nazir, H. A. Hussein, S. M. Shah, N. Azra, R. Iftikhar, M. S. Iqbal, Z. Qamar, Z. Ahmad, M. Afzaal, and A. D. Om, "New perspectives on biomass conversion and circular economy based on integrated algal-oil palm biorefinery framework for sustainable energy and bioproducts co-generation", *Ind. Crops Prod.*, vol. 213, p. 118452, 2024.

[16] J. Gaffey, G. Rajauria, H. McMahon, R. Ravindran, C. Dominguez, M. Ambye-Jensen, M.F. Souza, E. Meers, M.M. Aragonés, D. Skunca, and J.P.M. Sanders, "Green Biorefinery systems for the production of climate-smart sustainable products from grasses, legumes and green crop residues", *Biotechnol. Adv.*, vol. 66, p. 108168, 2023.
[http://dx.doi.org/10.1016/j.biotechadv.2023.108168] [PMID: 37146921]

[17] S. Kromus, B. Wachter, W. Koschuh, M. Mandl, C. Krotscheck, and M. Narodoslawsky, "The green biorefinery Austria-development of an integrated system for green biomass utilization", *Chem. Biochem. Eng. Q.*, vol. 18, no. 1, pp. 8-12, 2004.

[18] Y. Zhou, J. Remón, J. Gracia, Z. Jiang, J.L. Pinilla, C. Hu, and I. Suelves, "Toward developing more sustainable marine biorefineries: A novel 'sea-thermal' process for biofuels production from microalgae", *Energy Convers. Manage.*, vol. 270, p. 116201, 2022.
[http://dx.doi.org/10.1016/j.enconman.2022.116201]

[19] A.T. Ubando, E. Anderson S Ng, W.H. Chen, A.B. Culaba, and E.E. Kwon, "Life cycle assessment of microalgal biorefinery: A state-of-the-art review", *Bioresour. Technol.*, vol. 360, p. 127615, 2022.
[http://dx.doi.org/10.1016/j.biortech.2022.127615] [PMID: 35840032]

[20] D. C. Makepa and C. H. Chihobo, "Barriers to commercial deployment of biorefineries: A multi-faceted review of obstacles across the innovation chain", *Heliyon*, vol. 10, no. 12, 2024.

[21] A. Chauhan, and H. Chandra Joshi, "Energetic Efficiency of a Biofuels Production System Mathematical Modeling", *Trends in Mathematics,* no. Part F3197, pp. 337-357, 2024.
[http://dx.doi.org/10.1007/978-981-99-7250-0_8]

[22] S.Y. Pan, C.Y. Tsai, C.W. Liu, S.W. Wang, H. Kim, and C. Fan, "Anaerobic co-digestion of agricultural wastes toward circular bioeconomy", *iScience,* vol. 24, no. 7, p. 102704, 2021.
[http://dx.doi.org/10.1016/j.isci.2021.102704] [PMID: 34258548]

[23] A. Gouda, N. Merhi, M. Hmadeh, T. Cecchi, C. Santato, and M. Sain, "Sustainable strategies for converting organic, electronic, and plastic waste from municipal solid waste into functional materials", *Glob. Challenges*, vol. 9, no. 4, p. 2400240, 2025.

[24] K. Chojnacka and K. Moustakas, "Anaerobic digestate management for carbon neutrality and fertilizer use: A review of current practices and future opportunities", *Biomass Bioenergy*, vol. 180, p. 106991, 2024.

[25] J.A. Shukor, M.F. Omar, M.M. Kasim, M.H. Jamaludin, and M.A. Naim, "Assessment of composting technologies for organic waste management J", *Assessment,* vol. 9, no. 8, pp. 1579-1587, 2018.

[26] Z. Duan, Z. Kang, X. Kong, G. Qiu, Q. Wang, T. Wang, X. Yang, G. Zhu, J. Yue, X. Han, and H. Yu, "Synergistic hyperthermophilic microbial consortia in self-elevating ultra-high temperature composting: Mechanism and application investigation for sustainable organic waste upcycling", *Chem. Eng. J.*, vol. 512, p. 162364, 2025.

[27] M. Ramos-Suarez, Y. Zhang, and V. Outram, "Current perspectives on acidogenic fermentation to produce volatile fatty acids from waste", *Rev. Environ. Sci. Biotechnol.,* vol. 20, no. 2, pp. 439-478, 2021.
[http://dx.doi.org/10.1007/s11157-021-09566-0]

[28] Q. Wang, H. Li, K. Feng, and J. Liu, "Oriented fermentation of food waste towards high-value products: A review", *Energies,* vol. 13, no. 21, p. 5638, 2020.
[http://dx.doi.org/10.3390/en13215638]

[29] W.A. Wan Mahari, E. Azwar, S.Y. Foong, A. Ahmed, W. Peng, M. Tabatabaei, M. Aghbashlo, Y.K. Park, C. Sonne, and S.S. Lam, "Valorization of municipal wastes using co-pyrolysis for green energy production, energy security, and environmental sustainability: A review", *Chem. Eng. J.,* vol. 421, p. 129749, 2021.
[http://dx.doi.org/10.1016/j.cej.2021.129749]

[30] Rezvani, "Novel techniques in bio-oil production through catalytic pyrolysis of waste biomass: Effective parameters, innovations, and techno-economic analysis", *Can. J. Chem. Eng.*, vol. 103, p. 3531-3544, 2025.

[31] Y. Zhang, J. Yang, P. Yu, R. Ruan, L. Dai, J. Zhang, E. Huo, R. Zou, C. Wang, Y. Zhao, and Y. Wang, "Catalytic co-pyrolysis of cotton stalks and ground film plastic using fishbone-based metal catalysts: Enhanced production of olefins and aromatics", *Chem. Eng. J.*, vol. 521, p. 166274, 2025.

[32] A. Chanthakett, M.T. Arif, M.M.K. Khan, and A.M.T. Oo, "Performance assessment of gasification reactors for sustainable management of municipal solid waste", *J. Environ. Manage.,* vol. 291, p. 112661, 2021.

[http://dx.doi.org/10.1016/j.jenvman.2021.112661] [PMID: 33962284]

[33]   S.I. Hussain Shah, T.H. Seehar, M. Raashid, R. Nawaz, Z. Masood, S. Mukhtar, T.A. Al Johani, A. Doyle, M.N. Bashir, M.M. Ali, and M.A. Kalam, "Biocrude from hydrothermal liquefaction of indigenous municipal solid waste for green energy generation and contribution towards circular economy: A case study of urban Pakistan", *Heliyon,* vol. 10, no. 17, p. e36758, 2024. [http://dx.doi.org/10.1016/j.heliyon.2024.e36758] [PMID: 39281648]

[34]   M. Scarsella, B. de Caprariis, M. Damizia, and P. De Filippis, "Heterogeneous catalysts for hydrothermal liquefaction of lignocellulosic biomass: A review", *Biomass Bioenergy,* vol. 140, p. 105662, 2020. [http://dx.doi.org/10.1016/j.biombioe.2020.105662]

[35]   S. Mazhkoo, S. Soltanian, H. O. Odebiyi, O. Norouzi, M. Ubene, A. Hayder, O. Pourali, R. M. Santos, R. C. Brown, and A. Dutta, "Process intensification in hydrothermal liquefaction of biomass: A review", *J. Environ. Chem. Eng.,* vol. 13, no. 2, p. 115722, 2025.

[36]   C. Yang, H. Shang, J. Li, X. Fan, J. Sun, and A. Duan, "A Review on the Microwave-Assisted Pyrolysis of Waste Plastics", *Processes (Basel),* vol. 11, no. 5, p. 1487, 2023. [http://dx.doi.org/10.3390/pr11051487]

[37]   Y. Shen, "Microwave-assisted pyrolysis of biomass and plastic wastes for hydrogen production", *Green Chem.,* 2025.

[38]   A. Kumar, N. Kumar, P. Baredar, and A. Shukla, *A review on biomass energy resources, potential, conversion and policy in India,* 2015. [http://dx.doi.org/10.1016/j.rser.2015.02.007]

[39]   G. Singh, B. D. Schreiner, X. Sun, and V. Sethi, "A review of hydrogen micromix combustion technologies for gas turbine applications", *Int. J. Hydrogen Energy,* vol. 127, pp. 295-310, 2025.

[40]   M. Zeng, H. Gao, Y. Wu, L. Fan, T. Zheng, and D. Zhou, "Effects of Ultra-sonification Assisting Polyethylene Glycol Pre-treatment on the Crystallinity and Accessibility of Cellulose Fiber," Journal of Macromolecular Science, Part A, vol. 47, no. 10, pp. 1042-1049, 2010/08/31, 2010. [http://dx.doi.org/10.1080/10601325.2010.508016]

[41]   Y.H. Chai, S. Yusup, W.N. Kadir, C.Y. Wong, S.S. Rosli, M.S. Ruslan, B.L. Chin, and C.L. Yiin, "Valorization of Tropical Biomass Waste by Supercritical Fluid Extraction Technology", In: *Sustainability* vol. 13. , 2021.

[42]   T. Wang, Y. Zhai, Y. Zhu, C. Li, and G. Zeng, *A review of the hydrothermal carbonization of biomass waste for hydrochar formation: Process conditions, fundamentals, and physicochemical properties,* 2018.

[43]   N. Pachauri, C. W. Ahn, and T. J. Choi, "Biochar energy prediction from different biomass feedstocks for clean energy generation", *Environ. Technol. Innov.,* vol. 37, p. 104012, 2025.

[44]   A. S. Elgharbawy, M. Farghali, A. I. Osman, M. A. Hanafy, and A. A. Al-Muhtaseb, "Innovative biodiesel production for sustainable energy: Advances in feedstocks, transesterification, and cost efficiency", *Biomass Bioenergy,* vol. 201, p. 108114, 2025.

[45]   M. Hamza, M. Ayoub, R. B. Shamsuddin, A. Mukhtar, S. Saqib, I. Zahid, M. Ameen, S. Ullah, A. G. Al-Sehemi, and M. Ibrahim, "A review on the waste biomass derived catalysts for biodiesel production," Environmental Technology & Innovation, vol. 21, pp. 101200, 2021/02/01/, 2021. [http://dx.doi.org/10.1016/j.eti.2020.101200]

[46]   R. Goswami, S. Singh, P. Narasimhappa, P.C. Ramamurthy, A. Mishra, P.K. Mishra, H.C. Joshi, G. Pant, J. Singh, G. Kumar, N.A. Khan, and M. Yousefi, "Nanocellulose: A comprehensive review investigating its potential as an innovative material for water remediation", *Int. J. Biol. Macromol.,* vol. 254, no. Pt 3, p. 127465, 2024. [http://dx.doi.org/10.1016/j.ijbiomac.2023.127465] [PMID: 37866583]

[47]   A. T. Hoang, S. Nizetic, H. C. Ong, C. T. Chong, A. E. Atabani, and V. V. Pham, "Acid-based

lignocellulosic biomass biorefinery for bioenergy production: Advantages, application constraints, and perspectives," Journal of Environmental Management, vol. 296, pp. 113194, 2021/10/15/, 2021.
[http://dx.doi.org/10.1016/j.jenvman.2021.113194]

[48]   F. W. S. Lucas, R. G. Grim, S. A. Tacey, C. A. Downes, J. Hasse, A. M. Roman, C. A. Farberow, J. A. Schaidle, and A. Holewinski, "Electrochemical Routes for the Valorization of Biomass-Derived Feedstocks: From Chemistry to Application," ACS Energy Letters, vol. 6, no. 4, pp. 1205-1270, 2021/04/09, 2021.

[49]   S. Sun, Z. Liu, Z. J. Xu, and T. Wu, "Opportunities and challenges in biomass electrocatalysis and valorization," Applied Catalysis B: Environment and Energy, vol. 358, pp. 124404, 2024/12/05/, 2024.
[http://dx.doi.org/10.1016/j.apcatb.2024.124404]

[50]   E.U. Khan, and Å. Nordberg, "Thermal integration of membrane distillation in an anaerobic digestion biogas plant – A techno-economic assessment", *Appl. Energy,* vol. 239, pp. 1163-1174, 2019.
[http://dx.doi.org/10.1016/j.apenergy.2019.02.023]

[51]   D. Hidalgo, A. Urueña, J. M. Martín-Marroquín, and D. Díez, "Integrated approach for biomass conversion using thermochemical routes with anaerobic digestion and syngas fermentation", *Sustainability*, vol. 17, no. 8, p. 3615, 2025.

[52]   A. Ebrahimi and E. Houshfar, "A comprehensive exergoeconomic analysis of pyrolysis, anaerobic digestion, and integrated Py-AD plants for sustainable energy and waste management", *Fuel*, vol. 384, p. 133928, 2025.

[53]   Y. Liang, "A Critical Review of Challenges Faced by Converting Food Waste to Bioenergy Through Anaerobic Digestion and Hydrothermal Liquefaction", *Waste Biomass Valoriz.,* vol. 13, no. 2, pp. 781-796, 2022.
[http://dx.doi.org/10.1007/s12649-021-01540-9]

[54]   F. Xu, Y. Li, X. Ge, L. Yang, and Y. Li, "Anaerobic digestion of food waste – Challenges and opportunities", *Bioresour. Technol.,* vol. 247, pp. 1047-1058, 2018.
[http://dx.doi.org/10.1016/j.biortech.2017.09.020] [PMID: 28965912]

[55]   A. Zayen, N. Gharsallah, M. Jraou, S. Loukil, I. E. Nikolaou, T. Tsalis, S. Sayadi, and S. Khoufi, "Potential valorization of fruits and vegetables waste from the wholesale market in Sfax (Tunisia) *via* anaerobic digestion: Long-term characterization and stakeholders' attitude", *Biomass Convers. Biorefin.,* pp. 1-3, 2025.

[56]   E. Hsu, K. Barmak, A.C. West, and A.H.A. Park, "Advancements in the treatment and processing of electronic waste with sustainability: a review of metal extraction and recovery technologies", *Green Chem.,* vol. 21, no. 5, pp. 919-936, 2019.
[http://dx.doi.org/10.1039/C8GC03688H]

[57]   D. A. Ferreira-Filipe, A. C. Duarte, A. S. Hursthouse, T. Rocha-Santos, and A. L. Silva, "Biobased strategies for e-waste metal recovery: A critical overview of recent advances", *Environments*, vol. 12, no. 1, p. 26, 2025.

[58]   A. Sharma, T. Kuthiala, K. Thakur, K. S. Thatai, G. Singh, P. Kumar, and S. K. Arya, "Kitchen waste: Sustainable bioconversion to value-added product and economic challenges", *Biomass Convers. Biorefin.*, vol. 15, no. 2, pp. 1749-1770, 2025.

[59]   H. Wang, G. Kaur, N. Pensupa, K. Uisan, C. Du, X. Yang, and C.S.K. Lin, "Textile waste valorization using submerged filamentous fungal fermentation", *Process Saf. Environ. Prot.,* vol. 118, pp. 143-151, 2018.
[http://dx.doi.org/10.1016/j.psep.2018.06.038]

[60]   J. Ghosh, M. R. Repon, N. S. Rupanty, T. R. Asif, M. I. Tamjid, and V. Reukov, "Chemical valorization of textile waste: advancing sustainable recycling for a circular economy", *ACS Omega*, vol. 10, no. 12, pp. 11697-11722, 2025.

[61]   A.C.C. Chang, H.F. Chang, F.J. Lin, K.H. Lin, and C.H. Chen, "Biomass gasification for hydrogen

production", *Int. J. Hydrogen Energy,* vol. 36, no. 21, pp. 14252-14260, 2011.
[http://dx.doi.org/10.1016/j.ijhydene.2011.05.105]

[62]   G. Lopez, M. Artetxe, M. Amutio, J. Alvarez, J. Bilbao, and M. Olazar, "Recent advances in the gasification of waste plastics. A critical overview", *Renew. Sustain. Energy Rev.,* vol. 82, pp. 576-596, 2018.
[http://dx.doi.org/10.1016/j.rser.2017.09.032]

[63]   A. Moustafa, K. Abdelrahman, A. Abdelhaleem, and I. S. Fahim, "Valorization of plastic waste *via* hydrothermal liquefaction and hydrothermal gasification: Review and bibliometric analysis", *J. Anal. Appl. Pyrolysis*, p. 107112, 2025.

[64]   O. Mohan, O. A. Fakayode, and A. Kumar, "Biofuel production from pipeline-transported lignocellulosic biomass *via* hydrothermal liquefaction: Process optimization and product characterization", *Biomass Bioenergy*, vol. 202, p. 108210, 2025.

[65]   A. Mlonka-Mędrala, P. Evangelopoulos, M. Sieradzka, M. Zajemska, and A. Magdziarz, "Pyrolysis of agricultural waste biomass towards production of gas fuel and high-quality char: Experimental and numerical investigations," Fuel, vol. 296, pp. 120611, 2021/07/15/, 2021.
[http://dx.doi.org/10.1016/j.fuel.2021.120611]

[66]   H.S. Lee, S. Jung, K.Y.A. Lin, E.E. Kwon, and J. Lee, "Upcycling textile waste using pyrolysis process", *Sci. Total Environ.,* vol. 859, no. Pt 2, p. 160393, 2023.
[http://dx.doi.org/10.1016/j.scitotenv.2022.160393] [PMID: 36423842]

[67]   M. A. Khan, N. A. Sheikh, K. Z. Jadoon, A. Ayub, T. W. Awotwe, and R. Tariq, "Utilization of locally sourced waste fats for biodiesel production: Experimental characterization and environmental life cycle assessment", *Biomass Bioenergy*, vol. 194, p. 107692, 2025.

[68]   J. Poudel, T.-I. Ohm, and S. C. Oh, "A study on torrefaction of food waste," Fuel, vol. 140, pp. 275-281, 2015/01/15/, 2015.
[http://dx.doi.org/10.1016/j.fuel.2014.09.120]

[69]   C.-W. Huang, Y.-H. Huang, D.-E. Wu, P.-C. Chen, Y.-T. Lin, F.-H. Wu, and G.-B. Chen, "Optimization of torrefied black liquor and its combustion characteristics with pulverized coal," Journal of the Taiwan Institute of Chemical Engineers, pp. 105112, 2023/08/31/, 2023.

[70]   J. Kruyeniski, P.J.T. Ferreira, M.G. Videira Sousa Carvalho, M.E. Vallejos, F.E. Felissia, and M.C. Area, "Physical and chemical characteristics of pretreated slash pine sawdust influence its enzymatic hydrolysis", *Ind. Crops Prod.,* vol. 130, pp. 528-536, 2019.
[http://dx.doi.org/10.1016/j.indcrop.2018.12.075]

[71]   Z.H. Liu, L. Qin, J.Q. Zhu, B.Z. Li, and Y.J. Yuan, "Simultaneous saccharification and fermentation of steam-exploded corn stover at high glucan loading and high temperature", *Biotechnol. Biofuels,* vol. 7, no. 1, p. 167, 2014.
[http://dx.doi.org/10.1186/s13068-014-0167-x] [PMID: 25516770]

[72]   A.B. Díaz, I. Caro, I. de Ory, and A. Blandino, "Evaluation of the conditions for the extraction of hydrolitic enzymes obtained by solid state fermentation from grape pomace", *Enzyme Microb. Technol.,* vol. 41, no. 3, pp. 302-306, 2007.
[http://dx.doi.org/10.1016/j.enzmictec.2007.02.006]

[73]   K.K. Cheng, Q. Liu, J.A. Zhang, J.P. Li, J.M. Xu, and G.H. Wang, "Improved 2,3-butanediol production from corncob acid hydrolysate by fed-batch fermentation using Klebsiella oxytoca", *Process Biochem.,* vol. 45, no. 4, pp. 613-616, 2010.
[http://dx.doi.org/10.1016/j.procbio.2009.12.009]

[74]   Y. Gupta, B. Barrett, and D.G. Vlachos, "Understanding microwave-assisted extraction of phenolic compounds from diverse food waste feedstocks", *Chem. Eng. Process.,* vol. 203, p. 109870, 2024.
[http://dx.doi.org/10.1016/j.cep.2024.109870]

[75]   C. Mellinas, I. Solaberrieta, C.J. Pelegrín, A. Jiménez, and M.C. Garrigós, *Valorization of Agro-*

Industrial Wastes by Ultrasound-Assisted Extraction as a Source of Proteins, Antioxidants and Cutin: A Cascade Approach. vol. 11. Antioxidants, 2022.

[76] M.T. Afraz, X. Xu, M. Adil, M.F. Manzoor, X-A. Zeng, Z. Han, and R.M. Aadil, *Subcritical and Supercritical Fluids to Valorize Industrial Fruit and Vegetable Waste.* vol. 12. Foods, 2023.

[77] A.C. Fărcaş, S.A. Socaci, S.A. Nemeş, L.C. Salanţă, M.S. Chiş, C.R. Pop, A. Borşa, Z. Diaconeasa, and D.C. Vodnar, *Cereal Waste Valorization through Conventional and Current Extraction Techniques—An Up-to-Date Overview.* vol. 11. Foods, 2022.

[78] M. Jayakumar, G.T. Gindaba, K.B. Gebeyehu, S. Periyasamy, A. Jabesa, G. Baskar, B.I. John, and A. Pugazhendhi, "Bioethanol production from agricultural residues as lignocellulosic biomass feedstock's waste valorization approach: A comprehensive review", *Sci. Total Environ.,* vol. 879, p. 163158, 2023. [http://dx.doi.org/10.1016/j.scitotenv.2023.163158] [PMID: 37001650]

[79] K. Kusmiyati, H. Hadiyanto, and A. Fudholi, "Treatment updates of microalgae biomass for bioethanol production: A comparative study", *J. Clean. Prod.,* vol. 383, p. 135236, 2023. [http://dx.doi.org/10.1016/j.jclepro.2022.135236]

[80] K. Velusamy, J. Devanand, P. Senthil Kumar, K. Soundarajan, V. Sivasubramanian, J. Sindhu, and D.V.N. Vo, "A review on nano-catalysts and biochar-based catalysts for biofuel production", *Fuel,* vol. 306, p. 121632, 2021. [http://dx.doi.org/10.1016/j.fuel.2021.121632]

[81] M. Mulyatun, J. Prameswari, I. Istadi, and W. Widayat, "Production of non-food feedstock based biodiesel using acid-base bifunctional heterogeneous catalysts: A review", *Fuel,* vol. 314, p. 122749, 2022. [http://dx.doi.org/10.1016/j.fuel.2021.122749]

[82] C.R. Chilakamarry, A.M. Mimi Sakinah, A.W. Zularisam, A. Pandey, and D.V.N. Vo, "Technological perspectives for utilisation of waste glycerol for the production of biofuels: A review", *Environmental Technology & Innovation,* vol. 24, p. 101902, 2021. [http://dx.doi.org/10.1016/j.eti.2021.101902]

[83] G.P. Chutia, and K. Phukan, "Biomass derived heterogeneous catalysts used for sustainable biodiesel production: a systematic review", *Braz. J. Chem. Eng.,* vol. 41, no. 1, pp. 23-48, 2024. [http://dx.doi.org/10.1007/s43153-023-00371-6]

[84] F. Moazeni, Y.C. Chen, and G. Zhang, "Enzymatic transesterification for biodiesel production from used cooking oil, a review", *J. Clean. Prod.,* vol. 216, pp. 117-128, 2019. [http://dx.doi.org/10.1016/j.jclepro.2019.01.181]

[86] F. Toldrá-Reig, L. Mora, and F. Toldrá, "Developments in the use of lipase transesterification for biodiesel production from animal fat waste", *Appl. Sci. (Basel),* vol. 10, no. 15, p. 5085, 2020. [http://dx.doi.org/10.3390/app10155085]

[87] B. Kayode, and A. Hart, "An overview of transesterification methods for producing biodiesel from waste vegetable oils", *Biofuels,* vol. 10, no. 3, pp. 419-437, 2019. [http://dx.doi.org/10.1080/17597269.2017.1306683]

[88] H.C. Joshi, N. Grag, S. Kumar, and W. Ahmad, "Influence of the catalytic activity of MgO catalyst on the comparative studies of Schlichera oleosa, Michelia champaca and Putranjiva based biodiesel and its blend with ethanol-diesel", *Mater. Today Proc.,* vol. 38, no. 1, pp. 18-23, 2021. [http://dx.doi.org/10.1016/j.matpr.2020.05.432]

[89] Z. Ali, B. Subeshan, M.A. Alam, E. Asmatulu, and J. Xu, "Recent progress in extraction/transesterification techniques for the recovery of oil from algae biomass", *Biomass Convers. Biorefin.,* vol. •••, pp. 1-17, 2021.

[90] A. Nisar, K. Hashum, M. Bashir, N. Mubeen, S. Younus, S. Mehmood, F. Haq, and M. Haroon, *Advancing sustainable biofuel production from agricultural residues: a comprehensive mini-review.* Sustainable Chemical Engineering, 2024, pp. 115-128.

[91]   A. Ahmad, F. Banat, H. Alsafar, and S.W. Hasan, "An overview of biodegradable poly (lactic acid) production from fermentative lactic acid for biomedical and bioplastic applications", *Biomass Convers. Biorefin.,* vol. 14, no. 3, pp. 3057-3076, 2024.
[http://dx.doi.org/10.1007/s13399-022-02581-3]

[92]   Y.J. Sohn, H.T. Kim, K.A. Baritugo, S.Y. Jo, H.M. Song, S.Y. Park, S.K. Park, J. Pyo, H.G. Cha, H. Kim, J.G. Na, C. Park, J.I. Choi, J.C. Joo, and S.J. Park, "Recent advances in sustainable plastic upcycling and biopolymers", *Biotechnol. J.,* vol. 15, no. 6, p. 1900489, 2020.
[http://dx.doi.org/10.1002/biot.201900489] [PMID: 32162832]

[93]   S. Chavan, B. Yadav, R.D. Tyagi, J.W.C. Wong, and P. Drogui, "Trends and challenges in the valorization of kitchen waste to polyhydroxyalkanoates", *Bioresour. Technol.,* vol. 369, p. 128323, 2023.
[http://dx.doi.org/10.1016/j.biortech.2022.128323] [PMID: 36400275]

[94]   J.X. Chan, J.F. Wong, A. Hassan, and Z. Zakaria, *Bioplastics from agricultural waste,* 2021.
[http://dx.doi.org/10.1016/B978-0-12-819953-4.00005-7]

[95]   Y.F. Tsang, V. Kumar, P. Samadar, Y. Yang, J. Lee, Y.S. Ok, H. Song, K.H. Kim, E.E. Kwon, and Y.J. Jeon, "Production of bioplastic through food waste valorization", *Environ. Int.,* vol. 127, pp. 625-644, 2019.
[http://dx.doi.org/10.1016/j.envint.2019.03.076] [PMID: 30991219]

[96]   R. Reshmy, D. Thomas, E. Philip, S.A. Paul, A. Madhavan, R. Sindhu, R. Sirohi, S. Varjani, A. Pugazhendhi, A. Pandey, and P. Binod, "Bioplastic production from renewable lignocellulosic feedstocks: a review", *Rev. Environ. Sci. Biotechnol.,* vol. 20, no. 1, pp. 167-187, 2021.
[http://dx.doi.org/10.1007/s11157-021-09565-1]

[97]   F. Ebrahimian, K. Karimi, and R. Kumar, "Sustainable biofuels and bioplastic production from the organic fraction of municipal solid waste", *Waste Manag.,* vol. 116, pp. 40-48, 2020.
[http://dx.doi.org/10.1016/j.wasman.2020.07.049] [PMID: 32784120]

[98]   F. Liu, J. Li, and X.L. Zhang, "Bioplastic production from wastewater sludge and application",
[http://dx.doi.org/10.1088/1755-1315/344/1/012071]

[99]   S.H. Hamin, S.H.Y. Sayid Abdullah, F. Lananan, S.H. Abdul Hamid, N.A. Kasan, N.N. Mohamed, and A. Endut, "Effect of chemical treatment on the structural, thermal, and mechanical properties of sugarcane bagasse as filler for starch□based bioplastic", *J. Chem. Technol. Biotechnol.,* vol. 98, no. 3, pp. 625-632, 2023.
[http://dx.doi.org/10.1002/jctb.7218]

[100] Y.S. Jang, B. Kim, J.H. Shin, Y.J. Choi, S. Choi, C.W. Song, J. Lee, H.G. Park, and S.Y. Lee, "Bio□based production of C2–C6 platform chemicals", *Biotechnol. Bioeng.,* vol. 109, no. 10, pp. 2437-2459, 2012.
[http://dx.doi.org/10.1002/bit.24599] [PMID: 22766912]

[101] R. Naraian, and S. Kumari, "Microbial Production of Organic Acids",
[http://dx.doi.org/10.1002/9781119048961.ch5]

[102] M.V. Guettler, D. Rumler, and M.K. Jain, "Actinobacillus succinogenes sp. nov., a novel succinic-acid-producing strain from the bovine rumen", *Int. J. Syst. Evol. Microbiol.,* vol. 49, no. 1, pp. 207-216, 1999.
[http://dx.doi.org/10.1099/00207713-49-1-207] [PMID: 10028265]

[103] P.C. Lee, S.Y. Lee, S.H. Hong, and H.N. Chang, "Batch and continuous cultures of Mannheimia succiniciproducens MBEL55E for the production of succinic acid from whey and corn steep liquor", *Bioprocess Biosyst. Eng.,* vol. 26, no. 1, pp. 63-67, 2003.
[http://dx.doi.org/10.1007/s00449-003-0341-1] [PMID: 14530958]

[104] E. Stylianou, C. Pateraki, D. Ladakis, M. Cruz-Fernández, M. Latorre-Sánchez, C. Coll, and A. Koutinas, "Evaluation of organic fractions of municipal solid waste as renewable feedstock for

succinic acid production", *Biotechnol. Biofuels,* vol. 13, no. 1, p. 72, 2020.
[http://dx.doi.org/10.1186/s13068-020-01708-w] [PMID: 32322302]

[105] A. Djukić-Vuković, D. Mladenović, J. Ivanović, J. Pejin, and L. Mojović, "Towards sustainability of lactic acid and poly-lactic acid polymers production", *Renew. Sustain. Energy Rev.,* vol. 108, pp. 238-252, 2019.
[http://dx.doi.org/10.1016/j.rser.2019.03.050]

[106] M.A. Abdel-Rahman, Y. Tashiro, and K. Sonomoto, "Recent advances in lactic acid production by microbial fermentation processes", *Biotechnol. Adv.,* vol. 31, no. 6, pp. 877-902, 2013.
[http://dx.doi.org/10.1016/j.biotechadv.2013.04.002] [PMID: 23624242]

[107] H. Azaizeh, H.N. Abu Tayeh, R. Schneider, A. Klongklaew, and J. Venus, "Production of lactic acid from carob, banana and sugarcane lignocellulose biomass", *Molecules,* vol. 25, no. 13, p. 2956, 2020.
[http://dx.doi.org/10.3390/molecules25132956] [PMID: 32605022]

[108] H. Azaizeh, H.N. Abu Tayeh, R. Schneider, and J. Venus, "Pilot scale for production and purification of lactic acid from Ceratonia siliqua L.(carob) bagasse", *Fermentation (Basel),* vol. 8, no. 9, p. 424, 2022.
[http://dx.doi.org/10.3390/fermentation8090424]

[109] S. Costa, D. Summa, B. Semeraro, F. Zappaterra, I. Rugiero, and E. Tamburini, "Fermentation as a strategy for bio-transforming waste into resources: lactic acid production from agri-food residues", *Fermentation (Basel),* vol. 7, no. 1, p. 3, 2020.
[http://dx.doi.org/10.3390/fermentation7010003]

[110] H.J. Brownlee, and C.S. Miner, "Industrial development of furfural", *Ind. Eng. Chem.,* vol. 40, no. 2, pp. 201-204, 1948.
[http://dx.doi.org/10.1021/ie50458a005]

[111] C.B.T.L. Lee, and T.Y. Wu, "A review on solvent systems for furfural production from lignocellulosic biomass", *Renew. Sustain. Energy Rev.,* vol. 137, p. 110172, 2021.
[http://dx.doi.org/10.1016/j.rser.2020.110172]

[112] L. Zhang, G. Xi, J. Zhang, H. Yu, and X. Wang, "Efficient catalytic system for the direct transformation of lignocellulosic biomass to furfural and 5-hydroxymethylfurfural", *Bioresour. Technol.,* vol. 224, pp. 656-661, 2017.
[http://dx.doi.org/10.1016/j.biortech.2016.11.097] [PMID: 27913172]

[113] S. Maiti, G. Gallastegui, G. Suresh, V.L. Pachapur, S.K. Brar, Y. Le Bihan, P. Drogui, G. Buelna, M. Verma, and R. Galvez-Cloutier, "Microwave-assisted one-pot conversion of agro-industrial wastes into levulinic acid: An alternate approach", *Bioresour. Technol.,* vol. 265, pp. 471-479, 2018.
[http://dx.doi.org/10.1016/j.biortech.2018.06.012] [PMID: 29936351]

[114] J.M. Tukacs, A.T. Holló, N. Rétfalvi, E. Cséfalvay, G. Dibó, D. Havasi, and L.T. Mika, "Microwave□assisted valorization of biowastes to levulinic acid", *ChemistrySelect,* vol. 2, no. 4, pp. 1375-1380, 2017.
[http://dx.doi.org/10.1002/slct.201700037]

[115] J.E. Manson, N.R. Cook, I.M. Lee, W. Christen, S.S. Bassuk, S. Mora, H. Gibson, C.M. Albert, D. Gordon, T. Copeland, D. D'Agostino, G. Friedenberg, C. Ridge, V. Bubes, E.L. Giovannucci, W.C. Willett, and J.E. Buring, "Marine n− 3 fatty acids and prevention of cardiovascular disease and cancer", *N. Engl. J. Med.,* vol. 380, no. 1, pp. 23-32, 2019.
[http://dx.doi.org/10.1056/NEJMoa1811403] [PMID: 30415637]

[116] A. Paz, A. Karnaouri, C.C. Templis, N. Papayannakos, and E. Topakas, "Valorization of exhausted olive pomace for the production of omega-3 fatty acids by Crypthecodinium cohnii", *Waste Manag.,* vol. 118, pp. 435-444, 2020.
[http://dx.doi.org/10.1016/j.wasman.2020.09.011] [PMID: 32971378]

[117] P. Velho, L.R. Barroca, and E.A. Macedo, "Partition of antioxidants available in biowaste using a green aqueous biphasic system", *Separ. Purif. Tech.,* vol. 307, p. 122707, 2023.

[http://dx.doi.org/10.1016/j.seppur.2022.122707]

[118] R.D. Patria, S. Rehman, C.B. Yuen, D.J. Lee, A.K. Vuppaladadiyam, and S.Y. Leu, "Energy-environment-economic (3E) hub for sustainable plastic management – Upgraded recycling, chemical valorization, and bioplastics", *Appl. Energy,* vol. 357, p. 122543, 2024.
[http://dx.doi.org/10.1016/j.apenergy.2023.122543]

[119] K.Q. Tan, M.A. Ahmad, W.D. Oh, and S.C. Low, "Valorization of hazardous plastic wastes into value-added resources by catalytic pyrolysis-gasification: A review of techno-economic analysis", *Renew. Sustain. Energy Rev.,* vol. 182, p. 113346, 2023.
[http://dx.doi.org/10.1016/j.rser.2023.113346]

[120] X. Zheng, X. Chen, A. Qu, W. Yang, L. Tao, F. Li, J. Huang, X. Xu, J. Tang, P. Hou, and W. Han, "Valorisation of food waste for valuable by-products generation with economic assessment", *J. Environ. Manage.,* vol. 338, p. 117762, 2023.
[http://dx.doi.org/10.1016/j.jenvman.2023.117762] [PMID: 37003224]

[121] S.K. Malyan, S.S. Kumar, R.K. Fagodiya, P. Ghosh, A. Kumar, R. Singh, and L. Singh, "Biochar for environmental sustainability in the energy-water-agroecosystem nexus", *Renew. Sustain. Energy Rev.,* vol. 149, p. 111379, 2021.
[http://dx.doi.org/10.1016/j.rser.2021.111379]

[122] G. Atiwesh, A. Mikhael, C.C. Parrish, J. Banoub, and T.A.T. Le, "Environmental impact of bioplastic use: A review", *Heliyon,* vol. 7, no. 9, p. e07918, 2021.
[http://dx.doi.org/10.1016/j.heliyon.2021.e07918] [PMID: 34522811]

[123] Y. Pujara, P. Pathak, A. Sharma, and J. Govani, "Review on Indian Municipal Solid Waste Management practices for reduction of environmental impacts to achieve sustainable development goals", *J. Environ. Manage.,* vol. 248, p. 109238, 2019.
[http://dx.doi.org/10.1016/j.jenvman.2019.07.009] [PMID: 31319199]

[124] L.P. Pinheiro, A.A. Longati, A.M. Elias, C.L. Perez, L.P.R.C. Pereira, T.C. Zangirolami, F.F. Furlan, R.C. Giordano, and T.S. Milessi, "Improving the Feasibility of 2G Ethanol Production from Lignocellulosic Hydrolysate Using Immobilized Recombinant Yeast: A Technical–Economic Analysis and Life Cycle Assessment", *Fermentation (Basel),* vol. 11, no. 3, p. 116, 2025.
[http://dx.doi.org/10.3390/fermentation11030116]

[125] N. Mohan, "₹900-crore 2G ethanol plant in Panipat will generate fuel from crop waste", Hindustan Times, August 10, 2022. [Online]. Available from: https://www.hindustantimes.com/cities/chandigarh-news/900crore-2g-ethanol-plant-in-panipat-will-generate-fuel-from-crop-waste-101660079764635.html

[126] M.D. Gupta, "When farmers give you stubble, make ethanol — how IOCL Panipat plant can cut farm fires, emissions", ThePrint, November 22, 2022. [Online]. Available from:https://theprint.in/economy/when-farmers-give-you-stubble-make-ethanol-how-iocl-panipat-plant-can-cut-fa zrm-fire--emissions/1227511/

CHAPTER 3

Biomass-Derived Biochar Materials as Sustainable Energy Resources and Their Assessment

Sanskriti Thapliyal[1], Deepshikha Kothari[1], Alka N. Choudhary[2], Neetu Sharma[3,*], Bhawana[3], Harish Chandra Joshi[3] and Promila Sharma[4]

[1] *Department of Biotechnology, Graphic Era (Deemed to be) University, Dehradun, Uttarakhand, India*

[2] *Department of Pharmaceutical Science, ICFAI University, Dehradun, Uttarakhand, India*

[3] *Department of Chemistry, Graphic Era (Deemed to be) University, Dehradun, Uttarakhand, India*

[4] *Department of Microbiology, Graphic Era (Deemed to be) University, Dehradun, Uttarakhand, India*

Abstract: The rise in global warming, environmental pollution, and population growth has significantly increased energy consumption, leading to greater dependence on fossil fuels. Biomass and its derivatives offer substantial potential as renewable energy sources, making them promising alternatives for sustainable energy conversion, storage, and production. Through pyrolysis, biomass produces a porous, carbon-rich solid material called biochar. A green and sustainable platform for preparing various functional carbon materials is provided by biochar materials derived from biomass. The various types of biomass exhibit distinct physical and chemical properties. In this chapter, a summary of various methods used for preparing different types of biomass-derived biochar materials is provided. An investigation into the numerous potential applications of biomass in environmental reduction and its mechanisms has been conducted. Further assessment of these materials, including their properties and the challenges involved, is also discussed.

Keywords: Biomass, Biochar, Cellulose, Hemicellulose, Hydrothermal carbonization, Lignin, Organometallic compounds.

INTRODUCTION

Oil, natural gas, and coal are among the primary fossil fuels used to power various sectors, and they have significantly contributed to meeting the growing global

* **Corresponding author Neetu Sharma:** Department of Chemistry, Graphic Era (Deemed to be) University, Dehradun, Uttarakhand, India; E-mail: neetu.cherry@gmail.com

Harish Chandra Joshi, Anand Chauhan, Mikhail Vlaskin & Maulin P. Shah (Eds.)

energy demand. However, the rapid depletion of these resources, along with their environmental impact and the escalating demand for energy, underscores the urgent need for clean and sustainable alternatives [1]. Renewable energy systems—such as wind, hydrothermal, geothermal, and solar—often rely on intermittent sources, making efficient energy storage devices essential for storing surplus energy during peak production periods and supplying it when demand increases [2, 3]. Among all the elements, carbon is the fourth most plentiful element in the cosmos and the second highest component in the biosphere. Carbon is a fundamental element that underpins all life on Earth, including plants and animals, and plays a vital role in the energy supply chain through the formation of carbohydrates, which consist primarily of carbon, hydrogen, and oxygen atoms. In addition to the usable portions of plants and animals, the remaining or unused parts constitute biomass—a term that refers to natural organic materials, including plant matter, animal matter, and their wastes [4, 5]. Biomass primarily comes from forests and crops, as well as their post-harvest residues, which are among the most abundant resources, including plants and their residues. Generally speaking, these wastes are frequently burned in areas that pose a significant environmental threat due to pollution and greenhouse gas emissions. However, when biomass is processed into bioethanol, biogas, or solid pellets, it serves as an effective feedstock for energy generation and a valuable source of carbon-based raw materials for producing various commodities [6]. Biomass, used as a renewable source for the production of biochar, (*i.e.* carbon and advanced carbon materials) has recently attracted a lot of research interest in the development of materials that may be important in a variety of applications, including dye degradation, environmental remediation, energy storage devices, hydrogen storage, photovoltaics and carbon (CO_2) capture and storage [7 - 13]. Rapid population growth, along with economic and industrial development in recent years, has contributed to significant environmental and energy-related challenges. Over the past few decades, fossil fuel usage has increased due to the outdated waste management system and the growing energy demands of industry and communities. Even while fossil fuels have many benefits, including affordability, compatibility, and ease of access, they are non-renewable resources that could potentially run out in the future and produce significant greenhouse gas emissions [14]. Therefore, turning waste into eco-friendly and efficient products, and creating eco-friendly and renewable technologies for the production of energy, its storage, and use, are crucial worldwide concerns that must be addressed. However, throughout the pollutant's degradation process, the chemical energy present in pollutants is disregarded and wasted, while traditional waste treatment procedures primarily concentrate on eliminating or destroying the pollutant molecules [15]. To achieve carbon neutrality, the wasted energy is manifested in increased power consumption, higher costs in physicochemical processes or biomass generation, and additional costs throughout the life cycle. Finding appropriate technology to meet the growing demand for sustainable energy and waste treatment on a worldwide scale is therefore crucial. The heated disintegration of biomass, obtained from plants or animals, decomposed thermally without oxygen, produces biochar —a carbonaceous solid substance that is porous, with aromatization on the higher side, and is resistant to degradation [16]. Due to its diverse applications in various industrial and agricultural processes, it has garnered considerable attention recently. The numerous physicochemical properties of biochar have a substantial influence on its diverse range of uses [17]. According to the latest data, the material and production process have a significant impact on the properties of biochar, including density, porosity, pH, and concentrations of elemental ingredients. These properties collectively determine the suitability of biochar for various applications. It is widely used in waste treatment across multiple sectors to remove organic and inorganic contaminants, including pigments and dyes from textile effluents [18, 19].

It is used in agriculture to improve the soil's quality. It improves the quality of the soil by slowing down the rate at which nutrients degrade [20, 21]. Due to its high carbon content, biochar can be advantageous for generating power, as it can be used as a fuel [22]. It has been determined that biomass is a good source of minerals, chemicals, and renewable energy [23]. Biomass of any kind might be used to manufacture bio-oil and biochar [24]. However, the abundance of feedstock is reduced by the high cost of manufacturing and regulatory restrictions [25].

Sources of Biomass

The primary sources of biomass feedstock include digestate, energy crops, crop residues, animal waste, agricultural waste, algae biomass, and activated sludge [26, 27]. Biomass can be categorized based on its composition, source, and properties, as outlined below:

Forestry

Forest biomass has historically played a vital role in the development of human society through the utilization of various forest components, including timber, bark, branches, leaves, and roots. Traditional forest biofuels, including firewood and charcoal, are still widely used in many regions around the world.

One of the primary sources of forest biomass is the material obtained from harvesting trees and their by-products. Additionally, energy crops represent another significant source. These crops are cultivated specifically for energy production and are considered cost-effective and low-maintenance. Energy crops are utilized to produce various forms of bioenergy, including liquid bioethanol, solid fuel, and biogas, which are used to generate heat and electricity. To improve their usability and efficiency, these biomasses are often densified into pellets or briquettes.

These plant-based biomasses, composed primarily of cellulose, hemicellulose, and lignin, are collectively referred to as lignocellulosic biomass. Research has shown that carbon derived from poplar wood contains a notable amount of nitrogen heteroatoms, which enhances its overall electrical conductivity. Furthermore, Chen *et al.* developed porous activated carbon from bamboo, demonstrating its potential use in supercapacitor electrodes.

Agricultural Residues

Agricultural residue refers to the leftover plant materials that remain after agricultural activities, such as crop harvesting or processing. These residues are typically categorized into two groups.

- **Crop residues** include materials left on the field after harvest, such as stems, stalks, straws, leaves, and bagasse.
- **Processing residues** are by-products generated during the processing of harvested crops, including bran, husks, and shells.

These residues are produced in large quantities annually from crops such as rice, maize, sugarcane, groundnuts, cotton, beans, jute, and others. They are primarily composed of **cellulose, hemicellulose, and lignin**, though the proportions vary by biomass type. These components are essential for bioenergy production and other industrial applications.

Industrial and Food Residues

Industrial and food residues refer to the by-products or waste materials generated during manufacturing and food processing activities. They are often rich in organic or mineral content and can be repurposed for various sustainable uses. Both types of residues can be valuable resources for producing bioenergy, biochar, fertilizer, or biochemicals, helping reduce environmental impact and support circular economy practices.

Industries generate various types of biowaste from resources such as food waste, wood processing, and the pulp industry. For example, fruit pulp processing generates waste, including seeds, peels, and sacks, which are rich in minerals such as magnesium, phosphorus, potassium, calcium, and phenolic compounds.

Additionally, shellfish waste (*e.g.*, shrimp, lobster, crab shells) and butterfly wings contain chitin—a nitrogen-rich, structural polysaccharide that is the second most abundant organic material on earth. These chitin-rich materials can serve as natural templates for producing biochar, a carbon-rich material used in energy storage, soil improvement, and environmental applications.

Municipal Waste

Municipal waste—also referred to as garbage—is categorized into two main types based on their biotic and abiotic characteristics: organic and inorganic waste [28]. This waste can be further classified into materials such as textiles, wood debris, glass, metal scraps, food residues, and paper-based residues. Municipal waste is generated from a wide range of human activities. In line with the waste-to-energy concept, organic solid waste can be converted into energy through various processes, including pyrolysis and incineration.

BIOMASS CLASSIFICATION

Due to its diverse nature and composition, biomass is classified based on origin, function, and end use. It is generally grouped into two categories:

  i. **Natural Type-Based Classification** (*e.g.*, plant type, ecology)
 ii. **Usage-Based Classification** (*e.g.*, feedstock application)

According to the most widely accepted classification, biomass is classified as follows:

Wood and woody biomass

Herbaceous biomass

Aquatic biomass

Animal and human waste biomass

Wood and Woody Biomass

Lignin and biomass are key components of woody biomass, a category of plant-based material derived primarily from trees and shrubs. Woody biomass is particularly rich in lignin, which provides structural support to plant cell walls and contributes to the material's high energy content. Lignin is a complex organic polymer with having high molecular weight.

This type of biomass encompasses a diverse range of forest-derived materials, including tree trunks, branches, bark, roots, and woody shrubs, as well as their above- and below-ground components. It also encompasses tree remnants left behind after logging or land-clearing operations.

Woody biomass can be converted into energy through several processes, including direct combustion, gasification, pyrolysis, and biochemical conversion methods. These processes transform biomass into useful forms of energy, such as heat, electricity, bio-oil, or synthetic gas (syngas). Due to its high lignin content, woody biomass is particularly suitable for thermochemical conversion, offering a reliable and renewable alternative to fossil fuels in both industrial and residential energy applications.

Herbaceous Biomass

According to European Standard EN 14961-1, herbaceous biomass is categorized as solid biofuels [29]. It is derived from plants with non-woody stems that die back after the growing season. This category includes cereal straw and other by-products from grain and seed crops used in the food processing industry.

Aquatic Biomass

According to Di Benedetto, the term "aquatic biomass" refers to organic material derived from water-based organisms, such as microalgae, macroalgae (seaweed), and emergent aquatic plants [30]. These organisms thrive in various aquatic environments.

There are an estimated 55,000 species and over 100,000 strains of algae—both freshwater and terrestrial—that naturally occur in ecosystems worldwide. A key advantage of using these organisms lies in their remarkable ability to photosynthesize, converting sunlight, carbon dioxide (CO_2), and water into a wide range of valuable metabolites and compounds. These include lipids, proteins, carbohydrates, and pigments, which collectively make up the algal biomass [31].

This efficient conversion process, combined with their fast growth rates and ability to thrive in non-arable land and wastewater, makes aquatic biomass a promising and sustainable resource for applications in bioenergy, pharmaceuticals, agriculture, and environmental remediation.

Animal and Human Waste Biomass

Animal and human waste biomass refers to organic materials derived from the excreta and by-products of animals and humans that can be used as a renewable energy source or for other sustainable applications.

Animal waste biomass comprises organic materials such as manure, urine, by-products from slaughterhouses, and other residues generated through livestock and poultry farming.

Human waste biomass consists of sewage sludge, fecal matter, and wastewater from domestic or municipal sources.

These waste materials are rich in organic content and nutrients, making them suitable for biogas production (*via* anaerobic digestion), composting, and even fertilizer production. The utilization of animal and human waste biomass contributes to sustainable waste management, reduces greenhouse gas emissions, and supports circular economy practices.

Methods for the Synthesis of Biochar from Biomass

Various synthesis methods are used to convert biomass into carbon-based materials, including physical, thermochemical [32], and biochemical processes. Common techniques include pyrolysis, hydrothermal carbonization, chemical activation, and physical activation. Each technique differs in terms of temperature, pressure, and the use of activating agents, all of which influence the structure, surface area, and porosity of the resulting carbon material. For efficient biochar production, the moisture content of the biomass should be below 30%. Reducing moisture through methods such as sun-drying or air-drying lowers energy consumption and helps alleviate pressure on energy resources [33, 34].

Biochar from the Pyrolysis of Biomass

Pyrolysis refers to the thermal breakdown of organic matter in an oxygen-free environment, typically occurring at temperatures between 250°C and 900°C. This process is an alternative method for converting waste biomass into value-added products, including biochar, bio-oil, and syngas. Biomass, such as wood, agricultural waste, and coconut shells, is the source of the majority of carbon-based materials utilized in daily life. Biomass consists of three primary constituents: cellulose, hemicellulose, and lignin. These materials are heated to high temperatures and processed in an inert gas environment. During the initial phase of pyrolysis (below 100°C), moisture is primarily volatilized. At elevated temperatures (220°C–315°C), hemicellulose decomposes readily, while cellulose undergoes pyrolysis mainly between 315°C and 400°C. Lignin, in contrast, decomposes over a broader temperature range [35]. Pyrolysis produces solid, liquid, and gaseous products. The solid and liquid outputs include biochar and bio-oil, while the gaseous products consist of carbon dioxide, carbon monoxide, hydrogen, and syngas.

The biochar obtained by the pyrolysis process has a limited Specific Surface Area (SSA) and poorly developed pore structures, which restrict its effectiveness in applications like energy storage and catalysis. To improve these properties, activation processes are employed, and they are of two types:

i. **Physical Activation:** This involves a two-step process where the biomass is first carbonized at moderate temperatures (usually below 800°C) to produce a char. The char is then activated at higher temperatures using agents such as carbon dioxide (CO_2), steam, or air. This process creates a network of pores, increasing the SSA and enhancing the material's performance.

ii. **Chemical Activation:** In this method, the biomass is impregnated with a chemical activating agent like potassium hydroxide (KOH) or phosphoric acid (H_3PO_4). The mixture is then heated at lower temperatures (typically between 400 °C and 700°C). The chemical agent facilitates the breakdown of the biomass structure, leading to the formation of a highly porous carbon material with a high SSA [36]. Among various activating agents, KOH is particularly effective due to its ability to produce carbons with ultrahigh SSA (up to 3000 m^2/g) and well-defined micropore structures. These enhanced properties make the resulting carbons suitable for applications such as supercapacitors, batteries, and other energy storage devices.

The KOH activation of carbon involves three key steps:

i. During calcination, potassium-based compounds such as KOH, K_2CO_3, and K_2O interact with carbon, promoting the development of pores in the material.
ii. Intermediate gases (H_2O, CO_2) further enhance porosity through carbon gasification.
iii. Metallic potassium intercalates into the carbon lattice, expanding it, and upon removal, leaves behind a porous structure.

Molten Salt Carbonization (MSC)

Molten Salt Carbonization enables the synthesis of porous carbon materials with high conductivity and well-defined pore structures, eliminating the need for additional high-temperature treatment. Unlike solid-state reactions, MSC benefits from improved mass transfer in molten salt media, enhancing carbon formation. Biomass-derived precursors, such as glucose or peanut shells, can produce amorphous carbons with tunable porosity, especially when using salt systems like LiCl–KCl or Na_2CO_3–K_2CO_3. Adding activators (*e.g.*, KOH, K_2SO_3) can achieve ultra-high surface areas (up to 3200 m^2/g) [37]. However, challenges remain in controlling porosity when working with complex natural polymers (cellulose, lignin, hemicellulose) due to their insolubility and structural complexity. Developing suitable additives and optimized processes is key to advancing biomass-based carbon synthesis.

Hydrothermal Carbonization

Hydrothermal carbonization is a cost-effective technique for producing biochar, typically conducted at relatively low temperatures, ranging from 180°C to 250°C (ref). Most biomass resources contain up to 95% moisture. For pyrolysis, biomass with a moisture content exceeding 30% must be dried prior to processing [38]. The process known as hydrothermal carbonization (HTC) converts biomass into carbonaceous biofuel in the presence of water by applying pressure and heat. The hydrolyzed product proceeds through a series of reactions, such as dehydration, fragmentation, and isomerization, to form intermediate products and their derivatives. This method of converting wet biomass into biofuels without the need for drying is potentially promising [39]. An HTC method produces primary solids, known as hydrochar, when liquid-containing solid biomass is heated to lower temperatures (<200 °C) in a confined chamber under endogenous pressure [40].

Hydrothermal liquefaction (HTL) converts biomass fuel primarily into a liquid at temperatures of 200°C and 350 °C. At critical temperatures and pressures comparable to water (374°C and 22.1MPa), biomass is essentially converted into a gaseous medium (H_2, CO, CH_4, and CO_2) by the process of hydrothermal gasification (HTG). Hydrochar specifications feature a higher H/C ratio than those of biochar [41, 42].

Torrefaction

This method represents a recent advancement in biochar production technology. This process operates at a low heating rate and is therefore referred to as mild pyrolysis. The material must be heated gradually, from 200 °C to 300 °C, at atmospheric pressure in the absence of oxygen, to carry out torrefaction, a medium-temperature pyrolysis [43]. Torrefaction partially decomposes cellulose, hemicellulose, lignin, biopolymers, and removes excess moisture and volatile compounds through the production of organic volatiles. Compared to liquid or gaseous products, torrefaction predominantly yields solid products. For biochar, the molar oxygen-to-carbon (O/C) ratio should be less than 0.4. Notably, torrefied material typically achieves a more favorable O/C ratio than the standard set by the European Commission for biochar [20].

Characterization of Biochar

For evaluating the effectiveness of biomass in pollutant removal and other applications, its characterization is essential. The interaction between metal and biochar is influenced by pH. These properties underscore the strong potential of biochar as an effective adsorbent for removing various contaminants from soil. To understand its properties, biochar is characterized through methods that focus on its structural features, elemental makeup, and surface functional groups. Different types of advanced techniques, such as X-ray Diffraction (XRD), Nuclear Magnetic Resonance (NMR), Scanning Electron Microscopy (SEM), Fourier Transform Infrared Spectroscopy (FTIR), Thermogravimetric Analysis (TGA), Brunauer-Emmett-Teller (BET) surface area analysis, proximate and ultimate analysis, and Raman Spectroscopy is currently used for their characterization.

Applications of Biochar Derived from Biomass

Carbon Materials Derived (biochar) from biomass encapsulates an approach to renewable energy wherein carbon-based materials, produced from organic matter (biomass), are used to generate or store energy in environmentally friendly ways. Some of them are as follows:

Biochar used as a Catalyst

Biochar has emerged as a versatile catalyst with wide-ranging applications across various sectors, including agriculture, environmental remediation, and energy systems [44]. Its effectiveness as a catalyst is primarily attributed to its unique physicochemical properties. One of the most critical features is its large surface area, which gives numerous active sites for catalytic reactions due to the abundance of surface functional groups. For instance, hydroxyl groups (–OH) present on the biochar surface are responsible for the adsorption of contaminants, thereby enhancing its catalytic performance. Biochar is also used as a catalyst in the production of biodiesel, generation of energy, tar removal, waste management, syngas production, serving as electrode material in microbial fuel cells, chemical synthesis, and the removal of pollutants from the environmental.

Energy Production

In biomass gasification, the formation of tar is considered problematic, as its condensation can lead to the fouling and blockage of downstream equipment, ultimately decreasing the energy efficiency of the process. Catalytic conversion of tar offers a solution by transforming it into valuable gases, such as carbon monoxide (CO) and hydrogen (H_2), which are key components of synthesis gas (syngas). The type of char obtained from various biomass sources (rice straw) plays a significant role in tar removal efficiency.

Biofuels

Biofuel serves as an excellent alternative to petroleum-based fuels due to its biodegradability, non-toxicity, renewability, and comparable performance to fossil fuels. It is commonly synthesized by converting vegetable oils through transesterification or by reacting Free Fatty Acids (FFAs) with alcohols *via* the esterification process. The catalysts based on biochar are increasingly employed in these processes, enhancing the efficiency of both transesterification and esterification reactions.

Waste Management

Many synthetic chemical compounds are highly resistant to biological decay, making them bio-recalcitrant. These substances pose serious environmental and health risks, as they can be carcinogenic to humans, microorganisms, plants, and other living organisms. One practical approach for breaking down these persistent pollutants is the Catalytic Ozonation Process (COP). In this method, biochar derived from biomass is employed as a cost-effective catalyst due to its porous structure and the presence of functional groups such as phenolic and hydroxyl groups. This biochar has been successfully used to degrade resilient organic contaminants.

Energy Storage and Supercapacitors

It plays a crucial role in the functionality of electrical and electronic devices, directly affecting consumer usage and performance. Supercapacitors have garnered significant attention as energy storage systems due to their high power density, rapid charge-discharge capability, and long cycle life. In contrast, rechargeable batteries (Li batteries) have higher energy density but slower charge and discharge rates. The performance of these energy storage devices is primarily influenced by the electrode materials used, which typically require a high surface area and more porosity in their structure, providing active sites for electrochemical reactions. Mainly used carbon-based materials for electrodes include graphene, activated carbon, and carbon nanotubes. However, the production of these materials is costly, which limits their widespread application. To address this limitation, biochar has emerged as a promising alternative due to its comparable structural properties, such as high surface area and porosity, combined with its lower cost. Biochar is being increasingly explored as an electrode material in technologies such as microbial fuel cells and supercapacitors.

Control of Air Pollutants

Biochar has been explored as a potential catalyst for low-temperature Selective Catalytic Reduction (SCR) processes, aiming to control air pollutants. Research has demonstrated the use of biomass sources such as sewage sludge and rice straw to produce biochar, which is then applied as a catalyst in systems where ammonia serves as the reducing agent. These biochars are typically modified through either physical or chemical activation, with studies showing that chemically activated biochar achieves higher pollutant removal efficiency. During the catalytic process, biochar facilitates the generation of sulfate species and free radicals, which contribute to the degradation of pollutants. Additionally, the oxygen-containing functional groups on the surface of biochar form complexes that improve the overall catalytic activity, further enhancing the removal of air pollutants.

Soil Enhancement and Rehabilitation

Biochar is used to remove heavy metals and organic contaminants from soil. Biochar remediates soil by a primary method known as adsorption [45]. Interactions, including π-π interactions, electrostatic attractions, acid-base interactions, surface complexation, and hydrogen bonding, are all components of the adsorption mechanism of biochar [45].

Decontamination of Water and Wastewater

Numerous studies have demonstrated that biochar can effectively remove contaminants from water and wastewater, including both organic and inorganic pollutants, through adsorption processes [45]. Among emerging environmental pollutants, antibiotics are increasingly common organic contaminants [46]. Biochar derived from slurry has been demonstrated to be a cost-effective and efficient adsorbent for removing antibacterial drugs [45].

Carbon Sequestration

Climate change has intensified global efforts to reduce carbon dioxide (CO_2) emissions into the atmosphere. Soils play a vital role in the carbon cycle, which has a direct influence on climate regulation. Among various strategies, carbon sequestration in soil has emerged as a promising approach to mitigate CO_2 emissions [45]. Biochar contributes positively to this process due to its aromatic and stable structure, which makes it highly resistant to microbial degradation. Numerous studies have investigated the role of biochar in carbon sequestration. However, results have been mixed, with both increases and decreases in soil carbon emissions being reported [45].

The extent of carbon mineralization, where organic carbon is converted to CO_2, varies depending on soil fertility and organic carbon content. Soils with lower fertility tend to exhibit higher rates of mineralization, and similarly, soils with higher existing carbon content also show increased carbon turnover [47]. The carbon in biochar is of two types: labile (easily degradable) and recalcitrant (highly stable). Labile carbon is readily metabolized by soil microbes during the initial stages of biochar application, often resulting in a temporary increase in mineralization of carbon. In contrast, recalcitrant carbon remains stable in the soil over extended periods, contributing to long-term carbon fixation [48]. Overall, the net carbon sequestration potential of biochar depends on the balance between carbon losses from labile fractions and gains from the stability of recalcitrant carbon.

Gaps and Future Perspectives

Biochar is recognized as a renewable and sustainable material with significant potential to address a wide range of environmental challenges, including the remediation of pollutants in soil, water, and air. One area of growing interest is the activation of biochar to enhance its capacity for removing targeted contaminants. However, further research is necessary to develop novel activation techniques and better understand the mechanisms behind pollutant adsorption and desorption.

The interaction between biochar and soil microbial communities remains an area that requires further investigation. It is important to understand how biochar influences microbial growth, activity, and community composition. When incorporated into soil, biochar not only aids in contaminant removal and enhances soil fertility but also serves as a source of essential micro- and macronutrients. Further studies are needed to explore microbial dynamics during the mineralization process and their role in soil remediation. Additionally, the interaction of biochar with soil, particularly in terms of binding mechanisms and long-term stability, should be studied more comprehensively.

In wastewater treatment, the exact mechanisms by which biochar removes contaminants are still not fully understood. Emerging research suggests the potential of electrochemical conversion of solid carbon materials, such as biochar, into electricity through direct carbon fuel cells. However, key challenges remain in understanding the chemical kinetics of the reaction and the oxidation processes occurring at the anode/electrolyte interface. Limited knowledge exists about the interactions between solid carbon and electrolyte interfaces, highlighting a need for further investigation.

Despite the numerous advantages of biochar, particular concerns persist. Depending on the feedstock and pyrolysis conditions, biochar may contain harmful substances, including dioxins, chlorinated hydrocarbons, and polycyclic aromatic hydrocarbons. In applications like supercapacitors, the performance of biochar-based electrodes requires further optimization and study.

To fully assess the economic and environmental potential of biochar, comprehensive life-cycle assessments are essential. Advances in biochar characterization techniques have enhanced our understanding of its properties; however, optimizing these properties through activation remains crucial for maximizing its effectiveness. The adoption of advanced technologies for biochar production and analysis is influenced by both cost and accessibility. As biochar continues to gain prominence as a viable alternative material, the establishment of standardized characterization methods will be vital for consistent evaluation and application across various fields.

CONCLUSION

The production of biochar involves a diverse range of biomass feedstocks, which are pyrolyzed using various techniques to address water pollution. The properties of the resulting biochar are heavily influenced by factors such as pyrolysis temperature, feedstock type, and the specific pyrolysis technology employed. Biochar has proven to be a highly effective material for removing toxic pollutants, primarily due to the presence of functional groups, such as hydroxyl and carboxyl groups, on its surface. While the efficiency of biochar for pollutant removal depends on both the type of biomass and pyrolysis conditions, future developments in biochar production will likely focus on fine-tuning its properties to enhance performance.

Biochar holds significant promise as a sustainable solution for pollutant removal; however, its economic feasibility and recyclability must be carefully considered when developing biochar for broader environmental applications. The relationship between various waste management strategies and energy production varies based on several parameters, including the production techniques and the economic, social, and ecological constraints associated with each approach. By transitioning from a linear to a circular waste management model, higher energy recovery potential can be realized. This circular economy framework emphasizes the importance of sustainability and resource recovery.

This review paper provides a comprehensive summary of the current state of biochar research, offering valuable insights into new opportunities for scientific innovation in the field.

REFERENCES

[1] B. Hussain, Komal Batool, A. Naqvi, A. A. Nassani, and S. Ali, "Racing towards environmental sustainability by lowering fossil resources in the energy mix during era of global boiling," *Applied Energy*, pp. 124847–124847, 2025.
[http://dx.doi.org/10.1016/j.apenergy.2024.124847]

[2] E. Enasel and G. Dumitrascu, "Storage solutions for renewable energy: A review," *Energy Nexus*, pp. 100391–100391, 2025.
[http://dx.doi.org/10.1016/j.nexus.2025.100391]

[3] C. Raghutla and Y. Kolati, "Public-private partnerships investment in energy as new determinant of renewable energy: The role of political cooperation in China and India," *Energy Reports*, vol. 10, pp. 3092–3101, 2023.
[http://dx.doi.org/10.1016/j.egyr.2023.09.139]

[4] J.J. Chew, and V. Doshi, "Recent advances in biomass pretreatment – Torrefaction fundamentals and technology", *Renew. Sustain. Energy Rev.,* vol. 15, no. 8, pp. 4212-4222, 2011.
[http://dx.doi.org/10.1016/j.rser.2011.09.017]

[5] T. Wang, Y. Zhai, Y. Zhu, C. Li, and G. Zeng, "A review of the hydrothermal carbonization of biomass waste for hydrochar formation: Process conditions, fundamentals, and physicochemical properties", *Renew. Sustain. Energy Rev.,* vol. 90, pp. 223-247, 2018.
[http://dx.doi.org/10.1016/j.rser.2018.03.071]

[6] S. Mahapatra, D. Kumar, B. Singh, and P. K. Sachan, "Biofuels and their sources of production: A review on cleaner sustainable alternative against conventional fuel, in the framework of the food and energy nexus," Energy Nexus, vol. 4, no. 100036, p. 100036, 2021.
[http://dx.doi.org/10.1016/j.nexus.2021.100036]

[7] L. Li, F. Sun, J. Gao, L. Wang, X. Pi, and G. Zhao, "Broadening the pore size of coal-based activated carbon *via* a washing-free chem-physical activation method for high-capacity dye adsorption", *RSC Advances,* vol. 8, no. 26, pp. 14488-14499, 2018.
[http://dx.doi.org/10.1039/C8RA02127A] [PMID: 35540785]

[8] W. Tian, H. Zhang, H. Sun, M.O. Tadé, and S. Wang, "One-step synthesis of flour-derived functional nanocarbons with hierarchical pores for versatile environmental applications", *Chem. Eng. J.,* vol. 347, pp. 432-439, 2018.
[http://dx.doi.org/10.1016/j.cej.2018.04.139]

[9] J. Tang, W. Zhu, R. Kookana, and A. Katayama, "Characteristics of biochar and its application in remediation of contaminated soil", *J. Biosci. Bioeng.,* vol. 116, no. 6, pp. 653-659, 2013.
[http://dx.doi.org/10.1016/j.jbiosc.2013.05.035] [PMID: 23810668]

[10] L.S. Blankenship II, N. Balahmar, and R. Mokaya, "Oxygen-rich microporous carbons with exceptional hydrogen storage capacity", *Nat. Commun.,* vol. 8, no. 1, p. 1545, 2017.
[http://dx.doi.org/10.1038/s41467-017-01633-x] [PMID: 29146978]

[11] L. Wang, Y. Shi, Y. Wang, H. Zhang, H. Zhou, Y. Wei, S. Tao, and T. Ma, "Composite catalyst of rosin carbon/Fe3O4: highly efficient counter electrode for dye-sensitized solar cells", *Chem. Commun. (Camb.),* vol. 50, no. 14, pp. 1701-1703, 2014.
[http://dx.doi.org/10.1039/c3cc47163b] [PMID: 24396856]

[12] H. Jing, Y. Shi, D. Wu, S. Liang, X. Song, Y. An, and C. Hao, "Well-defined heteroatom-rich porous carbon electrocatalyst derived from biowaste for high-performance counter electrode in dye-sensitized solar cells", *Electrochim. Acta,* vol. 281, pp. 646-653, 2018.
[http://dx.doi.org/10.1016/j.electacta.2018.06.020]

[13] N. Balahmar, A.C. Mitchell, and R. Mokaya, "Generalized mechanochemical synthesis of biomass-derived sustainable carbons for high performance CO_2 storage", *Adv. Energy Mater.,* vol. 5, no. 22, p. 1500867, 2015.
[http://dx.doi.org/10.1002/aenm.201500867]

[14] E. Ravishankar, R.E. Booth, J.A. Hollingsworth, H. Ade, H. Sederoff, J.F. DeCarolis, and B.T. O'Connor, "Organic solar powered greenhouse performance optimization and global economic opportunity", *Energy Environ. Sci.,* vol. 15, no. 4, pp. 1659-1671, 2022.
[http://dx.doi.org/10.1039/D1EE03474J]

[15] D. Yuan, Z. Zhai, E. Zhu, H. Liu, T. Jiao, and S. Tang, "Humic acid removal in water *via* UV activated sodium perborate process", *Coatings,* vol. 12, no. 7, pp. 885-885, 2022.
[http://dx.doi.org/10.3390/coatings12070885]

[16] S. M. Derdag and N. Ouazzani, "Advancements in sustainable biochar production from waste: pathways for renewable energy generation and environmental remediation," Biomass, vol. 5, no. 2, p. 32, 2025.
[http://dx.doi.org/10.3390/biomass5020032]

[17] F. Amalina, A.S.A. Razak, S. Krishnan, A.W. Zularisam, and M. Nasrullah, "Water hyacinth (*Eichhornia crassipes*) for organic contaminants removal in water – A review", *Journal of Hazardous Materials Advances,* vol. 7, p. 100092, 2022.
[http://dx.doi.org/10.1016/j.hazadv.2022.100092]

[18] P.V. Nidheesh, A. Gopinath, N. Ranjith, A. Praveen Akre, V. Sreedharan, and M. Suresh Kumar, "Potential role of biochar in advanced oxidation processes: A sustainable approach", *Chem. Eng. J.,* vol. 405, p. 126582, 2021.
[http://dx.doi.org/10.1016/j.cej.2020.126582]

[19] A.K. Sakhiya, A. Anand, and P. Kaushal, "Production, activation, and applications of biochar in recent times", *Biochar,* vol. 2, no. 3, pp. 253-285, 2020.
[http://dx.doi.org/10.1007/s42773-020-00047-1]

[20] A.G. Daful, and M.R. Chandraratne, *Biochar Production From Biomass Waste-Derived Material,* 2020.
[http://dx.doi.org/10.1016/B978-0-12-803581-8.11249-4]

[21] *IFAD, UNICEF, WFP, and WHO, State Of Food Security And Nutrition In The World 2021. S.L.* Food & Agriculture Org, 2021.

[22] F. Amalina, A. S. A. Razak, S. Krishnan, H. Sulaiman, A. W. Zularisam, and M. Nasrullah, "Biochar production techniques utilizing biomass waste-derived materials and environmental applications – A review," Journal of Hazardous Materials Advances, vol. 7, no. 2772–4166, p. 100134, 2022.
[http://dx.doi.org/10.1016/j.heliyon.2022.e08905]

[23] P.K. Sadh, S. Duhan, and J.S. Duhan, "Agro-industrial wastes and their utilization using solid state fermentation: A review", *Bioresour. Bioprocess.,* vol. 5, no. 1, p. 1, 2018.
[http://dx.doi.org/10.1186/s40643-017-0187-z]

[24] W. N. R. W. Isahak, M. W. M. Hisham, M. A. Yarmo, and T. Yun Hin, "A review on bio-oil production from biomass by using pyrolysis method," *Renewable and Sustainable Energy Reviews,* vol. 16, no. 8, pp. 5910–5923, 2012.
[http://dx.doi.org/10.1016/j.rser.2012.05.039]

[25] M. Tripathi, J.N. Sahu, and P. Ganesan, "Effect of process parameters on production of biochar from biomass waste through pyrolysis: A review", *Renew. Sustain. Energy Rev.,* vol. 55, pp. 467-481, 2016.
[http://dx.doi.org/10.1016/j.rser.2015.10.122]

[26] M. Raud, T. Kikas, O. Sippula, and N.J. Shurpali, "Potentials and challenges in lignocellulosic biofuel production technology", *Renew. Sustain. Energy Rev.,* vol. 111, pp. 44-56, 2019.
[http://dx.doi.org/10.1016/j.rser.2019.05.020]

[27] D. Formenti, F. Ferretti, F.K. Scharnagl, and M. Beller, "Reduction of nitro compounds using 3d-no-noble metal catalysts", *Chem. Rev.,* vol. 119, no. 4, pp. 2611-2680, 2019.
[http://dx.doi.org/10.1021/acs.chemrev.8b00547] [PMID: 30516963]

[28] P. Li, K. Lin, Z. Fang, and K. Wang, "Enhanced nitrate removal by novel bimetallic Fe/Ni

nanoparticles supported on biochar", *J. Clean. Prod.,* vol. 151, pp. 21-33, 2017.
[http://dx.doi.org/10.1016/j.jclepro.2017.03.042]

[29]   D. Revel, "IPCC Special Report on Renewable Energy Sources and Climate Change Mitigation," May 2011.

[30]   J. Milledge, B. Smith, P. Dyer, and P. Harvey, "Macroalgae-derived biofuel: A review of methods of energy extraction from seaweed biomass", *Energies,* vol. 7, no. 11, pp. 7194-7222, 2014.
[http://dx.doi.org/10.3390/en7117194]

[31]   S.Jha, R.Singh,B.K. Pandey, "Recent aspects of algal biomass for sustainable fuel production: A review," *Discover Sustainability,* vol. 5, no. 1, 2024.
[http://dx.doi.org/10.1007/s43621-024-00472-3]

[32]   A. Saravanakumar, P. Vijayakumar, A.T. Hoang, E.E. Kwon, and W.H. Chen, "Thermochemical conversion of large-size woody biomass for carbon neutrality: Principles, applications, and issues", *Bioresour. Technol.,* vol. 370, p. 128562, 2023.
[http://dx.doi.org/10.1016/j.biortech.2022.128562] [PMID: 36587772]

[33]   K.M. Bryden, and M.J. Hagge, "Modeling the combined impact of moisture and char shrinkage on the pyrolysis of a biomass particle", *Fuel,* vol. 82, no. 13, pp. 1633-1644, 2003.
[http://dx.doi.org/10.1016/S0016-2361(03)00108-X]

[34]   D. Lv, M. Xu, X. Liu, Z. Zhan, Z. Li, and H. Yao, "Effect of cellulose, lignin, alkali and alkaline earth metallic species on biomass pyrolysis and gasification", *Fuel Process. Technol.,* vol. 91, no. 8, pp. 903-909, 2010.
[http://dx.doi.org/10.1016/j.fuproc.2009.09.014]

[35]   H. Yang, R. Yan, H. Chen, D.H. Lee, and C. Zheng, "Characteristics of hemicellulose, cellulose and lignin pyrolysis", *Fuel,* vol. 86, no. 12-13, pp. 1781-1788, 2007.
[http://dx.doi.org/10.1016/j.fuel.2006.12.013]

[36]   L.L. Zhang, and X.S. Zhao, "Carbon-based materials as supercapacitor electrodes", *Chem. Soc. Rev.,* vol. 38, no. 9, pp. 2520-2531, 2009.
[http://dx.doi.org/10.1039/b813846j] [PMID: 19690733]

[37]   N. Díez, A.B. Fuertes, and M. Sevilla, "Molten salt strategies towards carbon materials for energy storage and conversion", *Energy Storage Mater.,* vol. 38, pp. 50-69, 2021.
[http://dx.doi.org/10.1016/j.ensm.2021.02.048]

[38]   H.I. Abdel-Shafy, and M.S.M. Mansour, "Solid waste issue: Sources, composition, disposal, recycling, and valorization", *Egyptian Journal of Petroleum,* vol. 27, no. 4, pp. 1275-1290, 2018.
[http://dx.doi.org/10.1016/j.ejpe.2018.07.003]

[39]   H. Zhang, G. Xue, H. Chen, and X. Li, "Magnetic biochar catalyst derived from biological sludge and ferric sludge using hydrothermal carbonization: Preparation, characterization and its circulation in Fenton process for dyeing wastewater treatment", *Chemosphere,* vol. 191, pp. 64-71, 2018.
[http://dx.doi.org/10.1016/j.chemosphere.2017.10.026] [PMID: 29031054]

[40]   F. Mbarki, T. Selmi, A. Kesraoui, M. Seffen, P. Gadonneix, A. Celzard, and V. Fierro, "Hydrothermal pre-treatment, an efficient tool to improve activated carbon performances", *Ind. Crops Prod.,* vol. 140, p. 111717, 2019.
[http://dx.doi.org/10.1016/j.indcrop.2019.111717]

[41]   V.K. Ponnusamy, S. Nagappan, R.R. Bhosale, C.H. Lay, D. Duc Nguyen, A. Pugazhendhi, S.W. Chang, and G. Kumar, "Review on sustainable production of biochar through hydrothermal liquefaction: Physico-chemical properties and applications", *Bioresour. Technol.,* vol. 310, p. 123414, 2020.
[http://dx.doi.org/10.1016/j.biortech.2020.123414] [PMID: 32354676]

[42]   D. Castello, A. Kruse, and L. Fiori, "Biomass gasification in supercritical and subcritical water: The effect of the reactor material", *Chem. Eng. J.,* vol. 228, pp. 535-544, 2013.

[http://dx.doi.org/10.1016/j.cej.2013.04.119]

[43] J.J. Manyà, D. García-Morcate, and B. González, "Adsorption performance of physically activated biochars for postcombustion CO_2 capture from dry and humid flue gas", *Appl. Sci. (Basel)*, vol. 10, no. 1, p. 376, 2020.
[http://dx.doi.org/10.3390/app10010376]

[44] Diego, Rafael, R. C. Dutra, Paulo, and G. F. Ghesti, "Advancing biochar applications: A review of production processes, analytical methods, decision criteria, and pathways for scalability and certification," *Sustainability*, vol. 17, no. 6, 2025.
[http://dx.doi.org/10.1016/j.envpol.2011.07.023] [PMID: 21855187]

[45] K.A. Spokas, "Review of the stability of biochar in soils: predictability of O:C molar ratios", *Carbon Manag.*, vol. 1, no. 2, pp. 289-303, 2010.
[http://dx.doi.org/10.4155/cmt.10.32]

[46] P.R. Yaashikaa, P.S. Kumar, S. Varjani, and A. Saravanan, "A critical review on the biochar production techniques, characterization, stability and applications for circular bioeconomy", *Biotechnol. Rep. (Amst.)*, vol. 28, p. e00570, 2020.
[http://dx.doi.org/10.1016/j.btre.2020.e00570] [PMID: 33304842]

[47] A. El-Naggar, S.S. Lee, Y.M. Awad, X. Yang, C. Ryu, M. Rizwan, J. Rinklebe, D.C.W. Tsang, Y.S. Ok, and Y.S. Ok, "Corrigendum to "Influence of soil properties and feedstocks on biochar potential for carbon mineralization and improvement of infertile soils" [Geoderma 332 (2018) 100–108]", *Geoderma*, vol. 456, p. 117254, 2025.
[http://dx.doi.org/10.1016/j.geoderma.2025.117254]

[48] C.C. Figueiredo, J.K.M. Chagas, J. da Silva, and J. Paz-Ferreiro, "Short-term effects of a sewage sludge biochar amendment on total and available heavy metal content of a tropical soil", *Geoderma*, vol. 344, pp. 31-39, 2019.
[http://dx.doi.org/10.1016/j.geoderma.2019.01.052]

CHAPTER 4

Biomass-Derived Biochar for Supercapacitor Applications: A Sustainable Approach to Energy Storage

Biswajyoti Hazarika[1], Biplop Jyoti Hazarika[2] and Md. Juned K. Ahmed[1,*]

[1] Department of Chemistry, Arunachal University of Studies, Namsai-792103, Arunachal Pradesh, India

[2] Department of Chemical Sciences, IISER-Kolkata, Nadia-741246, West Bengal, India

Abstract: Biochar is a carbonaceous material derived from biomass through pyrolysis. It can be used to serve as a sustainable platform for producing various functional carbon materials, such as porous carbon, carbon nanotubes, heteroatom-doped biochar, carbon quantum dots, and graphene. These materials are synthesized through physical and chemical activation processes, taking advantage of biochar's versatile physicochemical properties for electrochemical energy storage applications, particularly supercapacitors. Biochar exhibits a high specific surface area (> 1000 m^2/g), large pore volume (> 0.5 cm^3/g), and high electrical conductivity (10-100 S/m). These characteristics enhance supercapacitor performance metrics, including capacitance, energy density, power density and cycling stability. Studies show that biochar obtained from various feedstocks like agricultural residues and municipal waste can boost supercapacitor's performance while also helping with waste management and carbon sequestration. However, challenges like scalability, cost and optimization of biochar properties persist. Future research will focus on the integration of biochar with advanced materials like graphene and MXenes to further improve its electrochemical performance, showing its potential as a viable and eco-friendly material for next-generation energy storage technologies.

Keywords: Biochar, Electrochemical performance, Energy storage, Supercapacitors, Sustainability.

INTRODUCTION

As global energy demands increase and the environmental impacts of fossil fuels become more evident, the need for sustainable and efficient energy storage solutions has gained unprecedented urgency. Among the promising materials under investigation, biomass-derived biochar has emerged as a compelling and

---

* **Corresponding author Md. Juned K. Ahmed:** Department of Chemistry, Arunachal University of Studies, Namsai-792103, Arunachal Pradesh, India; E-mail: juned.nits@gmail.com

Harish Chandra Joshi, Anand Chauhan, Mikhail Vlaskin & Maulin P. Shah (Eds.)

eco-friendly option for enhancing energy storage technologies, particularly in supercapacitors. This article provides an in-depth exploration of biochar, its production, and applications in supercapacitors, supported by recent research and practical examples that highlight its potential.

Biochar is a carbon-rich material produced through the pyrolysis of organic biomass, such as agricultural residues, forestry by-products, or municipal waste [1, 2]. Pyrolysis is a thermochemical decomposition process carried out in the absence of oxygen that converts biomass into a stable form of carbon. The resulting biochar has a high surface area, significant porosity, and excellent electrical conductivity, making it highly suitable for various applications, including energy storage. The production of biochar typically involves heating the biomass to temperatures ranging from 300-800°C [3 - 5]. During this process, volatile compounds are released, leaving behind a solid carbonaceous material. The specific pyrolysis conditions, such as temperature, heating rate, and duration, affect the properties of the resulting biochar [3]. For instance, higher temperatures generally lead to increased carbon content and enhanced electrical conductivity of biochar, while lower temperatures tend to increase surface area and porosity [6 - 8]. The production of biochar offers significant environmental advantages, converting waste materials into a valuable resource while simultaneously sequestering carbon and reducing greenhouse gas emissions. Research has shown that biochar can store carbon in a stable form for hundreds or even thousands of years, thereby playing a role in mitigating climate change [1]. Recent studies have focused on the doping of biochar with heteroatoms such as boron (B), phosphorus (P), sulfur (S), and nitrogen (N) to enhance its performance as a supercapacitor electrode material [9, 10]. Fig. (**1**) illustrates the preparation methods for heteroatom-doped biochar, which typically involve a two-step process: carbonization followed by pyrolysis in an inert atmosphere [6]. This approach allows for the incorporation of heteroatoms into the biochar structure, thereby improving its electrical conductivity, surface area, and electrochemical properties, ultimately leading to enhanced supercapacitor performance.

Supercapacitors and Their Energy Storage Mechanism

Supercapacitors, also known as ultracapacitors or Electric Double-layer Capacitors (EDLCs), are advanced energy storage devices characterized by high power density, rapid charge/discharge capabilities, and long cycle life [2]. Unlike conventional capacitors and batteries, supercapacitors store energy through electrostatic charge accumulation rather than relying on chemical reactions. A typical supercapacitor consists of two electrodes, an electrolyte, and a separator. When a voltage is applied, ions from the electrolyte accumulate on the electrode surfaces, forming an electric double layer that stores energy, which can be discharged rapidly when needed. The performance of supercapacitors is largely determined by the properties of the electrode material, including surface area, conductivity, and porosity [7, 9]. In comparison to batteries, supercapacitors offer higher power density and faster charge/discharge rates but generally have lower energy density. Batteries, conversely, provide higher energy density but are typically characterized by slower charge/discharge rates and shorter cycle lives. As a result, supercapacitors are particularly well-suited for applications that require rapid bursts of energy and frequent cycling, such as regenerative braking in electric vehicles and power backup systems.

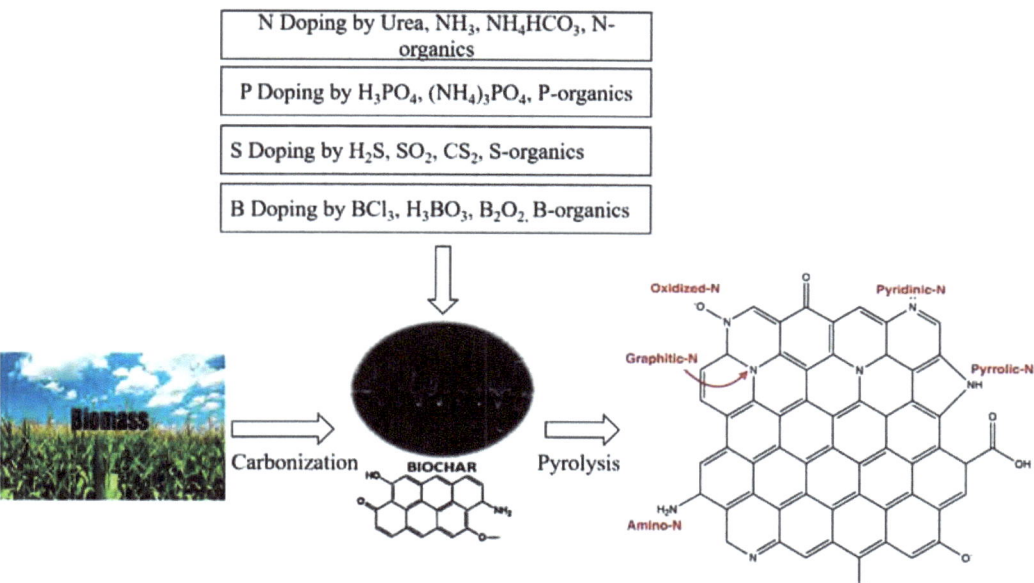

Fig. (1). Schematic diagram to prepare heteroatom doped biochar. Reprinted from ref [6]. Copyright 2020, with permission.

The unique physical and chemical properties of biochar make it an attractive candidate for use in supercapacitor electrodes (Fig. **2**) [1]. Its high surface area, porosity, and electrical conductivity significantly enhance its performance in energy storage applications. The porous structure and extensive surface area of biochar are crucial for increasing the capacitance of supercapacitors, as they enable a greater accumulation of charge. For instance, a study by Ray *et al.* demonstrated that hierarchical porous biochar derived from rice husks treated with NaOH exhibited a specific surface area (307.42 m^2/g) and high porosity, leading to a high capacitance value of 112 F/g at a current density of 0.5 A/g in supercapacitor applications [2]. Furthermore, the electrical conductivity of biochar is a critical factor in its performance as an electrode material. The carbon content in biochar plays a key role in enhancing its electrical conductivity, which is essential for efficient charge and discharge processes. In addition, studies by Xue *et al.* [3] revealed that self-graphenized biochar, derived from the biomass of *Ulva lactuca,* exhibited a specific surface area of 2296.7 m^2/g with a substantial pore volume of 4.59 m^3/g. This biochar also demonstrated a high electrical conductivity of 445.0 F/g at a current density of 0.5 A/g in 1.0 M H$_2$SO$_4$, along with a notable energy density of 27.17 Wh.kg^{-1} at a power density of 222.8 W.kg^{-1} in 1.0 M Na$_2$SO$_4$. These properties highlight the exceptional electrochemical performance of biochar, particularly in terms of energy density, making it a promising candidate for supercapacitor applications. Beyond performance, biochar offers significant sustainability advantages over conventional electrode materials, such as activated carbon or graphene. As a renewable resource derived from biomass, biochar contributes to waste management and carbon sequestration, making it an environmentally friendly alternative. This aligns with global sustainability goals, as biochar can mitigate climate change by sequestration of carbon in soil for extended periods. Additionally, biochar facilitates the recycling of organic materials and enhances soil health, promoting the principles of a circular economy. Shoudho *et al.* emphasized the environmental advantages of biochar in energy storage [4], highlighting its potential to reduce greenhouse gas (GHG) emissions and its role in advancing sustainable energy storage solutions.

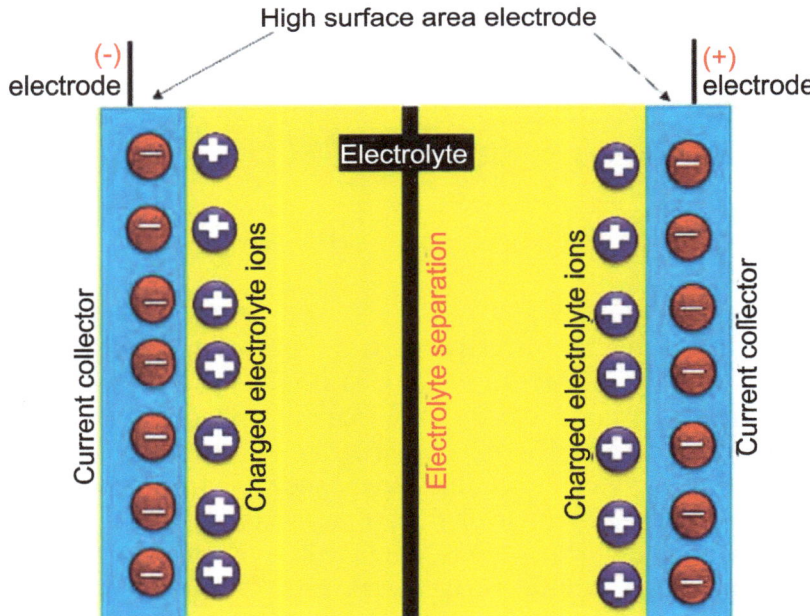

Fig. (2). Schematic representation of the supercapacitor [1].

The choice of biomass feedstocks is a critical factor in improving both the performance and sustainability of biochar-based supercapacitors. When converted into biochar, various biomass feedstocks provide an environmentally friendly alternative to traditional carbon materials while offering a wide range of chemical compositions and structural properties that can be optimized for supercapacitor applications. The production of biochar from feedstocks such as agricultural residues, forestry waste and other organic materials contributes to the creation of high-performance electrodes, improving the energy density, charge/discharge rates, and overall efficiency of supercapacitors. Understanding the relationship between feedstock selection and the electrochemical properties of biochar is essential for the continued development of renewable energy storage solutions.

Production of Biomass-Derived Biochar

The production of biochar involves several key steps, each of which influences the properties of the final material.

1. *Feedstock selection*

The choice of biomass feedstock significantly impacts the characteristics of the resulting biochar. Common feedstocks include agricultural residues (*e.g.,* rice husks, corn stalks), forestry by-products (*e.g.,* wood chips), and municipal waste (*e.g.,* paper, yard waste). Each feedstock has unique properties that affect the surface area, porosity, and chemical composition of the biochar (Fig. **3**).

Fig. (3). Common biomass feedstocks used for biochar production [1, 11, 12].

- ***Agricultural residues:*** Biochar derived from agricultural residues, including rice husks, corn stalks, etc., exhibits exceptional physicochemical properties, characterized by a high surface area and favorable electrochemical performance. For instance, Chen *et al.* demonstrated that rice husk-based hierarchical porous biochar [11] achieved a specific capacitance of 51.4 F/g at a current density of 0.5 A/g, along with remarkable cycle stability, retaining 99.6% of its capacitance after 20000 charge-discharge cycles in supercapacitor applications.

- ***Forestry by-products:*** Biochar produced from forestry by-products, such as wood chips, demonstrates significant potential for energy storage applications. Research conducted by Malhotra *et al.* indicates that biochar derived from wood pellets [12], primarily composed of 70-80% pine and the remainder spruce, exhibits a high surface area ranging from 1300-1500 m^2/g and substantial porosity. This biochar achieves a specific capacitance of 149-152 F/g at a current density of 50 mA/g, nearly doubling the performance of commercially available activated carbon, which has a specific resistance of 79 F/g at the same current density. These findings highlight the efficacy of biochar as an alternative material for supercapacitor applications in energy storage systems.

- ***Municipal waste:*** Municipal waste-derived biochar, although sometimes exhibiting lower surface areas, still demonstrates competitive performance in various applications. Research conducted by Wang *et al.* focused on a food waste-derived porous biochar [13] modified through a hydrogel template (FWH). The nitrogen-enriched variant (FWHB) was produced *via* pyrolysis at temperatures of 500, 700 and 900°C. Notably, the biochar obtained at 700°C (FWHB700) exhibited a substantial surface area of 693 m^2/g, which is approximately 20 times greater than that of untreated food waste biochar. This enhanced surface area contributed to a specific capacitance of 461 F/g at a current density of 1 A/g, along with an impressive 88.2% capacitance retention after 10000 cycles when employed in supercapacitor applications. These findings underscore the potential of food waste-derived biochar in advancing waste reduction strategies and promoting environmental sustainability.

1. *Pyrolysis conditions*

The properties of biochar produced through pyrolysis are significantly influenced by the conditions of the process, including temperature, heating rate, and duration [10, 11]. Specifically, lower pyrolysis temperatures tend to produce biochar with higher surface area and enhanced porosity due to the retention of volatile compounds. In contrast, elevated pyrolysis temperatures facilitate the conversion of biomass into char, resulting in higher carbon content and improved electrical conductivity. This relationship underscores the critical role of pyrolysis parameters in tailoring biochar characteristics for specific applications.

- ***Temperature effects:*** A study by Zhang *et al.* investigated the effects of pyrolysis temperature on the production of activated carbon derived from reed residue waste. The carbonization was conducted at four distinct temperatures: 500, 600, 700 and 800°C within the N_2 atmosphere, followed by activation with KOH. Notably, the sample carbonized at 600°C (referred to as C/600°C) exhibited a specific surface area of 2074.72 m^2/g and a substantial pore volume of 0.930 cm^3/g. This sample demonstrated a specific capacitance of 228 F/g at 1 A/g in 6M KOH electrolyte, with an impressive capacitance retention of 98.1% after 8000 cycles [14]. The research concluded that increased pyrolysis temperature correlated with enhanced carbon content and electrical conductivity, thereby improving the performance of biochar-based supercapacitors.

- ***Heating rate and duration***: The characteristics of biochar are significantly influenced by the heating rate and duration of the pyrolysis process. Babu *et al.* [15] investigated the production of biochar from low-value biomass, specifically Mixed Wood Waste (MWW) and Coconut Husk Waste (CNW), and utilized slow pyrolysis at various temperatures (400, 600 and 800°C) and holding times (30, 45 and 60 min) while maintaining a constant heating rate of 10°C/min. The results demonstrated that increasing the pyrolysis temperature from 400-800°C led to a reduction in biochar yield, accompanied by an increase in carbon content. The volatile content for MWW and CNW was found to be 84.08% and 71.38%, respectively, while the carbon content was recorded at 12.92% for MWS and 20.44% for CNW. Notably, biochar produced from MWW exhibited a maximum specific surface area of 171.52 m^2/g and a pore volume of 7.02 cm^3/g at a pyrolysis temperature of 600°C with a holding time of 60 min. Similarly, biochar derived from CNW revealed the highest specific surface area of 165.48 m^2/g and a pore volume of 4.51 cm^3/g under the same pyrolysis conditions. The combination of a slow heating rate and extended pyrolysis duration enhanced the surface area and porosity of the resulting biochar, thereby improving their potential performance in supercapacitor applications.

1. *Activation*

Activation processes, such as steam activation and chemical activation, are employed to significantly enhance the physicochemical properties of biochar [16]. These methods effectively increase the specific surface area and develop additional porosity, thereby optimizing biochar for applications in energy storage. The enhanced surface characteristics facilitate improved adsorption capacities and contribute to the material's overall performance in energy-related applications.

- ***Steam activation:*** Steam activation of biochar involves the treatment of biomass with high-temperature steam, leading to the development of additional porosity and an increased surface area. A study by Sun *et al.* [16] demonstrated that steam-activated lignocellulosic biochar achieved a high surface area ranging from 451-730 m^2/g, characterized by narrow micropores centered at 0.5-0.6 nm and broader small mesopores centered at 3.5-4 nm. This biochar exhibited an electrical conductivity of 90-608 S/m, correlating with significantly enhanced electrochemical capacitance. In contrast, research by Wang *et al.* [17] reported on non-activated biochar derived from banana biomass, which was co-doped with nitrogen and sulfur (referred to as CM-550). This biochar had a surface area of 138 m^2/g and exhibited a capacitance of 100.5 F/g at a current density of 1 A/g. These findings underscore the critical role of activation in improving the electrochemical performance of biochar for energy storage applications.

- *Chemical activation:* Chemical activation of biochar employs agents such as potassium hydroxide (KOH), sodium hydroxide (NaOH) and phosphoric acid (H_3PO_4) to enhance its physicochemical properties. Research conducted by Chen *et al.* demonstrated that biochar derived from pretreated rice husk [18] and stimulated with KOH exhibited a significantly increased specific surface area of 2556.8 m^2/g. Furthermore, this biochar showed superior electrochemical performance, achieving a specific capacitance of 224 F/g at a current density of 0.5 A/g with stability of 86% after 10000 cycles in supercapacitor applications.

The activation of biochar involves two steps: the pyrolysis of biomass, followed by surface modification achieved through activation techniques. These activation processes can be categorized as physical, chemical and physicochemical. For instance, Januszewicz *et al.* utilized both physical and chemical methods to synthesize activated carbon from chestnut shells [19], demonstrating effective surface area enhancement. Similarly, Taer *et al.* employed a combination of physical and chemical methods, specifically physicochemical techniques, to produce activated carbon from sugarcane bagasse [6, 20], which was subsequently applied in supercapacitor applications. Fig. (**4**) illustrates the SEM images of biochar produced through various activation methods, highlighting differences in morphology and porosity.

Characterization

Characterizing biochar is essential for evaluating its suitability in supercapacitor applications. Techniques like Brunauer-Emmett-Teller (BET) analysis and Scanning Electron Microscopy (SEM) are widely used to assess surface area, pore size distribution and structural properties (Fig. **5**). For instance, studies by Huang *et al.* [21] and Jin *et al.* [22] demonstrated the efficacy of BET analysis to characterize biochar derived from agricultural residues, reporting a high surface area of 2240.6 m^2/g and 3333 m^2/g, respectively, which correlate with enhanced supercapacitor performance. Additionally, Wang *et al.* utilized these characterization techniques to analyze biochar produced from forestry by-products, such as bamboo leaves [23]. Their findings indicated a complex morphology characterized by interconnected carbon layers forming a micro- and mesoporous network, underscoring its potential as an effective electrode material in supercapacitors.

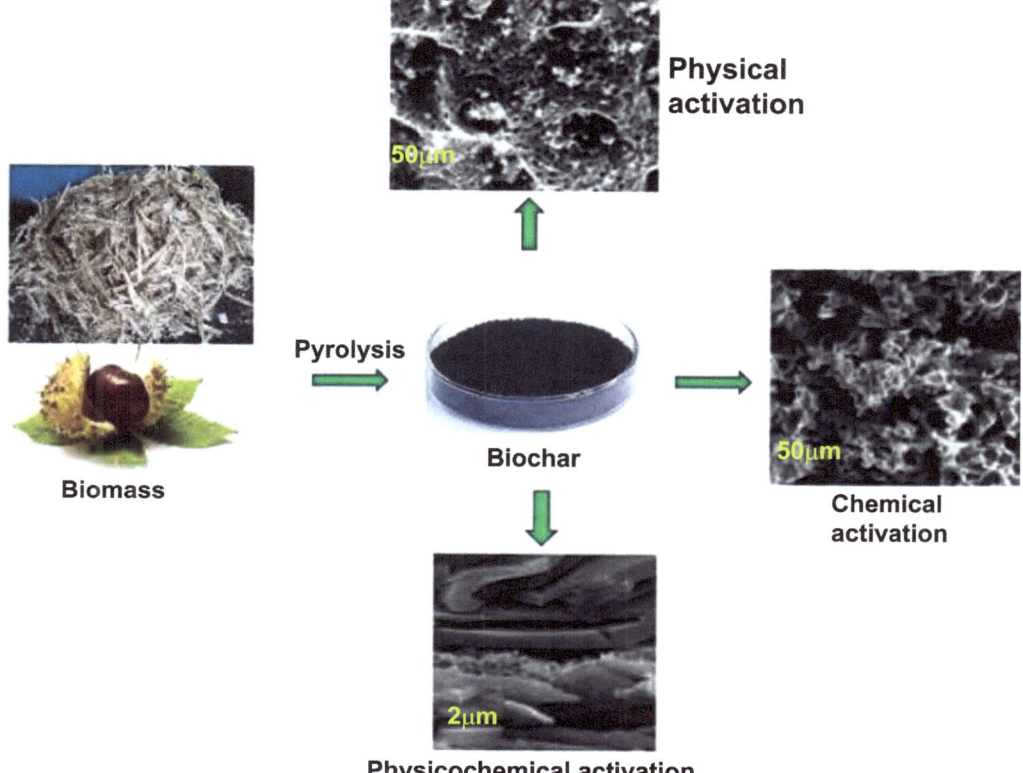

Fig. (4). SEM images of biomass-derived biochar produced *via* physical, chemical, and physicochemical activation methods. Reprinted from ref [18]. and [19]. Copyright 2021 and 2020, with permission.

Performance of Biochar in Supercapacitors

Biochar's performance in supercapacitors is evaluated based on several parameters, including capacitance, energy density, power density, and cycle life. Research has shown that biochar can achieve performance levels comparable to or even exceeding those of traditional materials.

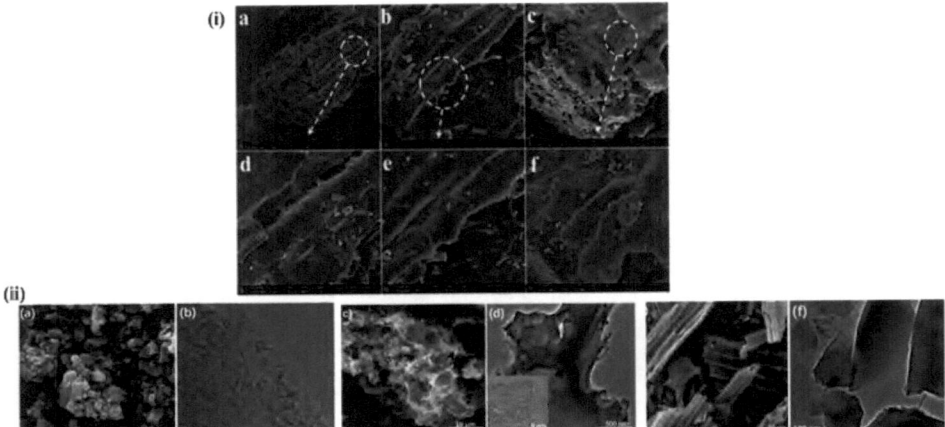

Fig. (5). SEM images showing the morphology of; **(i)** CuO/Cu$_2$O/hierarchical porous biomass-derived carbon hybrid composites and **(ii)** Corn stover-derived biomass. Reprinted from ref [23]. and [22]. Copyright 2019 and 2018, with permission.

1. *Capacitance*

Capacitance quantifies the amount of electric charge that a supercapacitor can store per unit voltage. The high surface area and porous structure of biochar significantly enhance its ability to accumulate charge, thereby improving the overall capacitance of the supercapacitor. For instance, research conducted by Senthil *et al.* [24] demonstrated that hollow tubular porous carbon derived from feather finger grass flowers, when activated with KOH, exhibited a specific surface area of 637.1 m^2/g. This material achieved a remarkable specific capacitance of 315 F/g at a current density of 1 A/g, with a retention of 96% capacitance stability after 50000 cycles. These findings underscore the potential of biochar-derived materials in supercapacitor applications, where higher specific capacitance correlates with increased energy storage capacity, thereby enhancing efficiency and performance in applications requiring rapid charge and discharge cycles.

2. *Energy density*

Energy density refers to the amount of energy that a supercapacitor can store per unit volume or mass. Biochar-based supercapacitors have demonstrated significant energy densities due to the material's high surface area, which facilitates efficient charge storage. Research by Choi *et al.* [25] developed biomass-derived activated carbon combined with reduced graphene oxide as a binder, achieving a specific capacitance of 440 F/g at a current density of 0.5 A/g in EMIMTFSI electrolyte. This configuration exhibited a high energy density of

187.3 Wh.kg^{-1} at power density 438 W.kg^{-1}, and retained an energy density of 153.8 Wh.kg^{-1} at power density 8750 W.kg^{-1}. These performance metrics are competitive with those reported by Dong *et al.* [26], who demonstrated a Zn^{+2}-ion hybrid supercapacitor with an energy density of 84 Wh.kg^{-1} at a power density of 14.9 kW.kg^{-1}, highlighting the potential of biochar-based supercapacitor as promising candidates for high-energy storage applications.

3. *Power density*

Power density is the rate at which a supercapacitor can deliver energy, a crucial factor for various applications. The incorporation of biochar, noted for its high electrical conductivity, enhances the efficiency of charge and discharge cycles, thereby contributing to elevated power density. A study by Xue *et al.* [27] investigated a porous biochar material derived from the carbonization and pyrolysis of *Suaeda Glauca Bunga* (SGB) at 700°C, which was co-doped with N, O, S and Cl heteroatoms. This biochar, referred to as SGB-700, demonstrated a commendable energy density of 30.2 Wh.kg^{-1} alongside a high power density of 164.0 W.kg^{-1}. Furthermore, it exhibited a specific capacitance of 638 F/g at 0.5 A/g in H_2SO_4. Such characteristics render SGB-700 particularly suitable for supercapacitor applications that require rapid energy discharge, such as regenerative braking systems in electric vehicles and energy storage solutions in renewable energy technologies. The high power density allows for swift energy delivery, essential for transient load demands in these applications.

4. *Cycle life*

Cycle life refers to the number of charge-discharge cycles that a supercapacitor can undergo before experiencing significant degradation in performance. Biochar-based supercapacitors have demonstrated excellent cycle stability, maintaining high capacitance over extended cycling periods. A study by Zhou *et al.* [28] investigated biochar derived from carbonized corn bract and activated with KOH for use as a supercapacitor. The result showed that the supercapacitor retained 86.7% of its initial capacitance after 120000 cycles using 6M KOH electrolyte and 70% after 30000 cycles with 1M $MeEt_3NBF_4$/PC electrolyte, at a current density of 4 A/g. These findings underscore the excellent cycle life and durability of biochar-based materials in supercapacitor applications.

Some Findings

Several case studies have explored the use of biochar in supercapacitors, highlighting its potential and performance across various applications. Table **1** represents various biomass-derived biochar materials and their applications as supercapacitor electrodes, detailing specific conductance (in F/g), current density (in A/g), and cyclic stability determined by cyclic voltammetry analysis. These parameters are crucial for evaluating the electrochemical performance of biochar in supercapacitor applications, providing insights into the material's charge storage capabilities, response to varying current loads and long-term stability under repeated cycling conditions.

Table 1. Recent studies on biomass-derived supercapacitor and their specific capacitance.

Biomass	Production reaction condition	Composite-based electrode	Electrolyte	Specific capacitance (F/g)	Cycling stability	Ref.
Wheat straw	Pyrolysis was performed at varying heat ramp rates to 800°C with N-dopant.	NPCM electrode	6.0 M KOH	223.9 (0.5 A/g)	91.4% after 10000 cycles	[9, 29]
Tea saponin	Pyrolysis with variable heat ramp rates to 800°C under N_2 atmosphere	AC_{TS}-1.0 electrode	6.0 M KOH	278 (1 A/g)	95.6% after 10000 cycles	[10]
Millet straw	KOH-assisted pyrolysis under N_2 atmosphere was conducted at varying heat ramp rates to 800°C.	MAC electrode	2.0 M KOH	144 (0.2 A/g)	125% after 10000 cycles	[30]
Rice straw	Pyrolysis was conducted with KOH (4.3 M) at varying heat ramps to 800°C under N_2 atmosphere.	IPGC electrode	6.0 M KOH	400 (0.1 A/g)	93.6% after 10000 cycles	[21]
Pomelo peel	Pyrolysis was carried out under N_2 atmosphere at different heat ramp rates up to 800°C using C and KOH (1:3)	HPC electrode	1.0 M KOH	285 (0.5 A/g)	100% after 10000 cycles	[31]
Lotus root	Pyrolysis was conducted with K_2CO_3 with various heating ramps up to 800°C under N_2 atmosphere.	BAC700 electrode	3.0 M KOH	390 (0.4 A/g)	99.2% after 10000 cycles	[28, 32]
Soft tissues of watermelon	Carbonized at 180°C in the presence of N_2 atmosphere and polypyrrole (PPy).	N-CA-600 electrode	6.0 M KOH	150.6 (0.2 A/g)	90.4% after 10000 cycles	[33]

(Table 1) cont.....

Biomass	Production reaction condition	Composite-based electrode	Electrolyte	Specific capacitance (F/g)	Cycling stability	Ref.
Corn stover	Pyrolysis was conducted at varying heating rates up to 800°C in the presence of KOH (1:3) under N_2 atmosphere.	CSB-800 electrode	1.0 M H_2SO_4	398 (0.5 A/g)	80% after 10000 cycles	[22]
Peanut shell	Pyrolyzed with Na_2CO_3-K_2CO_3 (1:4) with various heating rates up to 800°C under N_2 atmosphere.	HMC-24 electrode	1.0 M H_2SO_4	447 (0.2 A/g)	91.4% after 10000 cycles	[34]
Bamboo leaves	Carbonized at 600°C in the presence of N_2 atmosphere and hybrized with nanosized CuO/Cu_2O	CuO_x@C electrode	3.0 M KOH	147 (1 A/g)	93% after 10000 cycles	[23]

Comparative Analysis of Biochar-Based and Alternative Materials in Supercapacitor Applications

The increasing demand for efficient and sustainable energy storage systems has stimulated extensive research into the development of advanced materials for supercapacitors. Among these, biochar-based materials have been recognized as highly promising candidates for supercapacitor electrodes, as illustrated in Table **1**. This is primarily due to their exceptional physicochemical properties and minimal environmental impact, positioning them as a sustainable alternative to conventional carbon materials. In parallel, conducting polymers such as polyaniline (PANI), polypyrrole (PPy), polythiophene (PTh), and MXenes ($Ti_3C_2T_x$) have emerged as materials of interest due to their superior electrochemical properties. Conducting polymers are known for their high capacitance and flexibility with tunable properties, which enhance their charge storage capabilities and make them adaptable for various applications. MXenes, with their exceptional conductivity, high specific surface area, and excellent electrochemical stability, outperform biochar in terms of energy density and charge-discharge rates. A comparative analysis in Table **2** highlights the strengths and limitations of each material in terms of electrochemical performance and sustainability, providing key insights for optimizing high-performance, eco-friendly supercapacitor designs.

Table 2. Electrochemical performance of some conducting polymer and MXene-based electrode materials.

Material	Preparation method	Electrolyte	Specific capacitance (F/g)	Cycling stability	Ref.
PANI/rGN	Dilute polymerization	1.0 M H$_2$SO$_4$	740 (0.5 A/g)	87% after 1000 cycles	[35]
PANI/MnO$_2$ /MWCNT	*In-situ* polymerization	1.0 M KCl	517 ± 15.25 (1 mA)	90% after 1000 cycles	[36]
Pln/G	Chemical oxidative polymerization	1.0 M KOH	389.17 (1 mV/s)	98.6% after 1000 cycles at 0.1 V/s	[37]
MnCoS/CeO$_2$	Hydrothermal techniques	6.0 M KOH	772.5 (1 A/g)	89% after 2500 cycles	[38]
GO/CDs/PPy	*In-situ* polymerization	1.0 M LiCl	576 (0.5 A/g)	92.9% after 5000 cycles	[39]
SnO$_2$/CDs@MXene	Hydrothermal technique	3.0 M KOH	92 (2 A/g)	103% after 10000 cycles	[40]
SnO$_2$ QDs/GO/PPy	*In-situ* chemical synthesis and hydrothermal techniques	6.0 M KOH	928.56 (40 mV/s)	77.68% after 11000 cycles	[41]
RuCo$_2$O$_4$/Ti$_3$C$_2$T$_x$ @Ni foam	Electrophoretic deposition	1.0 M KOH	229 (3 A/g)	90% after 5000 cycles	[42]

Challenges and Future Directions

While biochar shows great promise as a supercapacitor electrode material, several challenges and opportunities for further development remain:

1. *Scalability and cost*

The scalability of biochar production and its associated costs are crucial factors for widespread adoption. Ongoing research aims to optimize production methodologies and minimize costs, enhancing biochar's economic viability. A study by Sun *et al.* [43] explored various production methods, demonstrating that the optimization of pyrolysis conditions and the strategic selection of feedstocks could significantly reduce production expenses. Additionally, Li *et al.* [44] examined the derivation of hierarchical micro- and mesoporous carbon from rice husk, a low-cost and eco-friendly agricultural residue, through hydrothermal pre-treatment followed by carbonization and activation. Their findings revealed an excellent specific capacitance of 302.2 F/g at 1 A/g at supercapacitor applications, highlighting the potential of cost-effective production methods to yield excellent performance outcomes.

2. *Material optimization*

Further research is essential to optimize the properties of biochar for specific supercapacitor applications, particularly in enhancing surface area, porosity and electrical conductivity through advanced activation techniques and material modifications. Current investigations are increasingly focusing on the development of 3D biochar architectures, as these structures facilitate continuous electron pathways and efficient ion diffusion due to their hierarchical pore systems. Methods for fabricating these interconnected 3D structures include soft-templating, hard-templating and template-free synthesis [45, 46]. Soft-templating employs surfactants and block copolymers as templates, while hard-templating utilizes specialized nanoparticles and porous inorganic materials. However, the complexity and associated costs of these methods have led to a growing interest in template-free synthesis within the electrochemical research community [47]. A notable study by Song *et al.* [48] introduced an innovative self-template method employing KOH pre-treatment followed by direct pyrolysis (a two-step KOH activation). This approach demonstrated the large-scale production of 3D hierarchical porous carbon derived from corn husks, achieving an excellent specific capacitance of 356 F/g at 1 A/g in 6M KOH and 300 F/g at 20 A/g in 1M Na_2SO_4. This illustrates the efficacy of advanced activation methods in enhancing the electrochemical performance of biochar-derived materials for supercapacitor applications.

3. *Integration with other materials*

The integration of biochar with advanced materials such as graphene or carbon nanotubes has the potential to significantly enhance its electrochemical performance in supercapacitor applications. This composite approach has been shown to improve capacitance and energy density relative to biochar alone. Through methods such as heteroatom doping, functionalization and composite fabrication, the porosity and surface area of biochar can be increased, thereby facilitating ion transport and enhancing energy storage capabilities as well as charge-discharge rates. Incorporating metal oxides like ZnO, TiO_2, Cu_2O, WO_3, etc. further augments the electrical conductivity of the composite, promoting efficient electron transport [49]. The utilization of monolithic biochar enables the fabrication of thicker electrodes, which contributes to an increase in energy density. Additionally, the production of a hierarchical structure featuring interconnected micro- and mesopores is critical for optimizing supercapacitor performance by providing enhanced pathways for ion and electron movement [50].

4. *Environmental impact*

A comprehensive lifecycle assessment (LCA) is essential to evaluate the environmental impact of biochar throughout its entire lifecycle, despite its widespread recognition as an environmentally beneficial material. This assessment should encompass the sustainability of feedstock sourcing, production processes, and end-of-life management. LCA is a robust methodology for systematically analyzing the environmental performance of products and technologies [51, 52]. Consequently, it has been extensively applied in the context of biochar production and its application as electrode material in lithium-ion or sodium-ion batteries [53], as well as supercapacitors. This systematic evaluation aids in identifying both the environmental effects and potential improvements associated with these emerging electrode materials.

CONCLUSION

Biomass-derived biochar has garnered increasing attention as a sustainable and efficient material for energy storage applications in supercapacitors. Its high surface area, porosity, and electrical conductivity make it a competitive electrode material compared to conventional alternatives. Additionally, biochar contributes to waste management and carbon sequestration, aligning with environmental sustainability goals. Recent advancements in research are expected to enhance its performance in supercapacitor applications, boosting its viability as an energy storage solution. Biochar's integration into supercapacitors has significant potential for various industries, aiding the transition to greener energy technologies. Future research should focus on optimization of heteroatom doping in biochar-based electrodes, specifically with nitrogen, sulfur, oxygen, and phosphorus, to improve electrochemical performance by increasing active sites and conductivity without compromising structural integrity. Investigating the effects of these dopants on double-layer capacitance and pseudocapacitance will be crucial to maximizing energy storage capacity. Another promising direction is the development of flexible biochar-based supercapacitors for wearable electronics. Enhancing electrode flexibility, potentially through conductive polymers or nanomaterials like graphene, will ensure high performance under mechanical strain. Scalable fabrication methods like 3D printing and roll-to-roll processing will be essential for producing lightweight, flexible supercapacitors. Finally, addressing long-term stability and performance under varying environmental conditions is critical for ensuring the reliability of biochar-based supercapacitors. By optimizing doping strategies and refining fabrication techniques, biochar-based supercapacitors could become a leading solution for sustainable, high-performance energy storage in next-generation electronic devices.

ACKNOWLEDGEMENTS

The authors are grateful to Arunachal University of Studies and World Education Mission (WEM) for providing financial assistance in the form of a Ph.D. Scholarship to Biswajyoti Hazarika.

REFERENCES

[1]   J. Lehmann, and S. Joseph, *Biochar for environmental management: Science, technology and implementation.* 2nd ed. Earthscan, 2011.

[2]   S.K. Ray, B. Pant, M. Park, and B.P. Bastakoti, "Rice husk-derived sodium hydroxide activated hierarchical porous biochar as an efficient electrode material for supercapacitors", *J. Anal. Appl. Pyrolysis,* vol. 175, p. 106207, 2023.
[http://dx.doi.org/10.1016/j.jaap.2023.106207]

[3]   C.F. Xue, X.Q. Li, J.Q. Du, L.F. Wang, X.H. Li, W.J. Yan, and X.G. Hao, "Self-graphenized biochar with huge pore volume prepared from pre-boiled *Ulva lactuca* for electrochemical supercapacitor with high energy density", *J. Energy Storage,* vol. 72, p. 108498, 2023.
[http://dx.doi.org/10.1016/j.est.2023.108498]

[4]   K.N. Shoudho, T.H. Khan, U.R. Ara, M.R. Khan, Z.B.Z. Shawon, and M.E. Hoque, "Biochar in global carbon cycle: Towards sustainable development goals", *Current Research in Green and Sustainable Chemistry,* vol. 8, p. 100409, 2024.
[http://dx.doi.org/10.1016/j.crgsc.2024.100409]

[5]   S.A. Kadam, K.P. Kadam, and N.R. Pradhan, "Advancements in 2D MXene-based supercapacitor electrodes: synthesis, mechanisms, electronic structure engineering, flexible wearable energy storage for real-world applications, and future prospects", *J. Mater. Chem. A Mater. Energy Sustain.,* vol. 12, no. 29, pp. 17992-18046, 2024.
[http://dx.doi.org/10.1039/D4TA00328D]

[6]   O. Norouzi, F.D. Maria, and A. Dutta, "Biochar-based composites as electrode active materials in hybrid supercapacitors with particular focus on surface topography and morphology", *J. Energy Storage,* vol. 29, p. 101291, 2020.
[http://dx.doi.org/10.1016/j.est.2020.101291]

[7]   E. Taer, and S.T. Iwantono, "Manik, R. Taslim, D. Dahlan and M. Deraman, "Preparation of activated carbon monolith electrodes from sugarcane bagasse by physical and physical-chemical activation process for supercapacitor application,"", *Adv. Mat. Res.,* vol. 896, pp. 179-182, 2014.

[8]   S.B. Srinivasan, S. Devendiran, K.V. Savunthari, P. Arumugam, and S. Mukerjee, "Multifarious heteroatom-doped/enriched carbon-based materials for energy storage prospectives: A crucial insight", *ChemRxiv,* .
[http://dx.doi.org/10.26434/chemrxiv-2024-qx0k8]

[9]   S. Zhang, K. Tian, B.H. Cheng, and H. Jiang, "Preparation of N-doped supercapacitor materials by integrated salt templating and silicon hard templating by pyrolysis of biomass wastes", *ACS Sustain. Chem.& Eng.,* vol. 5, no. 8, pp. 6682-6691, 2017.
[http://dx.doi.org/10.1021/acssuschemeng.7b00920]

[10]   Z.W. Ma, H.Q. Liu, and Q.F. Lü, "Porous biochar derived from tea saponin for supercapacitor electrode: Effect of preparation technique", *J. Energy Storage,* vol. 40, p. 102773, 2021.
[http://dx.doi.org/10.1016/j.est.2021.102773]

[11]   Z. Chen, X. Wang, B. Xue, W. Li, Z. Ding, X. Yang, J. Qiu, and Z. Wang, "Rice husk-based hierarchical porous carbon for high performance supercapacitors: The structure-performance relationship", *Carbon,* vol. 161, pp. 432-444, 2020.
[http://dx.doi.org/10.1016/j.carbon.2020.01.088]

[12]   J.S. Malhotra, R. Valiollahi, and H. Wiinikka, From wood to supercapacitor electrode material *via* fast pyrolysis., *J. Energy Storage,* vol. 57, p. 106179, 2023.
[http://dx.doi.org/10.1016/j.est.2022.106179]

[13]   S. Wang, Y. Shi, S. Chen, C. Zhu, X. Wang, T. Zhou, L. Su, C. Tan, L. Zhang, and H. Xiang, "Porous biochars with nitrogen defects prepared from hydrogel template-modified food waste for high-performance supercapacitors", *J. Energy Storage,* vol. 72, p. 108720, 2023.
[http://dx.doi.org/10.1016/j.est.2023.108720]

[14]   D. Zhang, Y. Zhang, H. Liu, Y. Xu, J. Wu, and P. Li, "Effect of pyrolysis temperature on carbon materials derived from reed residue waste biomass for use in supercapacitor electrodes", *J. Phys. Chem. Solids,* vol. 178, p. 111318, 2023.
[http://dx.doi.org/10.1016/j.jpcs.2023.111318]

[15]   K.K.B. Suresh Babu, M. Nataraj, M. Tayappa, Y. Vyas, R.K. Mishra, and B. Acharya, "Production of biochar from waste biomass using slow pyrolysis: Studies of the effect of pyrolysis temperature and holding time on biochar yield and properties", *Mater. Sci. Energy Technol.,* vol. 7, pp. 318-334, 2024.
[http://dx.doi.org/10.1016/j.mset.2024.05.002]

[16]   J. Sun, A. Jayakumar, C.G. Díaz-Maroto, I. Moreno, J. Fermoso, and O. Mašek, "The role of feedstock and activation process on supercapacitor performance of lignocellulosic biochar", *Biomass Bioenergy,* vol. 184, p. 107180, 2024.
[http://dx.doi.org/10.1016/j.biombioe.2024.107180]

[17]   L. Wang, X. Li, J. Ma, Q. Wu, and X. Duan, "Non-activated, N,S-co-doped biochar derived from banana with superior capacitive properties", *Sustainable Energy,* vol. 2, pp. 39-43, 2014.

[18]   J. Chen, J. Liu, D. Wu, X. Bai, Y. Lin, T. Wu, C. Zhang, D. Chen, and H. Li, "Improving the supercapacitor performance of activated carbon materials derived from pretreated rice husk", *J. Energy Storage,* vol. 44, p. 103432, 2021.
[http://dx.doi.org/10.1016/j.est.2021.103432]

[19]   K. Januszewicz, A. Cymann-Sachajdak, P. Kazimierski, M. Klein, J. Łuczak, and M. Wilamowska-Zawłocka, "Chestnut-derived activated carbon as a prospective material for energy storage", *Materials (Basel),* vol. 13, no. 20, p. 4658, 2020.
[http://dx.doi.org/10.3390/ma13204658] [PMID: 33086654]

[20]   A. Chauhan, and H.C. Joshi, "Recent developments and applications in bioconversion and biorefineries", *Trends in Mathematics,* no. Part F3197, pp. 247-307, 2024.
[http://dx.doi.org/10.1007/978-981-99-7250-0_6]

[21]   L. Huang, Q. Wu, S. Liu, S. Yu, and A.J. Ragauskas, "Solvent-free production of carbon materials with developed pore structure from biomass for high-performance supercapacitors", *Ind. Crops Prod.,* vol. 150, p. 112384, 2020.
[http://dx.doi.org/10.1016/j.indcrop.2020.112384]

[22]   H. Jin, J. Hu, S. Wu, X. Wang, H. Zhang, H. Xu, and K. Lian, "Three-dimensional interconnected porous graphitic carbon derived from rice straw for high performance supercapacitors", *J. Power Sources,* vol. 384, pp. 270-277, 2018.
[http://dx.doi.org/10.1016/j.jpowsour.2018.02.089]

[23]   Q. Wang, Y. Zhang, J. Xiao, H. Jiang, T. Hu, and C. Meng, "Copper oxide/cuprous oxide/hierarchical porous biomass-derived carbon hybrid composites for high-performance supercapacitor electrode", *J. Alloys Compd.,* vol. 782, pp. 1103-1113, 2019.
[http://dx.doi.org/10.1016/j.jallcom.2018.12.235]

[24]   R.A. Senthil, V. Yang, J. Pan, and Y. Sun, "A green and economical approach to derive biomass porous carbon from freely available feather finger grass flower for advanced symmetric supercapacitors", *J. Energy Storage,* vol. 35, p. 102287, 2021.
[http://dx.doi.org/10.1016/j.est.2021.102287]

[25]   J.H. Choi, C. Lee, S. Cho, G.D. Moon, B. kim, H. Chang, and H.D. Jang, "High capacitance and energy density supercapacitor based on biomass-derived activated carbons with reduced graphene oxide binder", *Carbon,* vol. 132, pp. 16-24, 2018.
[http://dx.doi.org/10.1016/j.carbon.2018.01.105]

[26]   L. Dong, X. Ma, Y. Li, L. Zhao, W. Liu, J. Cheng, C. Xu, B. Li, Q.H. Yang, and F. Kang, "Extremely safe, high-rate and ultralong-life zinc-ion hybrid supercapacitors", *Energy Storage Mater.,* vol. 13, pp. 96-102, 2018.
[http://dx.doi.org/10.1016/j.ensm.2018.01.003]

[27]   C.F. Xue, Y. Lin, W. Zhao, T. Wu, Y.Y. Wei, X.H. Li, W.J. Yan, and X.G. Hao, "Green preparation of high active biochar with tetra-heteroatom self-doped surface for aqueous electrochemical supercapacitor with boosted energy density", *J. Energy Storage,* vol. 90, p. 111872, 2024.
[http://dx.doi.org/10.1016/j.est.2024.111872]

[28]   J. Zhou, X. Ren, Z. Liu, and S. Yuan, "Improving the cycling stability of biochar electrodes by purification *via* ion exchange", *Materials Today Sustainability,* vol. 20, p. 100225, 2022.
[http://dx.doi.org/10.1016/j.mtsust.2022.100225]

[29]   A. Chauhan, and H.C. Joshi, "Lignocellulosic biomass for the conversion of bioethanol: production and optimization", *Trends in Mathematics,* no. Part F3197, pp. 187-214, 2024.
[http://dx.doi.org/10.1007/978-981-99-7250-0_4]

[30]   Y. Ding, T. Wang, D. Dong, and Y. Zhang, "Using biochar and coal as the electrode material for supercapacitor applications", *Front. Energy Res.,* vol. 7, p. 159, 2020.
[http://dx.doi.org/10.3389/fenrg.2019.00159]

[31]   W. Zhong-Yu, F. Lei, T. You-Rong, W. Wei, W. Xing-Cai, and Z. Jian-Wei, "Pomelo peel derived hierarchical porous carbon as electrode materials for high-performance supercapacitor", *Wuji Huaxue Xuebao,* vol. 34, pp. 1249-1260, 2018.

[32]   R. Rajendiran, M. Nallal, K.H. Park, O.L. Li, H.J. Kim, and K. Prabakar, "Mechanochemical assisted synthesis of heteroatoms inherited highly porous carbon from biomass for electrochemical capacitor and oxygen reduction reaction electrocatalysis", *Electrochim. Acta,* vol. 317, pp. 1-9, 2019.
[http://dx.doi.org/10.1016/j.electacta.2019.05.139]

[33]   Y. Ren, J. Zhang, Q. Xu, Z. Chen, D. Yang, B. Wang, and Z. Jiang, "Biomass-derived three-dimensional porous N-doped carbonaceous aerogel for efficient supercapacitor electrodes", *RSC Advances,* vol. 4, no. 45, p. 23412, 2014.
[http://dx.doi.org/10.1039/c4ra02109f]

[34]   W. Lei, B. Yang, Y. Sun, L. Xiao, D. Tang, K. Chen, J. Sun, J. Ke, and Y. Zhuang, "Self-sacrificial template synthesis of heteroatom doped porous biochar for enhanced electrochemical energy storage", *J. Power Sources,* vol. 488, p. 229455, 2021.
[http://dx.doi.org/10.1016/j.jpowsour.2021.229455]

[35]   S. Wang, L. Ma, M. Gan, S. Fu, W. Dai, T. Zhou, X. Sun, H. Wang, and H. Wang, "Free-standing 3D graphene/polyaniline composite film electrodes for high-performance supercapacitors", *J. Power Sources,* vol. 299, pp. 347-355, 2015.
[http://dx.doi.org/10.1016/j.jpowsour.2015.09.018]

[36]   M.M. Sk, C.Y. Yue, and R.K. Jena, "Non-covalent interactions and supercapacitance of pseudo-capacitive composite electrode materials (MWCNTCOOH/MnO_2/PANI)", *Synth. Met.,* vol. 208, pp. 2-12, 2015.
[http://dx.doi.org/10.1016/j.synthmet.2014.10.026]

[37] H. Mudila, S. Rana, and M.H. Zaidi, "Supercritical CO_2 aided polyindole-graphene nanocomposites for high power density electrode", *Adv. Mater. Lett.,* vol. 8, pp. 269-275, 2017.
[http://dx.doi.org/10.5185/amlett.2017.7018]

[38] Z. Molaei, A.A. Asgharinezhad, A. Larimi, C. Ghotbi, and F. Khorasheh, "Incorporation of CeO_2 nanosheets into $MnCoS_x$ hollow nanorods for next generation supercapacitors", *Energy Fuels,* vol. 39, no. 8, pp. 4047-4058, 2025.
[http://dx.doi.org/10.1021/acs.energyfuels.4c05048]

[39] X. Zhang, J. Wang, J. Liu, J. Wu, H. Chen, and H. Bi, "Design and preparation of a ternary composite of graphene oxide/carbon dots/polypyrrole for supercapacitor application: Importance and unique role of carbon dots", *Carbon,* vol. 115, pp. 134-146, 2017.
[http://dx.doi.org/10.1016/j.carbon.2017.01.005]

[40] M. Moniruzzaman, C.K. Maity, S. De, M.J. Kim, and J. Kim, "SnO_2 nanosphere/carbon dot-embedded $Ti_3C_2T_x$ MXene nanocomposites for high-performance binder-free asymmetric supercapacitor electrodes", *ACS Appl. Nano Mater.,* vol. 7, no. 6, pp. 6636-6649, 2024.
[http://dx.doi.org/10.1021/acsanm.4c00550]

[41] M. Vandana, S. Veeresh, H. Ganesh, Y.S. Nagaraju, H. Vijeth, M. Basappa, and H. Devendrappa, "Graphene oxide decorated SnO_2 quantum dots/polypyrrole ternary composites towards symmetric supercapacitor application", *J. Energy Storage,* vol. 46, p. 103904, 2022.
[http://dx.doi.org/10.1016/j.est.2021.103904]

[42] P. Asen, A. Esfandiar, and H. Mehdipour, "Urchin-like hierarchical ruthenium cobalt oxide nanosheets on $Ti_3C_2T_x$ MXene as a binder-free bifunctional electrode for overall water splitting and supercapacitors", *Nanoscale,* vol. 14, no. 4, pp. 1347-1362, 2022.
[http://dx.doi.org/10.1039/D1NR07145A] [PMID: 35014999]

[43] Y. Sun, P. Sun, J. Jia, Z. Liu, L. Huo, L. Zhao, Y. Zhao, W. Niu, and Z. Yao, "Machine learning in clarifying complex relationships: Biochar preparation procedures and capacitance characteristics", *Chem. Eng. J.,* vol. 485, p. 149975, 2024.
[http://dx.doi.org/10.1016/j.cej.2024.149975]

[44] C. Li, D. He, Z.H. Huang, and M.X. Wang, "Hierarchical micro-/mesoporous carbon derived from rice husk by hydrothermal pre-treatment for high performance supercapacitor", *J. Electrochem. Soc.,* vol. 165, no. 14, pp. A3334-A3341, 2018.
[http://dx.doi.org/10.1149/2.0121814jes]

[45] J. Deng, M. Li, and Y. Wang, "Biomass-derived carbon: synthesis and applications in energy storage and conversion", *Green Chem.,* vol. 18, no. 18, pp. 4824-4854, 2016.
[http://dx.doi.org/10.1039/C6GC01172A]

[46] A. Chauhan, and H. Chandra Joshi, "Energetic efficiency of a biofuels production system mathematical modeling", *Trends in Mathematics,* no. Part F3197, pp. 337-357, 2024.
[http://dx.doi.org/10.1007/978-981-99-7250-0_8]

[47] J. Lian, L. Xiong, R. Cheng, D. Pang, X. Tian, J. Lei, R. He, X. Yu, T. Duan, and W. Zhu, "Ultra-high nitrogen content biomass carbon supercapacitors and nitrogen forms analysis", *J. Alloys Compd.,* vol. 809, p. 151664, 2019.
[http://dx.doi.org/10.1016/j.jallcom.2019.151664]

[48] S. Song, F. Ma, G. Wu, D. Ma, W. Geng, and J. Wan, "Facile self-templating large scale preparation of biomass-derived 3D hierarchical porous carbon for advanced supercapacitors", *J. Mater. Chem. A Mater. Energy Sustain.,* vol. 3, no. 35, pp. 18154-18162, 2015.
[http://dx.doi.org/10.1039/C5TA04721H]

[49] A.P. Khedulkar, V.D. Dang, A. Thamilselvan, R. Doong, and B. Pandit, "Sustainable high-energy supercapacitors: Metal oxide-agricultural waste biochar composites paving the way for a greener future", *J. Energy Storage,* vol. 77, p. 109723, 2024.
[http://dx.doi.org/10.1016/j.est.2023.109723]

[50] N-A-S. Rafsan, S.F.B. Haque, S. Shah, J. Sagues, R. Ding, J. Ferraris, and P. Kolar, "Poultry litter-derived biochar for supercapacitor applications", *Next Energy,* vol. 5, p. 100171, 2024. [http://dx.doi.org/10.1016/j.nxener.2024.100171]

[51] Z. Jiang, Y. Zou, Y. Li, F. Kong, and D. Yang, "Environmental life cycle assessment of supercapacitor electrode production using algae derived biochar aerogel", *Biochar,* vol. 3, no. 4, pp. 701-714, 2021. [http://dx.doi.org/10.1007/s42773-021-00122-1]

[52] A. Malara, F. Pantò, S. Santangelo, P.L. Antonucci, M. Fiore, G. Longoni, R. Ruffo, and P. Frontera, "Comparative life cycle assessment of Fe_2O_3-based fibers as anode materials for sodium-ion batteries", *Environ. Dev. Sustain.,* vol. 23, no. 5, pp. 6786-6799, 2021. [http://dx.doi.org/10.1007/s10668-020-00891-y]

[53] J. Aslam, M.A. Waseem, X.M. Lu, W. Sun, and Y. Wang, "From biochar to battery electrodes: A pathway to green lithium and sodium-ion battery systems", *Chem. Eng. J.,* vol. 505, p. 159556, 2025. [http://dx.doi.org/10.1016/j.cej.2025.159556]

CHAPTER 5

Hydrogen Technology and Its Industrial Applications

Sujeet Kumar Pandey[1,*] and **Aash Mohammad**[1]

[1] *Department of Chemical and Biochemical Engineering, Rajiv Gandhi Institute of Petroleum Technology, Jais, Amethi, India*

Abstract: The potential of hydrogen technology to revolutionize different industrial applications and contribute considerably to a sustainable, low-carbon economy is gradually being acknowledged. Research in hydrogen production is crucial for the development of sustainable energy alternatives. Hydrogen is in high demand because of its significant heating value, which makes it well-suited for a variety of uses, such as aircraft fuel, steel manufacturing, and power storage in fuel cells. The progress in green hydrogen generation systems is highly encouraging, as they provide a means to fulfill future energy needs while reducing environmental harm. Hydrogen-generating technology can be classified into two primary categories. The methods used for producing hydrogen from fossil fuels include partial oxidation, steam reforming, and autothermal reforming. Renewable sources-based hydrogen production encompasses many biomass processes, including bio-photolysis, photo-fermentation, dark fermentation, pyrolysis, gasification, combustion, and liquefaction. Additionally, water splitting methods, such as thermolysis, photolysis, and electrolysis, are utilized. Hydrogen technology has favourable prospects for reducing carbon emissions. However, it is crucial to acknowledge that we should use hydrogen in conjunction with other renewable energy technologies rather than relying solely on it as a standalone solution. Its industrial uses are found in the chemical industry, metallurgical sector, transport, and energy storage. Although there are several positive attributes, there are also challenges associated with investment, infrastructure, and technological advancements.

This chapter discusses the production technologies for hydrogen, like electrolysis, reforming, and photocatalysis, and also briefly discusses various methods for storing hydrogen. The main focus of this chapter is its industrial applications across various domains, including power generation, aviation, and various industrial processes. Despite these advantages, hydrogen production and applications have some limitations, which we also discuss with a future perspective.

* **Corresponding author Sujeet Kumar Pandey:** Department of Chemical and Biochemical Engineering, Rajiv Gandhi Institute of Petroleum Technology, Jais, Amethi, India; E-mail: 20ce0012@rgipt.ac.in

Harish Chandra Joshi, Anand Chauhan, Mikhail Vlaskin & Maulin P. Shah (Eds.)

Keywords: Applications of H_2, Compressed hydrogen storage, Electrolysis, Hydrogen (H_2), Liquid hydrogen storage, Medical, Photocatalysis, Steam methane reforming, Storage of hydrogen, Solid-state hydrogen storage.

INTRODUCTION

Hydrogen technology signifies a promising advancement in the pursuit of sustainable and carbon-free energy. H_2 serves as an energy carrier in several forms, presenting the opportunity to decarbonize businesses and mitigate greenhouse gas emissions. This chapter concentrated on the generation, storage, and use of hydrogen, especially for industrial and energy purposes, as shown in Fig. (**1**). Hydrogen is produced through various means, like electrolysis, steam methane reforming, and photocatalysis, with electrolysis being the predominant technique.

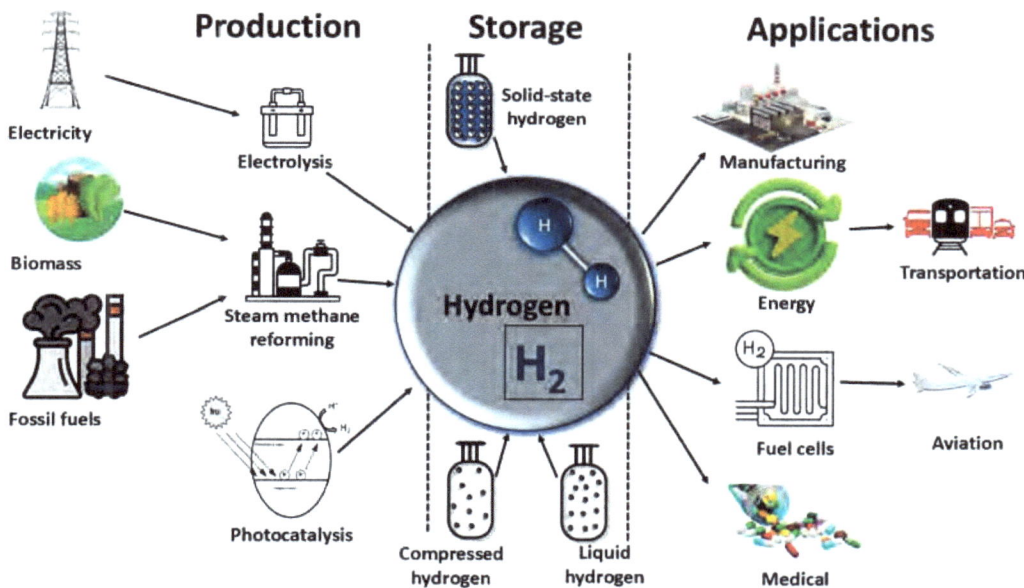

Fig. (1). Schematic diagram of hydrogen production, storage, and applications [14 - 16].

Production and Storage of Hydrogen

The production of H_2 is primarily divided into three technologies: electrolysis, steam methane reforming, and photocatalysis. Electrolysis is a viable method for generating carbon-free hydrogen. In electrolysis, electrical energy is used to decompose water into its constituent elements, oxygen and hydrogen. This process occurs within a device known as an electrolyzer. Electrolyzers have a cathode and an anode divided by an electrolyte. Various electrolyzers operate distinctly, depending on the electrolyte materials utilized and the type of ions they conduct.

Under elevated pressure and temperature conditions, SMR uses methane to generate hydrogen. Methane is a chemical molecule represented by the formula CH_4 and is a key component of natural gas. The conventional steam methane reforming process has four primary units: the desulfurization unit, the reforming unit, the shift reactor, and the separation unit.

Photocatalytic H_2 generation is a technique that utilizes photons to dissociate water molecules into oxygen and hydrogen. A photocatalyst, often a semiconductor, is suspended in water and exposed to light irradiation. The light generates electron-hole pairs that dissociate water into oxygen and hydrogen. Photocatalytic hydrogen generation is regarded as a promising, cost-effective, and eco-friendly method. Titanium dioxide (TiO_2) is a widely utilized photocatalyst due to its non-toxicity, chemical stability, and cost-effectiveness. Recent hydrogen-generating techniques include photoelectrochemical (PEC) water splitting and biological hydrogen synthesis. PEC water splitting electrolyzes water with semiconductors and solar energy. Solar energy generates hydrogen, making it a renewable energy source. To make PEC systems commercially viable, titanium dioxide (TiO_2) and other hybrid materials are being studied for their efficiency and scalability. Biochemical hydrogen production employs algae, bacteria, and fungi. This method is popular since it employs organic waste and operates under mild conditions. Scalability and efficiency concerns require additional investigation to enhance these systems for large-scale hydrogen production.

There are primarily three different kinds of hydrogen storage methods. Compressed hydrogen storage is the predominant process, entailing the containment of H_2 gas within high-pressure cylinders. It is economical and facilitates swift charging and discharging, although it possesses low volumetric density [1]. Liquid hydrogen storage, in which hydrogen is cooled to cryogenic temperatures to get a liquid state, hence providing excellent storage efficiency [2 - 4]. Compressed hydrogen storage is an essential technology for hydrogen fuel applications, especially in automotive contexts. This approach has shown notable improvements in safety, efficiency, and cost-effectiveness, rendering it a feasible candidate for commercialization [5]. The current hydrogen refueling infrastructure primarily utilizes compressed gas stations, thereby simplifying market entry. Material innovations comprise Contemporary tanks that are fabricated from high-strength carbon-composite materials, which improve safety and performance [6, 7].

Storage of liquid hydrogen (LH_2) is attracting interest owing to its elevated energy density and prospects for sustainable energy utilization. It functions as an efficient medium for the storage of energy, especially in the case of renewable energy systems, and has considerable ramifications for industries like aviation and cryogenics. LH_2 provides superior storage density compared to gaseous hydrogen, rendering it more cost-effective for extensive applications [8 - 10]. Storage of hydrogen in a solid state is becoming a more feasible alternative for effective hydrogen energy applications. This method uses various materials and technologies to securely and efficiently store hydrogen, tackling issues related to energy density and safety [11 - 13].

Applications of Hydrogen

Hydrogen is becoming a multifaceted element with substantial uses in several fields, including medicine, energy, and manufacturing. Hydrogen is a crucial energy carrier, facilitating the switch to renewable energy and decarbonizing transportation and power generation [17, 18]. Hydrogen is revolutionizing aviation, enabling decarbonization and sustainability [19]. Hydrogen gas demonstrates antioxidant, anti-inflammatory, and antiapoptotic properties, rendering it significant in medicinal applications [20].

Hydrogen Production Technologies

Hydrogen production technologies are diverse, encompassing various methods that leverage different feedstocks and processes. The transition to sustainable hydrogen production is crucial for achieving climate neutrality, as illustrated by several technologies in Fig. (**2**). The key hydrogen production methods are expressed below.

Electrolysis

Electrolysis for hydrogen generation is a crucial technique for sustainable energy, employing electrical energy to dissociate water into oxygen and hydrogen. This technique may be utilized for sources of renewable energy, rendering it a clean alternative to fossil fuels. Diverse electrolysis systems are available, each with distinct attributes and efficiency. There are primarily three types of electrolysis technology [21].

Polymer Electrolyte Membrane (PEM) electrolysis

In a PEM electrolysis, the electrolyte comprises a solid, specialized plastic substance. Water undergoes a reaction at the anode, producing oxygen and hydrogen ions, *i.e.*, protons. Electrons traverse an external circuit while protons preferentially migrate across the PEM to the cathode. At the cathode, protons react with electrons from the external circuit to produce H_2 gas [22, 23].

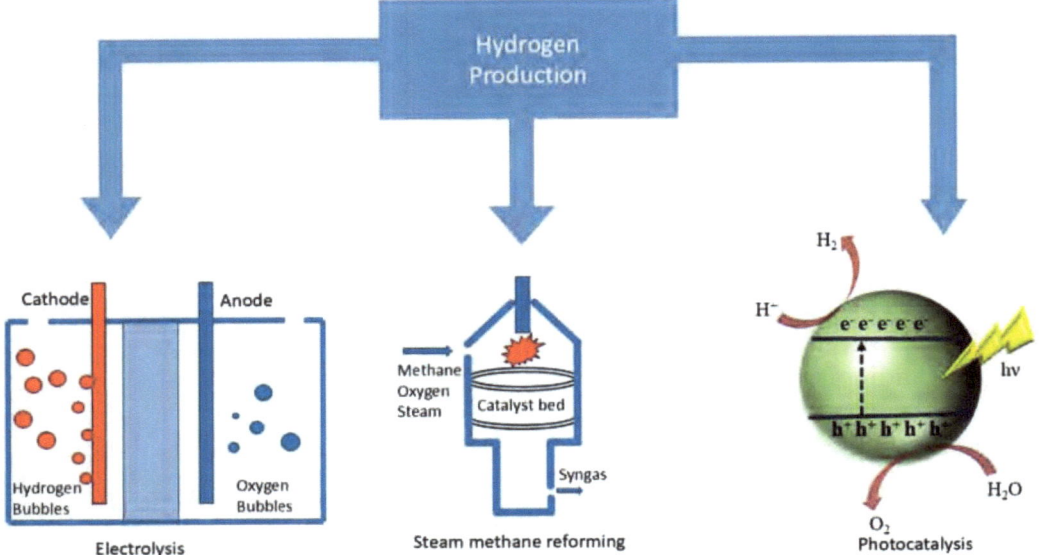

Fig. (2). Different methods of hydrogen production [38 - 40].

$2H_2O \rightarrow O_2 + 4H^+ + 4e^-$ (anodic reaction)

$4H+ + 4e^- \rightarrow 2H_2$ (cathodic reaction)

Alkaline Rlectrolysis

Alkaline electrolysis in which hydroxide ions move from the cathode to the anode through the electrolyte, resulting in H_2 production at the cathode. Electrolyzers utilizing a solution with a higher pH, like potassium or sodium hydroxide, as the electrolyte have been commercially accessible for several years. New methods that use solid Alkaline Exchange Membranes (AEMs) as the electrolyte have shown promising results on a small scale in the lab. Porous nickel foam electrodes speed up the release of hydrogen. The best parameters were found to be a pore density of 75 PPI and a thickness of 1000 μm [24]. The use of pulse current enhances efficiency, especially under low-load situations, by minimizing parasitic current losses [25]. Integrating alkaline electrolysis with solar systems can optimize energy efficiency and enhance hydrogen production rates, mitigating variations in renewable energy supply [26]. Combining alkaline electrolysis with oxyfuel combustion systems can improve overall efficiency, reaching thermodynamic efficiencies of up to 62.5% [27].

Solid Oxide Electrolysis

Solid oxide electrolysis uses a solid ceramic electrolyte that conducts negatively charged oxygen ions more efficiently at high temperatures to create H_2 differently. Steam at the cathode reacts with electrons to produce H_2 gas and oxygen ions. The oxygen ions traverse from the membrane of the solid ceramic and generate oxygen gas. Solid oxide electrolysis functions at high temperatures (generally 700-1000°C), improving energy efficiency relative to conventional electrolysis techniques [28]. Utilizing new materials and configurations can markedly enhance the electrochemical performance of solid oxide electrolysis [28]. Solid oxide electrolysis may be efficiently combined with sources of renewable energy, such as wind and solar, to generate green hydrogen, hence aiding decarbonization initiatives [28, 29]. Using Concentrating Solar Spectrum Splitters (CSSS) to split the solar spectrum more effectively makes Solid Oxide Electrolyzer Cells (SOECs) more efficient by providing them with both electricity and heat [29].

Steam Methane Reforming (SMR)

SMR is a prevalent technique for H_2 synthesis, utilizing methane as the fuel. Recent research has emphasized progress in catalyst development and process optimization, resulting in improved hydrogen output and minimized byproduct creation. The primary focus of steam methane reforming is advancements in catalyst technology. Nickel-based catalysts continue to be prevalent owing to their economical nature and superior activity. Research demonstrates that the morphology of ceria substantially affects hydrogen production and CO selectivity, with ceria nanocubes exhibiting superior performance compared to other forms at elevated temperatures [30]. Bimetallic catalysts, exemplified by Ni_3Cu_1/Al_2O_3, have demonstrated enhanced methane conversion and hydrogen purity, attaining up to 97% conversion at 700°C [31]. The process optimization derived from simulation models indicates ideal conditions for SMR, achieving hydrogen yields of 30% at 800°C with a steam-to-carbon ratio of 2 [32]. Industrial catalysts have been refined by response surface methods, attaining a peak hydrogen output of 78.60% with no coke generation [33]. Although SMR is efficient, other techniques, such as carbon dioxide sorption-enhanced reforming, are developing, providing sustainable avenues for hydrogen production from biogas [34, 35].

Photocatalysis

Photocatalysis offers a viable method for sustainable hydrogen generation, utilizing light to facilitate chemical processes that produce hydrogen from water or biomass. Recent breakthroughs have underscored the diverse range of photocatalytic materials and techniques that improve efficiency and scalability. Photocatalysis employs semiconductor materials, including TiO_2, enhanced with silver and graphene to augment light absorption and catalytic efficacy. For example, Ag-G-TiO$_2$ attained a hydrogen generation rate of 191 μmoles g^{-1} h^{-1} under visible light [36]. The method may also incorporate biomass, utilizing lignocellulosic waste as a substrate to enhance hydrogen generation and facilitate waste management [37].

Recently developed hydrogen generation systems include photoelectrochemical (PEC) water splitting and biological hydrogen synthesis.

PEC water splitting uses semiconductors and solar energy to electrolyze water. Hydrogen is produced directly from sunlight, making it a sustainable and renewable energy source. To make PEC systems economically feasible, titanium dioxide (TiO$_2$) and other hybrid materials are being investigated to increase efficiency and scalability.

Biological hydrogen generation uses algae, bacteria, and fungi to manufacture hydrogen biochemically. This approach is gaining popularity since it uses organic waste and works in moderate settings. Scalability and efficiency issues remain; thus, further study is needed to optimise these systems for large-scale hydrogen generation.

These technologies enhance the sustainability of hydrogen production by complementing established techniques.

Storage and Transportation of Hydrogen

The storage of H$_2$ is important for its efficient use as a carbon-free energy source. Diverse technologies in Fig. (**3**) have been developed, each presenting unique benefits and obstacles, to enable hydrogen storage for uses spanning from automobiles to marine transport.

Compressed Hydrogen Storage

Compressed hydrogen storage is an essential technique for improving the feasibility of hydrogen as a clean energy source. This approach basically includes the storage of hydrogen gas at elevated pressures, generally over 350 bar, to enhance volumetric density and enable efficient energy transmission. A three-stage compression system is advised for multi-stage compression to enhance energy efficiency and regulate temperature, maintaining it below 200°C while reducing energy expenses to around 10% of the fuel's calorific value [41]. Cryo-compressed storage is a technique that integrates cryogenic chilling with high-pressure storage, markedly enhancing mass and volume density while simultaneously mitigating safety and economic issues [42].

Storage of hydrogen

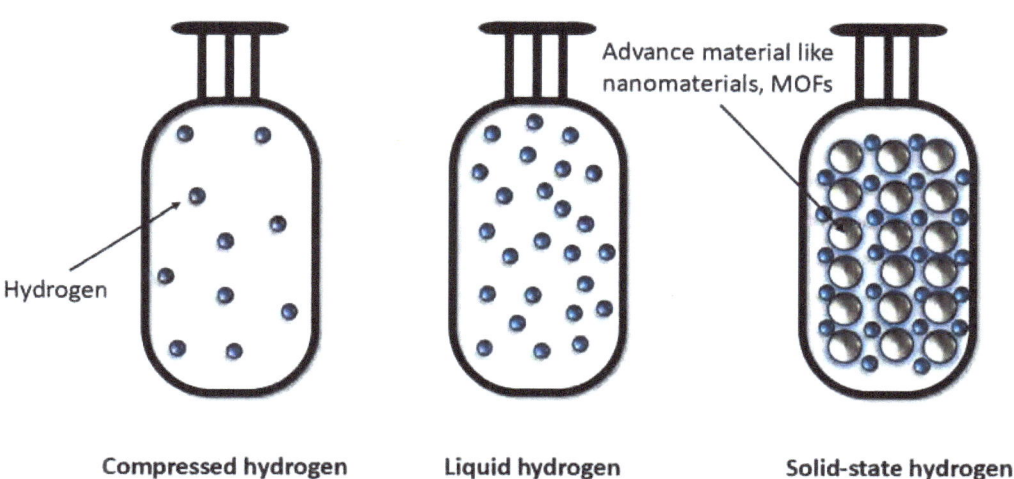

Fig. (3). Different methods of storage of hydrogen [53, 54].

Liquid Hydrogen Storage

Liquid hydrogen storage is an appropriate method for energy storage and transportation, primarily owing to its elevated energy density and efficiency. This strategy is becoming increasingly pertinent across several industries, including aviation and the integration of renewable energy. Liquid hydrogen possesses a density of 70.8 kg/m³, markedly surpassing that of gaseous hydrogen, thereby enhancing its efficiency for storage and transportation [43]. Liquid hydrogen functions as a cryogenic working medium, supplying both thermal energy and electricity; hence, it alleviates variations in sources of renewable energy like the sun and wind [44]. Efficient thermal insulation is crucial for a tank storing liquid hydrogen. Some methods are Vapor-cooled Shields (VCS) and multilayer insulation (MLI), which diminish heat flux and mitigate boil-off [10]. Recent research has refined insulating thickness, resulting in substantial decreases in heat transmission [45]. Advanced models forecast internal pressure and behavior in cryogenic tanks, facilitating the design and efficiency of LH_2 storage systems [46].

Hydrogen Storage in Solid-State

Solid-state hydrogen storage is becoming an attractive choice for efficient and secure hydrogen energy applications. Recent developments in materials, especially nanomaterials and Metal-organic Frameworks (MOFs), have demonstrated considerable promise in improving hydrogen storage efficiency. Nanomaterials, including palladium-graphene composites, demonstrate enhanced hydrogen absorption attributed to the hydrogen spillover effect, resulting in a 64% increase in storage capacity [47]. The dimensions and surface alterations of nanoparticles are pivotal in enhancing storage efficacy, as emphasized in recent studies [48]. Researchers studying rubidium-based metal hydrides have found that they can store hydrogen well, with 2.09 wt.% for $RbCrH_3$ and 1.64 wt.% for $RbZrH_3$ [49]. These materials have excellent thermal stability, rendering them attractive candidates for applications in hydrogen storage. Metal-organic frameworks (MOFs) are acknowledged for their exceptional adsorption capabilities and adjustable pore architectures, which improve hydrogen storage across diverse situations [50]. Their capacity to function as catalysts enhances the kinetics of hydrogen storage processes, in accordance with sustainability objectives [50].

Hydrogen Transportation

The conveyance of hydrogen is a pivotal element in its function as a clean energy carrier, presenting both challenges and opportunities. Recent research has identified several options for hydrogen transportation, including pipelines, road transport, and novel approaches such as blending with natural gas. Hydrogen can be conveyed as a compressed gas through pipes or specialized vehicles and as liquid hydrogen in cryogenic tanks, each method presenting unique benefits and obstacles. The use of existing pipelines for H_2 transport poses safety problems due to the distinct characteristics of hydrogen, requiring comprehensive risk evaluations [51]. Innovations in materials like metal-organic frameworks and nanoparticles are being investigated to improve storage and transit efficiency. Recent models that combine transportation route planning with hydrogen supply chains can markedly decrease expenses, as evidenced by case studies [52].

Applications of Hydrogen in Industrial Processes

H_2 is widely used in the industry, especially in the chemical sector. It is used in refining, ammonia synthesis, methanol production, hydrogenation, and methanation.

Hydrogen is needed for hydrotreating and hydrocracking, which improve oil quality. The reaction with hydrogen removes nitrogen, sulfur, and metals from crude oil fractions. Ammonia (NH_3) and hydrogen sulphide (H_2S) are produced, making fuels cleaner. Hydrocracking converts heavy hydrocarbons into lighter, more valuable products like gasoline, jet fuel, and diesel. Hydrogen cracks and saturates goods, preventing the synthesis of unsaturated molecules [55].

An iron-based catalyst synthesizes ammonia (NH_3) from nitrogen (N_2) and hydrogen (H_2) at high pressure and temperature in the Haber-Bosch process, a major commercial hydrogen production technique. Synthesising ammonia, a precursor to fertilizers, explosives, and chemicals, requires this process [56]:

$$N_2 + 3H_2 \rightarrow 2NH_3$$

Hydrogen is needed to synthesize methanol (CH_3OH), a feedstock for formaldehyde and acetic acid, and is a cleaner fuel. Methanol synthesis uses a copper-based catalyst to react H_2 with CO or CO_2 [57].

$$CO:CO + 2H_2 \rightarrow CH_3OH$$

$$CO_2: CO_2 + 3H_2 \rightarrow CH_3OH + H_2O$$

Hydrogenation adds hydrogen to unsaturated compounds using palladium or nickel catalysts. This method is widely used in food to hydrogenate unsaturated vegetable oils into semi-solid fats like margarine. These products are more stable and last longer after hydrogenation [58]. Hydrogenation is essential for fine compound, drug, and agrochemical production. Hydrogenation of unsaturated compounds like alkenes and alkynes creates saturated alkanes, making them more stable for chemical processing [59].

Methanation converts H_2 and CO or CO_2 to CH_4, the main component of natural gas. Converting CO_2 into methane helps produce natural gas and reduce industrial carbon emissions, creating a circular carbon economy [60 - 62].

$$CO + 3H_2 \rightarrow CH_4 + H_2O;$$

$$CO_2 + 4H_2 \rightarrow CH_4 + 2H_2O$$

Hydrogen-based direct reduced iron minimizes the environmental impact of steel production. Hydrogen reacts with iron ore (Fe_2O_3) to produce metallic iron (Fe) and water vapor (H_2O) [63]:

$$Fe_2O_3 + 3H_2 \rightarrow 2Fe + H_2O$$

Steelmaking and heat treatment need 20-30 million tons of H_2 annually. Chemicals for steelmaking produce 10–20 million tons of H_2 annually. This suggests that the steel sector will need 160-300 million tons of hydrogen annually, highlighting the sustainability of hydrogen [64].

Many steel companies are pioneering hydrogen-based steelmaking to reduce carbon emissions. The HYBRIT Project, a collaboration between SSAB, LKAB, and Vattenfall in Sweden, is forming HYBRIT to manufacture fossil-free steel utilizing hydrogen-based direct reduction iron. Sustainable hydrogen-powered steel manufacturing is the goal by 2026 [65]. Steel giant ArcelorMittal is investing heavily in hydrogen-direct reduced iron technology. The company aims to achieve carbon neutrality by 2050 and is exploring the potential applications of hydrogen in steel manufacturing [66]. SALCOS Projekt Salzgitter AG: This German project uses hydrogen in steelmaking to cut CO_2 emissions. Salzgitter AG uses hydrogen DRI technology to make steel more sustainable [67].

Hydrogen is essential for industrial decarbonization, especially as a clean replacement for high-temperature (over 1,800°C) heat generation. Hydrogen is a carbon-free option for high-temperature activities in energy-intensive industries, including steel, cement, glass, and ceramics. Hydrogen burning produces just water vapor and no CO_2, making it an attractive alternative for businesses where direct electrification of heat is problematic [68].

Fossil fuels heat cement and glass factories, contributing significantly to global greenhouse gas emissions. These enterprises may reduce their carbon footprints by using hydrogen instead of fossil fuels. In cement manufacture, hydrogen heats the kiln and reacts with clinker to make cement. Hydrogen's thermal energy can be used to melt raw materials for glass production, thereby reducing carbon emissions [69, 70].

Global fertilizer production requires much hydrogen. The IEA estimates that ammonia production uses 70% of the hydrogen used worldwide. Around 120 million tons of hydrogen were produced globally in 2022, with 31.5 million tons going to fertilizer ammonia. As the world population grows and food demand rises, nitrogen-based fertilizers and hydrogen will be in greater demand.

Applications of Hydrogen in Power Generation

Hydrogen Fuel Cells

H_2 fuel cells provide an up-and-coming technology for the generation of power, especially for the integration of renewable energy and decarbonization initiatives. A hydrogen fuel cell fundamentally comprises an anode, a cathode, and an electrolyte membrane. At the cathode, protons, electrons, and oxygen combine to produce water, the sole byproduct of the fuel cell. There are various types of fuel cells available for power production, such as Solid Oxide Fuel Cells (SOFCs) and Proton Exchange Membrane Fuel Cells (PEMFCs). PEMFCs are generally employed in smaller-scale applications, including domestic power generation and electric vehicles, owing to their high efficiency and comparatively low operating temperatures (60–100°C). In contrast, SOFCs, which function at elevated temperatures (500–1,000°C), are more appropriate for large-scale power generation, providing superior electrical efficiencies and the capability to employ diverse fuels, such as biogas, natural gas, and hydrogen [71 - 75].

Hydrogen for Grid Stabilization

Hydrogen is emerging as a viable solution for grid stabilization, due to its capacity to store substantial energy for prolonged durations. This method involves the electrolysis of water to separate it into its constituent elements, oxygen and hydrogen. Hydrogen is then stored in massive tanks or subterranean tunnels for future utilization. In times of elevated power demand or diminished renewable energy output, the stored hydrogen can be reconverted into energy using fuel cells or combustion turbines, therefore stabilizing the grid [76]. Hydrogen is optimal for mitigating seasonal variations in renewable energy output, as production may surpass demand during summer and be insufficient in winter [77]. Incorporation of hydrogen into the power grid facilitates interconnection of electricity, gas, and heat networks, a concept referred to as "sector coupling," which can improve total system flexibility and efficiency [78, 79].

Power-to-Hydrogen and Power-to-X Systems

Power-to-hydrogen (P2H) is a crucial component of hydrogen-centric grid stabilization techniques. P2H systems transform excess energy into hydrogen through electrolysis, functioning as a conduit between the electrical grid and the hydrogen industry. Power-to-X (P2X) denotes the comprehensive notion of transforming power into alternative energy sources, such as synthetic fuels or heat, hence enhancing the potential of hydrogen within an integrated energy system [80]. Although hydrogen has several benefits for grid stabilization, significant obstacles remain to its extensive use. The primary problem is the present expense of H_2 generation, particularly green hydrogen generated from sources of renewable energy. Electrolysis remains a costly method compared to alternative energy storage solutions, such as batteries. Nonetheless, continuous technical progress and economies of scale are anticipated to decrease expenses in the forthcoming years [81].

Hydrogen Combustion in Power Generation Facilities

Hydrogen combustion in power plants is an innovative technology that holds substantial promise for decarbonizing power generation. Hydrogen combustion happens when hydrogen interacts with oxygen, yielding heat and water vapor ($2H_2 + O_2 \rightarrow 2H_2O$). In a power plant, this thermal energy may be utilized to generate steam, which forces turbines to produce electricity [82, 83].

Combined Cycle Gas Turbine (CCGT) facilities are extensively employed for electricity production owing to their superior efficiency and adaptability. These facilities generally combust natural gas to create electricity *via* a gas turbine, using the resultant waste heat to provide supplementary power through a steam turbine [83]. Integrating hydrogen with natural gas provides a transitional route to complete hydrogen combustion [84].

The combustion of pure hydrogen, however, is more complex, signifying the paramount objective for zero-carbon power facilities. Specialized hydrogen turbines are being engineered to enhance the combustion process and accommodate the unique properties of hydrogen [85]. However, nitrogen in the atmosphere can react with oxygen at elevated combustion temperatures, resulting in the production of nitrogen oxides (NO_x), which are detrimental pollutants. Modern hydrogen turbines use techniques like dry low-NO_x combustion and exhaust gas recirculation to lower NO_x emissions. This makes hydrogen combustion less harmful to the environment [86 - 90].

Hydrogen in Distributed Energy Systems

Distributed energy systems are gaining significance in boosting energy resilience, augmenting local energy security, and facilitating the incorporation of renewable energy sources. Distributed energy systems denote small-scale power-producing installations situated near the consumption point, including residential, commercial, or industrial locations [91]. A significant benefit of hydrogen in distributed energy systems is its ability to be generated locally from renewable energy sources. Water electrolysis enables the utilization of surplus power from sources such as wind turbines or solar panels to dissociate water into oxygen and hydrogen [92].

Hybrid Systems: Hydrogen and Renewables

The combined use of hydrogen with sources of renewable energy, such as wind and solar, *via* hybrid systems is a highly efficient and sustainable approach to augmenting energy resilience and stability.

Hybrid systems that employ hydrogen and renewable resources generally have three essential components: renewable energy-generating units, electrolyzers, and fuel cells. Renewable systems, like solar panels and wind turbines, provide power that can be utilized immediately or sent to electrolyzers. The electrolyzers utilize the surplus energy to decompose water into oxygen and hydrogen, thereby storing hydrogen for future utilization [93].

Hydrogen for energy storage in hybrid systems facilitates the equilibrium of power supply, consequently augmenting grid stability and resilience. These devices can serve as a safeguard against energy shortages, particularly beneficial for areas with unreliable grid infrastructure [93]. Employing hydrogen sourced from renewable origins significantly reduces carbon emissions compared to traditional fossil fuels. The water output of hydrogen fuel cells contributes to environmental advantages, bolstering worldwide decarbonization initiatives [93].

Applications of Hydrogen in Aviation and Aerospace

H_2 is a new energy source in aviation and aerospace that can decarbonize. Its applications cover aircraft design, operation, propulsion, and onboard technology. Hydrogen has aviation and aerospace applications. A modified gas turbine engine may generate thrust from H_2. Hydrogen's high auto-ignition temperature increases compression ratios and thermal efficiency [94]. Hydrogen is an effective rocket propellant utilized for space exploration and satellite launches [95]. UAVs: Hydrogen fuel cells have three times the energy density of batteries, allowing UAVs to fly three times farther. Hydrogen fuel cells produce only water and heat, making them a greener alternative for UAVs. Hydrogen cylinders are faster than battery charging, as they may be replaced in minutes. Hydrogen fuel cells increase the payload capacity of UAVs [96].

Hydrogen fuel cells, or proton-exchange membranes, are being optimized for aircraft to reduce weight and increase efficiency [97]. Compared to traditional systems, these systems may save up to 10% in weight and minimize greenhouse gas emissions [97]. Metal hydrides (MHs) are being studied for hydrogen storage, cabin air conditioning, and hydrogen safety sensors [98]. Because they recover hydrogen boil-off and control temperature, hydrogen-powered aircraft are more efficient. Fuel cell efficiency needs intricate temperature control systems, although liquid hydrogen storage is favored due to its weight [99]. Additive manufacturing enables the creation of lightweight, efficient cryogenic heat exchangers for fuel cell applications [99].

Aviation must tackle these difficulties to utilize hydrogen technology effectively. Adapting planes and engines for hydrogen use might raise manufacturing and maintenance costs by 25% [100]. New airport refueling and storage facilities are needed. Hydrogen-powered aircraft will have 6% larger fuselages [101]. Bending and vibration resistance must be improved by reengineering the wings.

Challenges and Opportunities

Hydrogen technology encounters several obstacles that restrict its industrial utilization, despite its enormous potential as a clean energy resource. Principal concerns are manufacturing efficiency, storage safety, and material compatibility. Confronting these obstacles is essential for the extensive use of hydrogen across several industries.

Challenges in Production

Hydrogen technology encounters several production problems that limit its industrial uses, despite its promising potential as a clean energy source. Critical concerns encompass production technique efficiency, material compatibility, and safety measures. Hydrogen generation techniques, including hydro electrolysis and biomass conversion, show potential but necessitate enhancements in efficiency and cost-effectiveness [102]. The standardized price of hydrogen is anticipated to decline significantly; however, attaining large-scale commercial viability poses a hurdle [102]. Hydrogen embrittlement presents hazards to infrastructure, requiring investigation into material performance under elevated pressure and temperature circumstances [103, 104]. Ensuring the safe handling and storage of hydrogen is essential, with continuous endeavors to create new storage methods [105]. Comprehending hydrogen's interaction with metals is crucial for avoiding malfunctions in industrial applications [106]. Modifying current industrial systems to facilitate the utilization of hydrogen is crucial for its extensive adoption.

Storage and Transportation Challenges

Hydrogen technology faces substantial storage and transportation obstacles that hinder its industrial use. Significant challenges encompass poor storage density, elevated prices, and insufficient infrastructure, despite progress in storage techniques like compressed gas and cryogenic liquid hydrogen [107]. Moreover, the incorporation of hydrogen into current energy infrastructures is essential for its extensive acceptance. Inadequate storage density restricts the quantity of hydrogen that can be stored. Pressure vessels and cryogenic tanks incur significant costs. Concerns surround the safety of handling and storing hydrogen. The infrastructure needed for extensive hydrogen transportation is insufficient. Hydrogen transportation necessitates the use of modern materials in pipeline systems. Although hydrogen offers a promising alternative to fossil fuels, the economic viability and safety of its production, storage, and delivery pose considerable challenges that must be addressed to unlock its full potential in the energy sector.

Hydrogen fuel cells deliver efficient, zero-emission energy for transportation and industrial uses, improving energy efficiency [108]. Hydrogen extraction strategies enhance the production and quality of bioactive chemicals derived from agricultural and food waste, providing an environmentally sustainable alternative to conventional extraction methods [109]. This approach is scalable and economical, meeting the increasing need for natural food additives while reducing environmental impact. Green hydrogen, generated from sources of renewable energy, has the potential to mitigate the emission of greenhouse gases in the agro-livestock industry, a large contributor to world emissions [110]. It increases energy security and diminishes dependence on fossil fuels, fostering sustainable agricultural practices [110]. Hydrogen has demonstrated the ability to stimulate plant development, augment nutritional benefits, and improve livestock tolerance to infections [111]. This application can result in increased yields and enhanced quality of agricultural goods, hence contributing to food security [111].

CONCLUSION

Multiple methods are used to generate hydrogen, each with its own pros and cons. A water-gas shift reaction and hydrocarbon-steam contact produce hydrogen in SMR, a standard process. Photoelectrochemical and high-temperature steam electrolysis are becoming successful electrolysis methods. These devices split water into oxygen and hydrogen using renewable energy, minimizing carbon emissions. Hydrogen can be created from biomass and other renewable sources to achieve carbon neutrality. Feedstock choice significantly impacts production sustainability. Solar energy can break water into oxygen and hydrogen for photocatalytic hydrogen synthesis. Nanoparticles and plasmonic nanoparticles have been tailored to optimize light absorption and charge carrier generation, improving photocatalytic performance. Photocatalytic hydrogen production offers a sustainable alternative to fossil fuels. These technologies have potential, but issues include efficiency, cost, and scalability.

Due to the high pressures and low temperatures needed for liquefaction, compressed gas and liquid storage methods have density and safety limitations. Solid-state storage utilizes metal and complex hydrides, which offer higher gravimetric capacities and improved safety, but require temperature control for the release of hydrogen. Engineered carbon nanospaces store hydrogen *via* physisorption due to their high surface area and customized pore size. Hydrogen storage technologies for energy storage are uneconomical. In a hydrogen economy, where hydrogen is sold for transportation, they are more feasible. Hydrogen storage technology has improved, but safety, economic feasibility, and efficiency remain significant challenges that require further study.

Hydrogen's ability to decarbonize businesses and meet rising energy needs is driving its use in industrial operations. Hydrogen's energy carrier properties affect transportation, chemical manufacture, and energy storage. Hydrogen is needed to refine heavy crude oil into greener fuels to fulfill increased demand for low-sulfur goods. Hydrogen power generating may reduce carbon emissions and boost energy efficiency. Hydrogen-oxygen combustion turbines are efficient and emit less than fossil fuels. Hydrogen storage combined with wind energy systems allows electricity generation to fulfill load demands and reduce waste. Hydrogen in aviation and aerospace might reduce fossil fuel emissions and promote sustainable air travel. H_2's ability to produce just water reduces greenhouse gas emissions, making it a clean fuel. Hydrogen fuel cells may reduce carbon emissions and noise while improving airplane design with multifunctional systems.

REFERENCES

[1]   P. Lv, "Hydrogen storage technology", In: *Towards Hydrogen Infrastructure.* Elsevier, 2024, pp. 165-184.
[http://dx.doi.org/10.1016/B978-0-323-95553-9.00001-7]

[2]   G. AlZohbi, A. Almoaikel, and L. AlShuhail, "An overview on the technologies used to store hydrogen", *Energy Rep.,* vol. 9, pp. 28-34, 2023.
[http://dx.doi.org/10.1016/j.egyr.2023.08.072]

[3]   S. Zhu, X. Shi, C. Yang, Y. Li, H. Li, K. Yang, X. Wei, W. Bai, and X. Liu, "Hydrogen loss of salt cavern hydrogen storage", *Renew. Energy,* vol. 218, p. 119267, 2023.
[http://dx.doi.org/10.1016/j.renene.2023.119267]

[4]   P. Gabrielli, J. Garrison, S. Hässig, E. Raycheva, and G. Sansavini, "The role of hydrogen storage in an electricity system with large hydropower resources", *Energy Convers. Manage.,* vol. 302, p. 118130, 2024.
[http://dx.doi.org/10.1016/j.enconman.2024.118130]

[5]   N. Sirosh, "FUELS – HYDROGEN STORAGE | Compressed", In: *Encyclopedia of Electrochemical Power Sources.* Elsevier, 2009, pp. 414-420.
[http://dx.doi.org/10.1016/B978-044452745-5.00322-1]

[6]   D. Nash, D. Aklil, E. Johnson, R. Gazey, and V. Ortisi, "Hydrogen Storage", In: *Comprehensive Renewable Energy.* Elsevier, 2012, pp. 131-155.
[http://dx.doi.org/10.1016/B978-0-08-087872-0.00413-3]

[7]   T.Q. Hua, R.K. Ahluwalia, J-K. Peng, M. Kromer, S. Lasher, K. McKenney, K. Law, and J. Sinha, "Technical assessment of compressed hydrogen storage tank systems for automotive applications", *Int. J. Hydrogen Energy,* vol. 36, no. 4, pp. 3037-3049, 2011.
[http://dx.doi.org/10.1016/j.ijhydene.2010.11.090]

[8]   Z. Zhang, M. He, H. Zhou, and L. Gong, "Model establishment and process analysis of liquid hydrogen energy storage", *IOP Conf Ser Mater Sci Eng,* vol. vol. 1301, 2024p. 012057
[http://dx.doi.org/10.1088/1757-899X/1301/1/012057]

[9]   B. Liu, Y. Li, Y. Ma, and L. Wang, "Electrostatic characteristics analysis and risk assessments of liquid hydrogen storage system", *Int. J. Hydrogen Energy,* vol. 55, pp. 1322-1334, 2024.
[http://dx.doi.org/10.1016/j.ijhydene.2023.12.036]

[10]   L. Yin, H. Yang, and Y. Ju, "Review on the key technologies and future development of insulation structure for liquid hydrogen storage tanks", *Int. J. Hydrogen Energy,* vol. 57, pp. 1302-1315, 2024.
[http://dx.doi.org/10.1016/j.ijhydene.2024.01.093]

[11]   S. Park, W. L. Tae, and H. N. Dong, "Solid-state hydrogen storage device", U.S. Patent 20170244124A1, March 30, 2020.

[12]   C. Wu, P. Zhu, Y. Han, Y. Chen, W. Yao, and Y. Yigang, "Solid-state hydrogen storage device with

high heat exchange characteristics", China Patent 111188988, May 21, 2020.

[13]   Y. Xu, Y. Zhou, Y. Li, and Z. Ding, "Research progress and application prospects of solid-state hydrogen storage technology", *Molecules,* vol. 29, no. 8, p. 1767, 2024.
[http://dx.doi.org/10.3390/molecules29081767] [PMID: 38675587]

[14]   E.B. Agyekum, C. Nutakor, A.M. Agwa, and S. Kamel, "A critical review of renewable hydrogen production methods: Factors affecting their scale-up and its role in future energy generation", *Membranes (Basel),* vol. 12, no. 2, p. 173, 2022.
[http://dx.doi.org/10.3390/membranes12020173] [PMID: 35207094]

[15]   L. Ge, B. Zhang, W. Huang, Y. Li, L. Hou, J. Xiao, Z. Mao, and X. Li, "A review of hydrogen generation, storage, and applications in power system", *J. Energy Storage,* vol. 75, p. 109307, 2024.
[http://dx.doi.org/10.1016/j.est.2023.109307]

[16]   S. Ahmad, A. Ullah, A. Samreen, M. Qasim, K. Nawaz, W. Ahmad, A. Alnaser, A.M. Kannan, and M. Egilmez, "Hydrogen production, storage, transportation and utilization for energy sector: A current status review", *J. Energy Storage,* vol. 101, p. 113733, 2024.
[http://dx.doi.org/10.1016/j.est.2024.113733]

[17]   J.A. Okolie, B.R. Patra, A. Mukherjee, S. Nanda, A.K. Dalai, and J.A. Kozinski, "Futuristic applications of hydrogen in energy, biorefining, aerospace, pharmaceuticals and metallurgy", *Int. J. Hydrogen Energy,* vol. 46, no. 13, pp. 8885-8905, 2021.
[http://dx.doi.org/10.1016/j.ijhydene.2021.01.014]

[18]   A. Chauhan, and H.C. Joshi, "Technologies for the production of biogas and biodiesel in different scenario", *Trends in Mathematics,* no. Part F3197, pp. 127-185, 2024.
[http://dx.doi.org/10.1007/978-981-99-7250-0_3]

[19]   I. Vajdová, E. Jenčová, and P. Koščák, "Hydrogen as one of the future alternative fuels in aviation - review", In: *New Trends in Civil Aviation (NTCA).* IEEE, 2024, pp. 89-97.
[http://dx.doi.org/10.23919/NTCA60572.2024.10517808]

[20]   L. Qian, J. Shen, and X. Sun, "Methods of hydrogen application", In: *Hydrogen Molecular Biology and Medicine.* Springer Netherlands: Dordrecht, 2015, pp. 99-107.
[http://dx.doi.org/10.1007/978-94-017-9691-0_7]

[21]   K. Andreassen, "Hydrogen production by electrolysis", In: *Hydrogen Power: Theoretical and Engineering Solutions.* Springer Netherlands: Dordrecht, 1998, pp. 91-102.
[http://dx.doi.org/10.1007/978-94-015-9054-9_11]

[22]   J. Hnát, M. Paidar, and K. Bouzek, "Hydrogen production by electrolysis", In: *Current Trends and Future Developments on (Bio-) Membranes.* Elsevier, 2020, pp. 91-117.
[http://dx.doi.org/10.1016/B978-0-12-817384-8.00005-4]

[23]   H.C. Joshi, "Reetika, Waseem, Bhawana, Nishesh Sharma, "A review on carbonaceous materials for fuel cell technologies:An advanced approach"", *Vietnam J. Chem.,* pp. 1-10, 2024.

[24]   E. S. Akyüz, E. Telli, and M. Farsak, "Hydrogen generation electrolyzers: Paving the way for sustainable energy," *Int J Hydrogen Energy*, vol. 81, pp. 1338–1362, Sep. 2024.
[http://dx.doi.org/10.1016/J.IJHYDENE.2024.07.175]

[25]   H. Cheng, Y. Xia, Z. Hu, and W. Wei, "Optimum pulse electrolysis for efficiency enhancement of hydrogen production by alkaline water electrolyzers", *Appl. Energy,* vol. 358, p. 122510, 2024.
[http://dx.doi.org/10.1016/j.apenergy.2023.122510]

[26]   A. Tello, "Green hydrogen production by photovoltaic-assisted alkaline water electrolysis: A review on the conceptualization and advancements", *Int. J. Hydrogen Energy,* no. May, 2024.
[http://dx.doi.org/10.1016/j.ijhydene.2024.04.333]

[27]   J. Jeddizahed, P.A. Webley, and T.J. Hughes, "Integrating alkaline electrolysis with oxyfuel combustion for hydrogen and electricity production", *Appl. Energy,* vol. 361, p. 122856, 2024.
[http://dx.doi.org/10.1016/j.apenergy.2024.122856]

[28]   H. Liu, M. Yu, X. Tong, Q. Wang, and M. Chen, "High temperature solid oxide electrolysis for green hydrogen production", *Chem. Rev.,* vol. 124, no. 18, pp. 10509-10576, 2024.
[http://dx.doi.org/10.1021/acs.chemrev.3c00795] [PMID: 39167109]

[29]   S. Lang, J. Yuan, and H. Zhang, "Optimally splitting solar spectrums by concentrating solar spectrums splitter for hydrogen production *via* solid oxide electrolysis cell", *Energies,* vol. 17, no. 9, p. 2067, 2024.
[http://dx.doi.org/10.3390/en17092067]

[30]   A. Baudh, M. Garjola, R. Sharma, S. Sharma, and R.K. Upadhyay, "Effect of ceria morphology on hydrogen production *via* methane steam reforming for membrane reformer", *Can. J. Chem. Eng.,* vol. 102, no. 11, pp. 3803-3816, 2024.
[http://dx.doi.org/10.1002/cjce.25396]

[31]   S. Wang, Z. Shen, A. Osatiashtiani, S.A. Nabavi, and P.T. Clough, "Ni-based bimetallic catalysts for hydrogen production *via* (sorption-enhanced) steam methane reforming", *Chem. Eng. J.,* vol. 486, p. 150170, 2024.
[http://dx.doi.org/10.1016/j.cej.2024.150170]

[32]   R. Kumar, A. Kumar, and A. Pal, "Simulation modelling of hydrogen production from steam reforming of methane and biogas", *Fuel,* vol. 362, p. 130742, 2024.
[http://dx.doi.org/10.1016/j.fuel.2023.130742]

[33]   S. Zolghadri, M.R. Kiani, R. kamandi, and M.R. Rahimpour, "Enhanced hydrogen production in steam methane reforming: Comparative analysis of industrial catalysts and process optimization", *Journal of the Energy Institute,* vol. 113, p. 101541, 2024.
[http://dx.doi.org/10.1016/j.joei.2024.101541]

[34]   A.M. Chandole, and P.D. Vaidya, "Hydrogen production from the carbon dioxide sorption-enhanced steam reforming of biogas over a Ni-based hybrid material with a calcium oxide sorbent", *Ind. Eng. Chem. Res.,* vol. 63, no. 19, pp. 8581-8590, 2024.
[http://dx.doi.org/10.1021/acs.iecr.4c00593]

[35]   A. Chauhan, and H.C. Joshi, "Optimizing the biofuels infrastructure for the transportation network", *Trends in Mathematics,* no. Part F3197, pp. 309-335, 2024.
[http://dx.doi.org/10.1007/978-981-99-7250-0_7]

[36]   T. Ahasan, P. Xu, and H. Wang, "Dual-function photocatalysis in the visible spectrum: AG-G-TiO2 for simultaneous dye wastewater degradation and hydrogen production", *Catalysts,* vol. 14, no. 8, p. 530, 2024.
[http://dx.doi.org/10.3390/catal14080530]

[37]   M. Singh, "Photocatalysis-derived biomass conversion for green hydrogen production", *Towards Sustainable and Green Hydrogen Production by Photocatalysis: Insights into Design and Development of Efficient Materials,* vol. 2, pp. 47-78, 2024.
[http://dx.doi.org/10.1021/bk-2024-1468.ch003]

[38]   R.D. Tentu, and S. Basu, "Photocatalytic water splitting for hydrogen production", *Curr. Opin. Electrochem.,* vol. 5, no. 1, pp. 56-62, 2017.
[http://dx.doi.org/10.1016/j.coelec.2017.10.019]

[39]   C.H. Kim, J-Y. Han, S. Kim, B. Lee, H. Lim, K-Y. Lee, and S-K. Ryi, "Hydrogen production by steam methane reforming in a membrane reactor equipped with a Pd composite membrane deposited on a porous stainless steel", *Int. J. Hydrogen Energy,* vol. 43, no. 15, pp. 7684-7692, 2018.
[http://dx.doi.org/10.1016/j.ijhydene.2017.11.176]

[40]   J. Chi, and H. Yu, "Water electrolysis based on renewable energy for hydrogen production", *Chin. J. Catal.,* vol. 39, no. 3, pp. 390-394, 2018.
[http://dx.doi.org/10.1016/S1872-2067(17)62949-8]

[41]   A. Franco, and C. Giovannini, "Hydrogen gas compression for efficient storage: Balancing energy and

increasing density", *Hydrogen,* vol. 5, no. 2, pp. 293-311, 2024.
[http://dx.doi.org/10.3390/hydrogen5020017]

[42] M. He, "Design and development of vehicle cryo-compressed hydrogen storage vessel", *IOP Conf Ser Mater Sci Eng,* vol. vol. 1301, 2024p. 012056
[http://dx.doi.org/10.1088/1757-899X/1301/1/012056]

[43] D.H. Kang, J.H. An, and C.J. Lee, "Numerical modeling and optimization of thermal insulation for liquid hydrogen storage tanks", *Energy,* vol. 291, p. 130143, 2024.
[http://dx.doi.org/10.1016/j.energy.2023.130143]

[44] H. Mun, S. Park, and I. Lee, "Liquid hydrogen cold energy recovery to enhance sustainability: Optimal design of dual-stage power generation cycles," Energy, vol. 284, p. 129229, Dec. 2023.
[http://dx.doi.org/10.1016/J.ENERGY.2023.129229]

[45] M. Ozel, "Effect of insulation location on dynamic heat-transfer characteristics of building external walls and optimization of insulation thickness," Energy Build, vol. 72, pp. 288–295, Apr. 2014.
[http://dx.doi.org/10.1016/J.ENBUILD.2013.11.015]

[46] D. Choi, S. Lee, and S. Kim, "A thermodynamic model for cryogenic liquid hydrogen fuel tanks", *Appl. Sci. (Basel),* vol. 14, no. 9, p. 3786, 2024.
[http://dx.doi.org/10.3390/app14093786]

[47] Y. Sun, J. Hong, M. Zhu, Z. Liu, W. Qiao, S. Yan, Z. Li, Y. Wu, J. Wu, H. Zhang, and H. Bai, "Solid hydrogen storage with palladium-graphene composites: Synthesis, characterization, and mechanistic insights", *Mater. Chem. Phys.,* vol. 322, p. 129537, 2024.
[http://dx.doi.org/10.1016/j.matchemphys.2024.129537]

[48] Y. Xu, Y. Li, L. Gao, Y. Liu, and Z. Ding, "Advances and prospects of nanomaterials for solid-state hydrogen storage", *Nanomaterials (Basel),* vol. 14, no. 12, p. 1036, 2024.
[http://dx.doi.org/10.3390/nano14121036] [PMID: 38921912]

[49] Y. Didi, S. Bahhar, A. Tahiri, M. Naji, and A. Rjeb, "A computational study of metal hydrides based on rubidium for developing solid-state hydrogen storage", *ChemistrySelect,* vol. 9, no. 22, p. e202401444, 2024.
[http://dx.doi.org/10.1002/slct.202401444]

[50] Y. Li, Q. Guo, Z. Ding, H. Jiang, H. Yang, W. Du, Y. Zheng, K. Huo, and L.L. Shaw, "MOFs-based materials for solid-state hydrogen storage: strategies and perspectives", *Chem. Eng. J.,* vol. 485, p. 149665, 2024.
[http://dx.doi.org/10.1016/j.cej.2024.149665]

[51] M.O. Amer, S.M. Hoseyni, and J. Cordiner, "Fuelling the future with safe hydrogen transportation through natural gas pipelines: A quantitative risk assessment approach", *Trans. Indian Natl. Acad. Eng.,* vol. 9, no. 4, pp. 763-781, 2024.
[http://dx.doi.org/10.1007/s41403-024-00482-7]

[52] X. Wang, Y. Wu, Z. Wen, Z. Cui, and Y. Wang, "A new transportation route planning method for wind-based hydrogen supply chains", *ACS Sustain. Chem.& Eng.,* vol. 12, no. 22, pp. 8436-8452, 2024.
[http://dx.doi.org/10.1021/acssuschemeng.4c01352]

[53] L. Mulky, S. Srivastava, T. Lakshmi, E.R. Sandadi, S. Gour, N.A. Thomas, S. Shanmuga Priya, and K. Sudhakar, "An overview of hydrogen storage technologies – Key challenges and opportunities", *Mater. Chem. Phys.,* vol. 325, p. 129710, 2024.
[http://dx.doi.org/10.1016/j.matchemphys.2024.129710]

[54] M.R. Usman, "Hydrogen storage methods: Review and current status", *Renew. Sustain. Energy Rev.,* vol. 167, p. 112743, 2022.
[http://dx.doi.org/10.1016/j.rser.2022.112743]

[55] A. Lubis, J. Jeong, N. Giannetti, S. Yamaguchi, K. Saito, H. Yabase, M.I. Alhamid, and Nasruddin,

"Operation performance enhancement of single-double-effect absorption chiller", *Appl. Energy,* vol. 219, pp. 299-311, 2018.
[http://dx.doi.org/10.1016/j.apenergy.2018.03.046]

[56] Ž. Ponikvar, B. Likozar, and S. Gyergyek, "Electrification of catalytic ammonia production and decomposition reactions: from resistance, induction, and dielectric reactor heating to electrolysis", *ACS Appl. Energy Mater.,* vol. 5, no. 5, pp. 5457-5472, 2022.
[http://dx.doi.org/10.1021/acsaem.1c03045]

[57] B. Balopi, P. Agachi, and Danha, "Methanol synthesis chemistry and process engineering aspects- a review with consequence to botswana chemical industries", *Procedia Manuf.,* vol. 35, pp. 367-376, 2019.
[http://dx.doi.org/10.1016/j.promfg.2019.05.054]

[58] C. zhang, C. Guo, F. Liu, W. Kong, Y. He, and B. Lou, "Hyperspectral imaging analysis for ripeness evaluation of strawberry with support vector machine", *J. Food Eng.,* vol. 179, pp. 11-18, 2016.
[http://dx.doi.org/10.1016/j.jfoodeng.2016.01.002]

[59] W. Cheng, and R.G. Compton, "Oxygen reduction mediated by single nanodroplets containing attomoles of vitamin B12: electrocatalytic nano-impacts method", *Angew. Chem. Int. Ed.,* vol. 54, no. 24, pp. 7082-7085, 2015.
[http://dx.doi.org/10.1002/anie.201501820] [PMID: 25917121]

[60] X. Luo, X. Zhu, and E.G. Lim, "A parametric bootstrap algorithm for cluster number determination of load pattern categorization", *Energy,* vol. 180, pp. 50-60, 2019.
[http://dx.doi.org/10.1016/j.energy.2019.04.089]

[61] W. Zhang, F. Gu, F. Dai, X. Gu, F. Yue, and B. Bao, "Decision framework for feasibility analysis of introducing the steam turbine unit to recover industrial waste heat based on economic and environmental assessments", *J. Clean. Prod.,* vol. 137, pp. 1491-1502, 2016.
[http://dx.doi.org/10.1016/j.jclepro.2016.07.039]

[62] N. Rambhujun, M.S. Salman, T. Wang, C. Pratthana, P. Sapkota, M. Costalin, Q. Lai, and K.F. Aguey-Zinsou, "Renewable hydrogen for the chemical industry", *MRS Energy Sustain.,* vol. 7, no. 1, p. 33, 2020.
[http://dx.doi.org/10.1557/mre.2020.33] [PMID: 38624624]

[63] T. Jafary, A. Al-Mamun, H. Alhimali, M.S. Baawain, M.S. Rahman, S. Rahman, B.R. Dhar, M. Aghbashlo, and M. Tabatabaei, "Enhanced power generation and desalination rate in a novel quadruple microbial desalination cell with a single desalination chamber", *Renew. Sustain. Energy Rev.,* vol. 127, p. 109855, 2020.
[http://dx.doi.org/10.1016/j.rser.2020.109855]

[64] H. Sudrajat, S. Babel, S. Hartuti, J. Phanthuwongpakdee, K. Laohhasurayotin, T.K. Nguyen, and H.D. Tong, "Origin of the overall water splitting activity over Rh/Cr2O3@ anatase TiO_2 following UV-pretreatment", *Int. J. Hydrogen Energy,* vol. 46, no. 61, pp. 31228-31238, 2021.
[http://dx.doi.org/10.1016/j.ijhydene.2021.07.002]

[65] A. Lorente-Arevalo, M. Ladero, and J.M. Bolivar, "Intensification of oxygen-dependent biotransformations catalyzed by immobilized enzymes", *Curr. Opin. Green Sustain. Chem.,* vol. 32, p. 100544, 2021.
[http://dx.doi.org/10.1016/j.cogsc.2021.100544]

[66] C. Jia, P. Liu, and Z. Li, "Performance analysis of ceramic membrane tube modules for water and heat recovery in coal-fired power plants", *J. Clean. Prod.,* vol. 306, p. 127237, 2021.
[http://dx.doi.org/10.1016/j.jclepro.2021.127237]

[67] N.U. Blum, M. Haupt, and C.R. Bening, "Why "Circular" doesn't always mean "Sustainable"", *Resour. Conserv. Recycling,* vol. 162, p. 105042, 2020.
[http://dx.doi.org/10.1016/j.resconrec.2020.105042]

[68] Y. Ahn, K. Pandi, M. Lee, and J. Choi, "Removing hydrogen sulfide from a feed stream using suitable

adsorbent materials", *J. Clean. Prod.,* vol. 272, p. 122849, 2020.
[http://dx.doi.org/10.1016/j.jclepro.2020.122849]

[69]   M.S. Nazir, M. Bilal, H.M. Sohail, B. Liu, W. Chen, and H.M.N. Iqbal, "Impacts of renewable energy atlas: Reaping the benefits of renewables and biodiversity threats", *Int. J. Hydrogen Energy,* vol. 45, no. 41, pp. 22113-22124, 2020.
[http://dx.doi.org/10.1016/j.ijhydene.2020.05.195]

[70]   R. Sun, Q. Wu, J. Guo, T. Wang, Y. Wu, B. Qiu, Z. Luo, W. Yang, Z. Hu, J. Guo, M. Shi, C. Yang, F. Huang, Y. Li, and J. Min, "A layer-by-layer architecture for printable organic solar cells overcoming the scaling lag of module efficiency", *Joule,* vol. 4, no. 2, pp. 407-419, 2020.
[http://dx.doi.org/10.1016/j.joule.2019.12.004]

[71]   X. Zhang, M. Zeng, Y. She, X. Lin, D. Yang, Y. Qin, and X. Rui, "Enhanced low-temperature sodium storage kinetics in a $NaTi_2(PO_4)_3$@C nanocomposite", *J. Power Sources,* vol. 477, p. 228735, 2020.
[http://dx.doi.org/10.1016/j.jpowsour.2020.228735]

[72]   H. Gu, J. Ding, Q. Zhong, Y. Zeng, and F. Song, "Promotion of surface oxygen vacancies on the light olefins synthesis from catalytic CO_2 hydrogenation over Fe K/ZrO_2 catalysts", *Int. J. Hydrogen Energy,* vol. 44, no. 23, pp. 11808-11816, 2019.
[http://dx.doi.org/10.1016/j.ijhydene.2019.03.046]

[73]   N. Yasmin, and P. Grundmann, "Adoption and diffusion of renewable energy – The case of biogas as alternative fuel for cooking in Pakistan", *Renew. Sustain. Energy Rev.,* vol. 101, pp. 255-264, 2019.
[http://dx.doi.org/10.1016/j.rser.2018.10.011]

[74]   M.C. Tucker, "Corrigendum to Dynamic-temperature operation of metal-supported solid oxide fuel cells [J. Power Sources 395 (2018) 314–317]", *J. Power Sources,* vol. 492, p. 229647, 2021.
[http://dx.doi.org/10.1016/j.jpowsour.2021.229647]

[75]   P.K. Halder, "Potential and economic feasibility of solar home systems implementation in Bangladesh", *Renew. Sustain. Energy Rev.,* vol. 65, pp. 568-576, 2016.
[http://dx.doi.org/10.1016/j.rser.2016.07.062]

[76]   I. Staffell, D. Scamman, A. Velazquez Abad, P. Balcombe, P.E. Dodds, P. Ekins, N. Shah, and K.R. Ward, "The role of hydrogen and fuel cells in the global energy system", *Energy Environ. Sci.,* vol. 12, no. 2, pp. 463-491, 2019.
[http://dx.doi.org/10.1039/C8EE01157E]

[77]   Z. He, S. Xu, W. Shen, R. Long, and H. Yang, "Overview of the development of the Chinese Jiangsu coastal wind-power industry cluster", *Renew. Sustain. Energy Rev.,* vol. 57, pp. 59-71, 2016.
[http://dx.doi.org/10.1016/j.rser.2015.12.187]

[78]   M. Cavana, A. Mazza, G. Chicco, and P. Leone, "Electrical and gas networks coupling through hydrogen blending under increasing distributed photovoltaic generation", *Appl. Energy,* vol. 290, p. 116764, 2021.
[http://dx.doi.org/10.1016/j.apenergy.2021.116764]

[79]   A. Chauhan, and H. Chandra Joshi, "Energetic efficiency of a biofuels production system mathematical modeling", *Trends in Mathematics,* no. Part F3197, pp. 337-357, 2024.
[http://dx.doi.org/10.1007/978-981-99-7250-0_8]

[80]   A. Prinzhofer, C.S. Tahara Cissé, and A.B. Diallo, "Discovery of a large accumulation of natural hydrogen in Bourakebougou (Mali)", *Int. J. Hydrogen Energy,* vol. 43, no. 42, pp. 19315-19326, 2018.
[http://dx.doi.org/10.1016/j.ijhydene.2018.08.193]

[81]   H. Kim, and S. Song, "Concept design of a novel reformer producing hydrogen for internal combustion engines using fuel decomposition method: Performance evaluation of coated monolith suitable for on-board applications", *Int. J. Hydrogen Energy,* vol. 45, no. 16, pp. 9353-9367, 2020.
[http://dx.doi.org/10.1016/j.ijhydene.2020.01.227]

[82]   J.X. Flores-Lasluisa, F. Huerta, D. Cazorla-Amorós, and E. Morallón, "Manganese oxides/LaMnO3

perovskite materials and their application in the oxygen reduction reaction", *Energy,* vol. 247, p. 123456, 2022.
[http://dx.doi.org/10.1016/j.energy.2022.123456]

[83]   W.J. Zou, K.Y. Shen, S. Jung, and Y.B. Kim, "Application of thermoelectric devices in performance optimization of a domestic PEMFC-based CHP system", *Energy,* vol. 229, p. 120698, 2021.
[http://dx.doi.org/10.1016/j.energy.2021.120698]

[84]   Y. Du, K. Fujita, S. Shironita, Y. Sone, E. Hosono, D. Asakura, and M. Umeda, "Capacity fade characteristics of nickel-based lithium-ion secondary battery after calendar deterioration at 80°C", *J. Power Sources,* vol. 501, p. 230005, 2021.
[http://dx.doi.org/10.1016/j.jpowsour.2021.230005]

[85]   G. Yang, Y. Li, J. Sang, A. Wu, J. Yang, T. Liang, J. Xu, W. Guan, M. Chai, and S.C. Singhal, "*In-situ* analysis of anode atmosphere in a flat-tube solid oxide fuel cell operated with dry reforming of methane", *J. Power Sources,* vol. 533, p. 231246, 2022.
[http://dx.doi.org/10.1016/j.jpowsour.2022.231246]

[86]   Y. Yang, H. Meng, C. Kong, W. Ma, H. Zhu, F. Ma, C. Wang, and Z. Hu, "Template-free synthesis of 1D hollow Fe doped CoP nanoneedles as highly activity electrocatalysts for overall water splitting", *Int. J. Hydrogen Energy,* vol. 46, no. 55, pp. 28053-28063, 2021.
[http://dx.doi.org/10.1016/j.ijhydene.2021.06.047]

[87]   T.A. Jack, R. Pourazizi, E. Ohaeri, J. Szpunar, J. Zhang, and J. Qu, "Investigation of the hydrogen induced cracking behaviour of API 5L X65 pipeline steel", *Int. J. Hydrogen Energy,* vol. 45, no. 35, pp. 17671-17684, 2020.
[http://dx.doi.org/10.1016/j.ijhydene.2020.04.211]

[88]   R. Zhang, S. Lan, Z. Xu, D. Qiu, and L. Peng, "Investigation and optimization of the ultra-thin metallic bipolar plate multi-stage forming for proton exchange membrane fuel cell", *J. Power Sources,* vol. 484, p. 229298, 2021.
[http://dx.doi.org/10.1016/j.jpowsour.2020.229298]

[89]   A. C. Lewis, "Optimising air quality co-benefits in a hydrogen economy: a case for hydrogen-specific standards for NOx emissions," Environmental Science: Atmospheres, vol. 1, no. 5, pp. 201–207, Jul. 2021.
[http://dx.doi.org/10.1039/D1EA00037C]

[90]   G. Hou, J. Xiong, G. Zhou, L. Gong, C. Huang, and S. Wang, "Coordinated control system modeling of ultra-supercritical unit based on a new fuzzy neural network", *Energy,* vol. 234, p. 121231, 2021.
[http://dx.doi.org/10.1016/j.energy.2021.121231]

[91]   M. Kumar, and M.P. Sharma, "Assessment of potential of oils for biodiesel production", *Renew. Sustain. Energy Rev.,* vol. 44, pp. 814-823, 2015.
[http://dx.doi.org/10.1016/j.rser.2015.01.013]

[92]   P. Nussbaumer, M. Bazilian, and V. Modi, "Measuring energy poverty: Focusing on what matters", *Renew. Sustain. Energy Rev.,* vol. 16, no. 1, pp. 231-243, 2012.
[http://dx.doi.org/10.1016/j.rser.2011.07.150]

[93]   Z. Abdin, N. Al Khafaf, B. McGrath, K. Catchpole, and E. Gray, "A review of renewable hydrogen hybrid energy systems towards a sustainable energy value chain", *Sustain. Energy Fuels,* vol. 7, no. 9, pp. 2042-2062, 2023.
[http://dx.doi.org/10.1039/D3SE00099K]

[94]   A. Boretti, "Towards hydrogen gas turbine engines aviation: A review of production, infrastructure, storage, aircraft design and combustion technologies", *Int. J. Hydrogen Energy,* vol. 88, pp. 279-288, 2024.
[http://dx.doi.org/10.1016/j.ijhydene.2024.09.121]

[95]   A. Simonini, M. Dreyer, A. Urbano, F. Sanfedino, T. Himeno, P. Behruzi, M. Avila, J. Pinho, L. Peveroni, and J.B. Gouriet, "Cryogenic propellant management in space: open challenges and

perspectives", *NPJ Microgravity,* vol. 10, no. 1, p. 34, 2024.
[http://dx.doi.org/10.1038/s41526-024-00377-5] [PMID: 38509131]

[96]   X. Huang, Y. Li, H. Ma, P. Huang, J. Zheng, and K. Song, "Fuel cells for multirotor unmanned aerial vehicles: A comparative study of energy storage and performance analysis", *J. Power Sources,* vol. 613, p. 234860, 2024.
[http://dx.doi.org/10.1016/j.jpowsour.2024.234860]

[97]   M. Chiara Massaro, S. Pramotton, P. Marocco, A.H.A. Monteverde, and M. Santarelli, "Optimal design of a hydrogen-powered fuel cell system for aircraft applications", *Energy Convers. Manage.,* vol. 306, p. 118266, 2024.
[http://dx.doi.org/10.1016/j.enconman.2024.118266]

[98]   F. Franke, S. Kazula, and L. Enghardt, "Elaboration and outlook for metal hydride applications in future hydrogen-powered aviation", *Aeronaut. J.,* vol. 128, no. 1325, pp. 1501-1531, 2024.
[http://dx.doi.org/10.1017/aer.2024.53]

[99]   M. Vietze, and C. Evrim, "Development of additively manufactured cryogenic heat exchangers for hydrogen-electric aircraft propulsion", *IOP Conf Ser Mater Sci Eng,* vol. vol. 1301, 2024p. 012007
[http://dx.doi.org/10.1088/1757-899X/1301/1/012007]

[100]   D. Verstraete, "Long range transport aircraft using hydrogen fuel", *Int. J. Hydrogen Energy,* vol. 38, no. 34, pp. 14824-14831, 2013.
[http://dx.doi.org/10.1016/j.ijhydene.2013.09.021]

[101]   T. Yusaf, "Sustainable hydrogen energy in aviation – A narrative review", *Int. J. Hydrogen Energy,* vol. 52, pp. 1026-1045, 2024.
[http://dx.doi.org/10.1016/j.ijhydene.2023.02.086]

[102]   Y. Li, R. Lin, R. O'Shea, V. Thaore, D. Wall, and J.D. Murphy, "A perspective on three sustainable hydrogen production technologies with a focus on technology readiness level, cost of production and life cycle environmental impacts", *Heliyon,* vol. 10, no. 5, p. e26637, 2024.
[http://dx.doi.org/10.1016/j.heliyon.2024.e26637] [PMID: 38444498]

[103]   A. Ilyushechkin, L. Schoeman, L. Carter, and S.S. Hla, "Material challenges and hydrogen embrittlement assessment for hydrogen utilisation in industrial scale", *Hydrogen,* vol. 4, no. 3, pp. 599-619, 2023.
[http://dx.doi.org/10.3390/hydrogen4030039]

[104]   M. Ahad, M. Bhuiyan, A. Sakib, A. Becerril Corral, and Z. Siddique, "An overview of challenges for the future of hydrogen", *Materials (Basel),* vol. 16, no. 20, p. 6680, 2023.
[http://dx.doi.org/10.3390/ma16206680] [PMID: 37895667]

[105]   Y. S. Chen *et al.*, "Hydrogen trapping and embrittlement in metals – A review," Int J Hydrogen Energy, vol. 136, pp. 789–821, Jun. 2025.
[http://dx.doi.org/10.1016/J.IJHYDENE.2024.04.076]

[106]   A. Ilyushechkin, L. Schoeman, L. Carter, and S.S. Hla, "Material challenges and hydrogen embrittlement assessment for hydrogen utilisation in industrial scale", *Hydrogen,* vol. 4, no. 3, pp. 599-619, 2023.
[http://dx.doi.org/10.3390/hydrogen4030039]

[107]   Z. Xie, Q. Jin, G. Su, and W. Lu, "A review of hydrogen storage and transportation: Progresses and challenges", *Energies,* vol. 17, no. 16, p. 4070, 2024.
[http://dx.doi.org/10.3390/en17164070]

[108]   M. İnci, "Future vision of hydrogen fuel cells: A statistical review and research on applications, socio-economic impacts and forecasting prospects", *Sustain. Energy Technol. Assess.,* vol. 53, p. 102739, 2022.
[http://dx.doi.org/10.1016/j.seta.2022.102739]

[109]   D. Alwazeer, "Use of hydrogen extraction in the food industry", In: *Reference Module in Chemistry,*

Molecular Sciences and Chemical Engineering. Elsevier, 2024.

[110]   A. Maganza, A. Gabetti, P. Pastorino, A. Zanoli, B. Sicuro, D. Barcelò, A. Cesarani, A. Dondo, M. Prearo, and G. Esposito, "Toward sustainability: an overview of the use of green hydrogen in the agriculture and livestock sector", *Animals (Basel),* vol. 13, no. 16, p. 2561, 2023.
[http://dx.doi.org/10.3390/ani13162561] [PMID: 37627352]

[111]   L. Li, W. Lou, L. Kong, and W. Shen, "Hydrogen commonly applicable from medicine to agriculture: from molecular mechanisms to the field", *Curr. Pharm. Des.,* vol. 27, no. 5, pp. 747-759, 2021.
[http://dx.doi.org/10.2174/18734286MTEysMTQfx] [PMID: 33290194]

CHAPTER 6

State-of-the-Art Recycling Technologies for Composite Materials

Olesya A. Buryakovskaya[1] and **Mikhail S. Vlaskin**[1,*]

[1] *Joint Institute for High Temperatures of the Russian Academy of Sciences, Moscow, Russia*

Abstract: There is a lot of interest in the development of new technologies for composite materials recycling, as well as in the improvement of the existing procedures and their adaptation for alternative applications. Since numerous sorting techniques have been developed and tested to recover recyclable components from a waste stream, this chapter aims to elucidate the state-of-the-art materials identification technologies for waste sorting. The survey revealed that, aside from the commonly used set of identification methods for analyzing the quantitative and qualitative elemental compositions of the inspected materials (X-ray fluorescence, XRF) and their molecular structure (near-infrared spectroscopy, NIR) assisted by visual (VIS) spectroscopy, some of the less conventional methods, such as Raman spectroscopy, Laser-induced breakdown spectroscopy (LIBS), and even X-ray transmission (XRT) spectroscopy (traditionally employed for the mining and mineral processing industry), have been proposed either as individual or supplementary techniques for advanced hybrid technologies. The evolution of machine learning and artificial intelligence tools has already contributed to the improvement of material classification accuracy based on measurement data. This chapter contains a brief description of the adopted and prospective identification technologies with their principal illustrations and a presentation of the latest advances in their implementation for waste sorting.

Keywords: Composite materials, Materials identification, Plastics, Spectroscopy, Sorting, Waste recycling.

INTRODUCTION

Municipal Solid Waste (MSW) represents a mixture of paper, plastics, clothing, metals, glass, food, and other types of household waste. The world's annual production of MSW constitutes over 2 billion tons and, according to forecasts, will skyrocket up to about 4 billion tons by 2100. Thus, the increasing amount of Municipal Solid Waste (MSW) has gradually become one of the acute problems that hampers the global movement toward economic, environmental, and social

---

* **Corresponding author Mikhail S. Vlaskin:** Joint Institute for High Temperatures of the Russian Academy of Sciences, Moscow, Russia; E-mail: vlaskin@inbox.ru

Harish Chandra Joshi, Anand Chauhan, Mikhail Vlaskin & Maulin P. Shah (Eds.)

prosperity. To address this issue and promote the development of a circular economy and sustainable cities, it is essential to prioritize the sorting of recyclable materials, particularly plastics [1, 2].

At present, the predominant sorts of commonly used plastics include polyethylene terephthalate (PET), polypropylene (PP), high-density polyethylene (HDPE), and low-density polyethylene (LDPE)—altogether contributing over 50% of the plastic residues present in municipal solid waste, as well as polystyrene (PS), polyvinyl chloride (PVC), and some other sorts of plastic. The listed polymers are widely used as one-part materials as well as components of composite and multi-layer materials. The problems associated with Plastic Waste (PW) handling are that it is often represented by mixtures of different plastics with unknown compositions and can contain organic contaminants or other materials (paper, metal, *etc.*). Packaging waste accounts for over 61% of PW, with the following shares of individual plastics: PE > PP > PET> PS > PVC [3].

The current alternatives for PW handling include recycling, incineration, and burying at landfills (Fig. **1**). In 2020, over 29 million tons of post-consumer PW were collected in European Union countries. As much as 23.4% of that amount went to landfills, 34.6% was subjected to recycling, and 42% was utilized through incineration with energy recovery [4]. Recycling is the most rational approach to plastic waste handling, which employs a variety of chemical, mechanical, and thermochemical methods to convert PW into either raw materials with their properties close to those of the original ones, or into valuable substances (gaseous and liquid fuels, monomers, *etc.*), or some functional composite materials by their combining with different additives. However, a relatively low level of PW recycling has been observed so far, and the major reasons for this typically include the unsatisfactory quality of the resulting products, high energy demand, and the complexity of the processing methods.

Sorting

Effective waste recycling requires the execution of certain preparatory procedures, the first of which is the reliable sorting of the waste streams in order to extract recyclable components. Automated waste sorting encompasses both direct and indirect sorting methods. Direct sorting can be performed through air separation, magnetic separation, magnetic levitation, Eddy Current Separation (ECS), the dense medium method, and the electrostatic method. It is intended for direct solid waste separation using electromagnetic force, buoyancy, electric field force, and other methods. In the case of indirect sorting, the separation of the supplied stream into recyclable items and the remaining waste mass is preceded by the identification of the former ones by their material. Materials identification can be performed by a variety of methods, which comprise Delayed and Prompt Gamma Neutron Activation Analysis (DGNAA and PGNAA), Neutron Radiography (NR), X-ray fluorescence (XRF), dual-energy X-ray transmission (DE-XRT) scanning, dual-energy γ-ray transmission (DE-γRT), X-ray luminescence (XRL), Raman spectroscopy, laser-induced fluorescence spectroscopy (LIFS), Laser-induced breakdown spectroscopy (LIBS), visual (VIS) spectroscopy, infrared (IR) – visible to near-infrared (VNIR), Near-infrared (NIR), short-wavelength infrared (SWIR), Mid-wavelength infrared (MWIR), Long-wavelength infrared (LWIR), Far-infrared (FIR) – spectroscopy, terahertz (THz) spectroscopy, Microwave Imaging (MWI), radiofrequency radiation based techniques, electric conductivity, Thermal Imaging (TI), Impact Acoustic (IA) method, and their combinations [5 - 8]. Since existing techniques are continually updated and new ones are constantly introduced and adapted, it would be helpful to survey the current state of the art methods in this domain of knowledge.

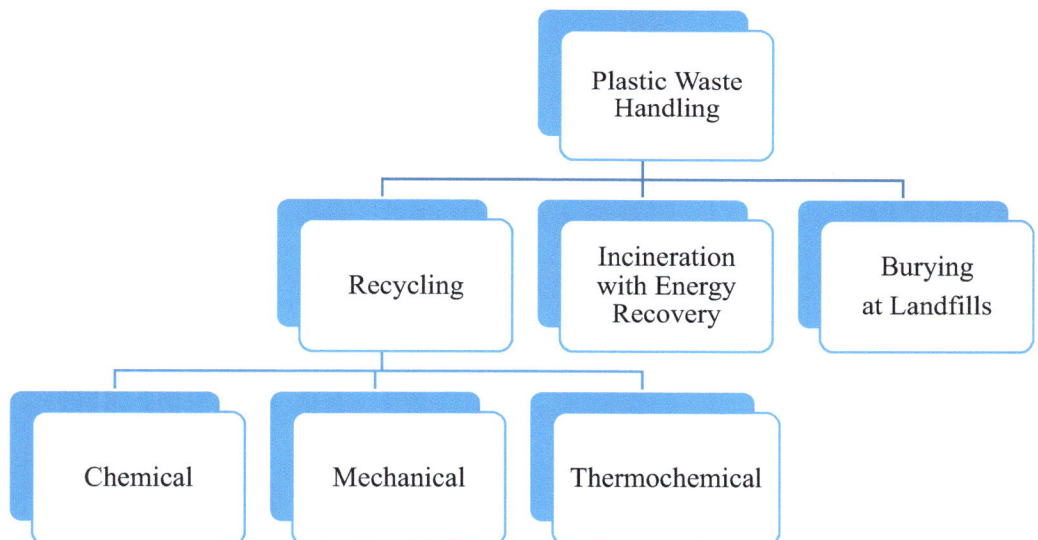

Fig. (1). Conventional approaches to waste handling.

Neutron and γ-Ray Based Methods

Delayed and Prompt Gamma Neutron Activation Analysis

Delayed and Prompt Gamma Neutron Activation Analysis (DGNAA and PGNAA) are the methods designed initially for scanning radioactive waste packages for non-radioactive substances, with the typical sample masses on the order of 0.001–1 g. PGNAA is based on the measurement of γ-radiation (wavelength, $\lambda < 10^{-11}$ m) emitted promptly (after 10^{-16}–10^{-12} s) due to the de-excitation of compound nuclei after capturing a thermal neutron (< 0.025 eV). Due to such a short emission delay, neutron irradiation and gamma spectrum acquisition are performed simultaneously. DGNAA employs the detection of γ-rays emitted from the radioactive (activated) nuclei, whose emission is delayed by nearly 10^{-6}–10^{-2} s (half-life of the activation product) [9, 10].

Neutron Radiography

Neutron Radiography (NR) is implemented by determining the neutron flux attenuation in the sample (neutron flux transmission). An important advantage of neutron-based techniques lies in the profound ability of neutrons (and resulting γ-rays) to penetrate materials (that is notable when dealing with high-density materials or heavy metals), making them promising for sorting waste electrical and electronic equipment (e-waste). A notable distinction of NR is that neutrons interact with the atom's nucleus rather than with its electron shell, which results in substantial variation in the attenuation coefficients even for similar materials, thus enhancing imaging contrast. The technique is ideal for detecting light elements (*e.g.*, hydrogen-containing plastics or lithium-containing battery materials), which are transparent in X-ray imaging. NR tests on e-waste sorting revealed that for thick metal items, clear images were obtained, while fine metal elements of printed electronic circuits (a few microns in thickness) were poorly visible or invisible. Another limitation was related to large amounts of hydrogen in plastic-containing e-waste, which caused excessive beam scattering and attenuation, thus resulting in blurred and insufficiently contrasted images [11].

X-Ray Based Methods

X-ray Transmission

X-ray transmission (XRT) scanning is a 2D imaging technique based on the emission of an X-ray beam (wavelength 10^{-11}–10^{-8} m) towards the inspected object and pixel-wise measurement of the local X-ray attenuation (reduction in the X-ray beam intensity from its passing through the sample) with a detector. The object material density and thickness define the local attenuation. The XRT scanning is a widely used method for baggage inspection at airports. Dual-energy XRT (DE-XRT) technology combines high- and low-energy X-ray levels for detection. Dual-energy γ-ray transmission (DE-γRT) employs the same principles, but utilizes a shorter wavelength, allowing it to penetrate materials more effectively. In combination with deep learning techniques, DE-XRT ensures reliable detection of batteries in e-waste. It can be adapted to recognize non-metals (such as ceramics, plastics, glass, and other types of solid waste). One of the technique's drawbacks is that similar materials may have only a marginal difference between their X-ray attenuation coefficients, and those of different materials can be nearly equal by coincidence. Other factors that can affect identification accuracy include overexposure, underexposure, measurement noise, and the fan-shaped effect [5, 12 - 14].

X-Ray Fluorescence

The X-ray fluorescence (XRF) analysis also utilizes X-ray beam emission. It focuses on measuring the wavelengths of the fluorescent X-rays emitted by different chemical elements with atomic numbers greater than 20. Unlike XRT imaging, which provides an image of the object's internal structure, XRF is limited to surface analysis. Despite its low penetration ability and inapplicability for the recognition of light elements, XRF is successfully implemented for the identification of precious, base, ferrous, and non-ferrous metals [5, 13, 14].

X-ray Luminescence

The X-ray luminescence (XRL) technique is based on photoluminescence, which occurs when an electron, impacted by an X-ray photon, receives extra energy and transitions from its ground state to an excited state. As in the case of fluorescence, the following relaxation results in the release of the energy margin between the conduction and valence bands as a photon, with the emitted energy being lower than that of the source. In contrast with an almost instant fluorescence (with photon absorption, internal conversion—a radiationless transition from a higher to a lower electronic state, and fluorescence lasting for 10^{-15}, 10^{-12}, and 10^{-9} s respectively), luminescence usually takes longer time (10^{-15} s for absorption, 10^{-9} s for an isoenergetic radiationless transition between the two electronic states with different spin multiplicity (also known as intersystem crossing), 10^{-12} s for internal conversion, and from 10^{-6}–10^{-3} s to several minutes/hours for photoluminescence). Its typical energy margins are higher so as the emitted photon energies fall within the visible light spectrum. This process also involves the origination and annihilation of electron-hole pairs. In diamonds, photoluminescence is associated with defects or impurities within the crystal lattice and is usually manifested as emissions in the visible spectrum (predominantly blue light, with peaks around 450 nm at room temperature). Different diamonds may vary in the intensity, color, duration, and distribution of their emissions, while some diamonds do not exhibit photoluminescence at all. Since XRL response is more consistent and less varied than photoluminescence using ultraviolet or other sources, it is suitable for the separation of diamonds with sizes ranging from 1.25 to 32 mm (self-absorption effects determine the size limit) [15, 16].

To date, the aforementioned X-ray techniques are primarily employed in the mining and mineral processing industries. Their application for waste sorting is currently limited to pilot testing on sorting non-ferrous metals (XRF and XRT) and discriminating between PET and PVC (XRF) [7, 8].

Ultraviolet, Visible, and Infrared Radiation-Based Methods

Laser-Induced Breakdown Spectroscopy

Laser-induced breakdown spectroscopy (LIBS), also known as Laser-induced plasma spectroscopy (LIPS), employs a brief, intense laser pulse to vaporize a small portion (a few micrograms) of the sample, which creates plasma by expelling electrons from the atom's outer shells. By detecting the characteristic emissions pulse (wavelength of approximately 250–760 nm), quantitative and qualitative elemental compositions of the sample are obtained. The method is used for sorting and differentiating metals and polymers in general. In combination with classification models, this method was claimed to identify e-waste polymers, such as PP, PE, PS, polyamide (PA), polycarbonate (PC), and acrylonitrile-butadiene-styrene (ABS) with 98% precision. However, since LIPS does not provide molecular structure data, this method is not suitable for distinguishing polymers with similar chemical formulas. For the reliable identification of PET, PP, PE, and PS, a hybrid method combining LIPS and Raman spectroscopy was proposed, which provides data on atomic emissions and molecular structure [2, 7, 17, 18].

Laser-Induced Fluorescence Spectroscopy

Laser-induced fluorescence spectroscopy (LIFS) utilizes ultraviolet (UV) lasers to excite the atoms in the sample. De-excitation of the atoms results in the emission of photons within the wavelength range of nearly 10–760 nm. The fluorescence wavelength region is slightly wider than the excitation interval and is observed in the ultraviolet-visual (UV-VIS) range by hyperspectral sensing. The fluorescence intensity is recorded over a timespan long enough for the decay tendency to be observed. The resulting time-dependent spectrum is used for materials characterization. Fluorescence, however, does not support elemental analysis and is affected by the overlap between the molecule's vibrational levels and its excited electronic energy levels. LIFS is used for plastic sorting, particularly for identifying carbon black plastics (HDPE, PP, PS, EPS) effectively [2, 7, 19].

Raman Spectroscopy

Raman spectroscopy focuses on measuring the rotational and vibrational frequencies of a molecule's chemical bonds. Photons emitted by a laser transmit their energy to vibrating molecules, thus exciting them to a higher energy level. The primary energy dissipating mechanism is elastic (Rayleigh) scattering, after which the photon wavelength remains unaffected. This part of the spectrum is eliminated through filtration. In Raman spectroscopy, the wavelength of inelastically scattered photons is measured. The emitted photons have energy either lower or higher than that of the laser photon. The typical wavelengths of monochromatic light emitted by lasers are 244, 405, 532, 633, 785, 830, and 1064 nm, which fall within the near-ultraviolet, visible, and near-infrared regions of the electromagnetic spectrum. Raman spectroscopy is a low-sensitivity method that is subject to interference from fluorescence. The performance of a dual spectroscopy system using LIBS and Raman spectroscopy techniques in classifying PP, PC, polylactide (PLA), Nylon-11, PMMA, PET, HDPE, and PP delivered promising results, since the elemental and molecular data sets acquired by the system complemented each other, thus resulting in a reduction in the number of drop-outs. The combination of Machine Learning (ML) algorithms and Raman spectral datasets, tested for the identification of plastic waste (PET, PP, HDPE, and LDPE), demonstrated an accuracy of 94.9% [20 - 24].

Visual Spectroscopy

Visual (VIS) spectroscopy provides sorting of waste components by color and applies to plastics, wood, paper, glass, and construction and demolition waste. The method employs electromagnetic radiation from the visible range (380–760 nm) to inspect the sample. The absorption of a specific wavelength corresponds to the valence electrons excitation followed by their transitions between orbitals of different energies. The resulting energy margins define the absorption of certain wavelengths, which increases as the energy margin decreases. A fraction of the source light undergoes diffuse reflection on the sample surface and falls onto the detector, which splits the light into its components. The resulting wavelength spectrum is a function of intensity. It should be noted that there is no actual color, but rather its complementary color that is observed, because the detected color is the reflected one, which is complementary to the absorbed one. The colors are defined either *via* the hue-saturation-brightness (HSB) or red-green-blue (RGB) method. The HSB method defines color by hue, saturation, and brightness, while the RGB method identifies it by the components of red, green, and blue. Color line-scan cameras (VIS cameras) record three channels (blue, green, and red) within the visible wavelength range. Their embodiment typically comprises either three individual sensor lines with optical filters or a single line, in which macropixels consist of three adjacent pixels, each sensitive to one color range. Color line-scan cameras are implemented for both transmittance and reflectance measurements (the former is effective for color glass sorting). They are used for sensor-based sorting applied to agricultural products, colored glass cullet, and processed minerals. Color line-scan cameras are a standard component of multi-sensor systems equipped with infrared cameras or four-channel VIS-NIR cameras, as they provide precise localization of individual particles due to their high spatial resolution. In area-scan cameras, Bayer sensors are usually used to obtain color information. Half of the sensor is equipped with equal numbers of red and blue filters, positioned in a checkerboard-like pattern, and the rest is equipped with green filters. Thus, each pixel of the resulting image initially contains information for only one color. The missing color values for all pixels are derived from interpolation. Due to the high frame rates provided by area-scan cameras, they can be used to observe the motion of individual particles. In combination with multi-object tracking algorithms, these cameras have a good potential for materials characterization based on motion-related features. Although the implementation of this camera type has been mentioned in numerous publications, no information about its industrial applications was found [7, 25, 26].

Infrared Spectroscopy

Infrared (IR) spectroscopy measures the IR radiation reflectance or absorption by the inspected materials. The IR wavelength range is associated with the vibrational and rotational movements of distinct chemical bonds within molecules. The molecules absorb infrared rays with the resonant frequencies, while those with non-resonant frequencies are transmitted. Fourier Transform Infrared Spectroscopy (FTIR) employs Fourier transformations to convert time-domain signals into the frequency dependence of radiation absorption and transmission. Rotational frequencies are observed within the FIR region, while fundamental vibrational frequencies are located within the MIR, and overtones of vibrational frequencies fall within the NIR. With its operational wavelengths region from 0.76 to 1.4 μm, NIR is capable of performing remote high-speed measurements. NIR radiation has a relatively high signal-to-noise ratio and penetration depth. This technique enables the detection of the PET, PVC, HDPE, PP, and PS waste. It was reported to differentiate between two groups of polymers, such as PE and PET, PE and PP, or polylactic acid (PLA) and PET, with a 97–100% classification accuracy. It is also sensitive to organic waste and oil-contaminated cardboard. The majority of studies employed variations of common plastics, including PVC, PP, PS, PET, LDPE, and HDPE. Actual plastic waste mixtures, however, are heterogeneous, which may affect the accuracy of chemometric models applied in conjunction with NIR. Its applicability to the analysis of polyolefins requires further clarification, as PE comprises both HDPE and LDPE. Mislabeling of some PP and PE samples was also reported. A significant drawback of NIR is its inability to differentiate between black plastics due to considerable radiation absorption within its operational interval. Also, NIR deals with overtones and combination bands of different functional groups, which results in less prominent spectral features and the associated challenges in obtaining distinctive identification. To improve NIR performance for heterogeneous plastic waste, its results can be supplemented with the data measured by MIR, Raman, and LIBS methods [20]. Absorbance spectroscopy is predominantly performed in the NIR range, while the SWIR–LWIR and VIS wavelength domains are less common for this method. Multi- or hyperspectral imaging in the UV–SWIR region can also be used for the detection and differentiation of certain materials due to the differences in their complex refractive indices, as well as absorption and scattering cross-sections. For waste characterization with the UV–SWIR imaging, the reflected light is spatially and spectrally resolved. Spectral imaging data can be effectively processed using machine learning methods to identify or quantify different materials. The UV–SWIR imaging techniques are generally based on hyperspectral (20–200 wavelength bands), multispectral (3–20 bands), and RGB (3 bands) or monochrome cameras. The monochrome and RGB color 3D imaging techniques are useful for estimating waste shape and volume and can be used for detecting particles, plastics, and metals. Human analysis of RGB color imaging data is applicable to the identification of various materials, including glass, organics, wood, paper, card plastic, dense plastic, film plastic, textiles, metals, and e-waste. In comparison to imaging in the UV–NIR region, that in the NIR–SWIR band employs more expensive detectors but also provides higher specificity in material detection due to the overtones and combination tones typical of this wavelength interval. A camera operating in the UV–VIS–NIR domain is capable of 'recognizing' polymethylmethacrylate (PMMA), ABS, PVC, PC, PP, PE, rubber, wood, and aluminum. VIS–NIR hyperspectral techniques are suitable for identifying fine particles, electronic waste, and metals (Cu, Zn, Al, Fe, and Ni). Hyperspectral imaging in the NIR–SWIR region applies to the detection of moisture, raw and colored cardboard, plastic and non-plastic waste, newspaper, and printed paper. Single-point waste monitoring *via* UV–LWIR spectroscopy is based on either absorbance spectroscopy or LIBS (the emitted wavelengths usually fall within the UV–NIR range). In the SWIR–LWIR region, PP, PE, PET, LDPE, HDPE, PS, and PLA can be differentiated. The MIR wavelength domain comprises the fundamental vibrational bands and is referred to as the 'molecular fingerprint' interval. In comparison to the NIR band, the MIR region is significantly farther from the VIS domain. For that reason, the MIR spectroscopy results are not affected by the plastic surface morphology and color (in particular, in the case of black plastics). The reduced sensitivity of MIR photodetectors (mercury cadmium telluride and deuterated triglycine sulfate) limits plastic recycling due to the slow spectral acquisition speed. A photon-up conversion technique was reported to be employed to convert MIR photons to a higher energy NIR signal that can be recorded by more sensitive InGaAs sensors. The MWIR–LWIR imaging is limited to the detection of thermal radiation. For various materials at the same temperature, their intensities of the emitted thermal radiation can be differentiated. Such a technique can be implemented with or without an illuminating source [27]. In the absence of an illuminating source, material characterization usually requires elevated sample temperatures. In the case of the FIR band, high absorption of light in water and its high reflection from metals are typical. Those effects are suitable for detecting these materials. Other materials demonstrate predominantly light scattering effects for the material particles, whose sizes are comparable to the respective wavelengths [7, 14, 25, 27 - 31].

Microwave/Terahertz Radiation-Based Methods

Terahertz Spectroscopy

Terahertz radiation relates to the frequency range of 0.1–1 THz (wavelength of 30–3,000 μm), which combines the straightness of light waves and the transparency of radio waves. That range corresponds to vibrations between molecular chains, while in the infrared region, local vibrational modes with small effective masses are typically detected in plastics. Spectroscopy measurements can be performed for single-point detection or imaging in either the frequency or time domain. In the frequency domain, contrast differences in the reflected and transmitted light intensities, attributed to specific materials such as water or metal, can be obtained. In the time domain, the reflected or transmitted pulse phase and amplitude are measured as a function of time. A fast Fourier transformation is applied to the amplitude–time measurements to convert them into the frequency domain. Since the reflectance and transmission of terahertz waves depend on the plastic material, its identification is based on measuring its dielectric constant at the characteristic terahertz frequency. PE and PP exhibit low reflectance, along with high transparency, over a wide terahertz frequency range. In contrast, PS is characterized by high reflectance and low transparency, while PET exhibits no transmittance above 1 THz; its reflectance is even greater than that of the listed plastics (1.0 THz, 1.8 THz, and 2.2 THz are effective for sorting PP, PE, PS, and PET). This technique is suitable for the identification of multi-layered plastics containing coloring, additives, and aluminum film, and is affected by plastic materials strain and degradation as well. The implementation of machine learning algorithms in conjunction with terahertz radiation has been shown to improve plastic sorting accuracy by up to 100%. Terahertz technology is still under research and development. It is slower than MWIR in point detection and throughput. The advantages of combining the THz and MWIR spectrums of black ABS, PS, and PE were that the MWIR was not affected by surface conditions, while THz in transmission mode was not affected by the sample thickness [31 - 33].

Microwave Imaging

Microwave imaging (MWI) is commonly used for sensor-based ore sorting. Compared to XRT and XRF, MWI radiation penetrates deeper into the rock particles because most gangue minerals expose a low dielectric loss in the microwave range. For that reason, a supplementary MWI method is used for sorting ores with high contrast in electromagnetic (EM) properties between gangue minerals and valuable minerals or metals. In contrast with an electromagnetic (inductive) sensor (EMS), which generates eddy current within conductive particles, thus inducing current in a receiving coil, MWI is based on receiving and processing the backscattered microwave signals from ore particles. It should be noted that this method is focused on the pure rotational motion of molecules with a permanent dipole moment. Rotation of such a molecule establishes an electric field that interacts with the microwave radiation's electric component. This results in energy absorption or emission through transitions between rotational energy levels. This technique is applicable for measuring the moisture content in construction and demolition materials, mixed solid waste, wood, cardboard, and textiles. It can be applied to materials with low dielectric constants (ceramics, glass, plastics, and composite materials) [14, 34, 35].

Radiofrequency Radiation-Based Methods

Radiofrequency radiation (RF) based techniques operate under frequencies below those for the FIR region, from 3 kHz to 300 GHz. Typical interactions between radio waves and matter are water absorption and metal reflection because of the high transmittance and low reflections of radio waves in most of the other materials. Their high electron densities cause high reflection in metals, while high attenuation in water is attributed to the induction of transitions in the molecules' rotational energy levels. Radiowave transmittance or reflectance can be used for single-point spectroscopic measurements of the moisture or metal contents in materials.

Nuclear Magnetic Resonance Spectroscopy

Nuclear Magnetic Resonance (NMR) is a radiofrequency spectroscopic technique used to identify the presence of specific nuclei in a material. It includes applying a powerful magnetic field to the sample, which induces a resonant precession of the nuclei about the magnetic field direction. The angular (Larmor) frequency of precession is proportional to the strength of the magnetic field and the nucleus's gyromagnetic ratio. Since each nucleus has its specific gyromagnetic ratio, its Larmor frequency can be probed using radiofrequency techniques to identify its presence in the sample. In a typical NMR spectrometer, the static magnetic field is commonly generated between the poles of a permanent magnet or electromagnet.

A sequence of radiofrequency current pulses (Hahn-Echo sequence) is applied to the solenoidal coil, encompassing the sample, at the Larmor frequency. The pulses create an alternating magnetic field at the same frequency, which, in the case of the target nucleus's presence, generates an 'echo' signal that is then detected *via* a radio receiver. The signal strength can be calibrated to determine the quantitative amount of nuclei, thereby obtaining the mass of the nuclei. The technique has been tested for bulk quantitative analysis of 7Li in spodumene and the characterization of cements. It is considered for the identification and quantification of microplastics, *e.g.*, by analyzing dilution series of PS, polyisoprene (PI), polybutadiene (PB), PLA, PVC, and polyurethane (PU) [36 - 39].

The techniques described above (except for neutron radiography) are based on the implementation of different wavelength regions of electromagnetic radiation, which are summarized in Fig. (**2**). The known approaches to waste sorting, however, are not limited to those based on the interactions between electromagnetic radiation and materials.

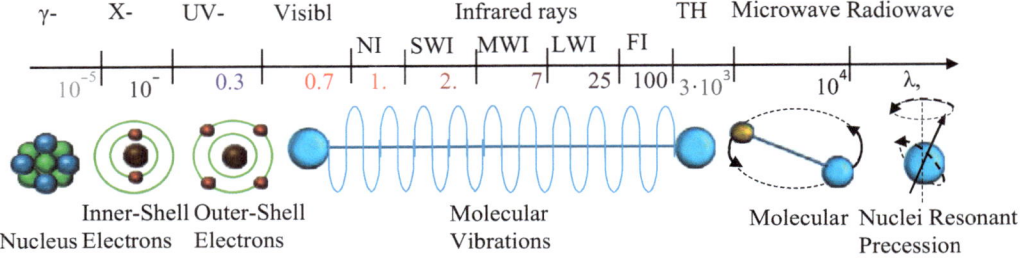

Fig. (2). The electromagnetic spectrum wavelength regions for different optical technologies.

Other Methods

Impact Acoustic Method

In recent years, the impact acoustic (IA) method has been intensively studied. Acoustic data demonstrate a high correlation with the type of waste material, as sound signals of objects depend on their material properties. Acoustic signals emitted from external actions (*e.g.*, striking, free-falling) can reveal certain intrinsic properties of waste, such as elasticity and internal friction. The elasticity influences the sound frequency, while internal friction determines the temporal attenuation of the signal and provides its shape-variant acoustic features, which are useful for waste sorting. A typical installation for this method includes a plate against which a free-falling waste item impacts, and a sound acquisition device to record the sounds generated by the impacts. The application of deep learning to image and sound processing has attracted considerable interest. Although reports on audition-based waste sorting are limited, efforts have been made to solve acoustic classification problems for agricultural products and end-of-life vehicle plastics. Object recognition using the raw acoustic data was reported to achieve an accuracy of 91.50%. The implementation of deep 2D convolutional neural networks (CNNs) involving the initial conversion of 1D signals into 2D images has been tested, while other approaches suggested direct 1D signal processing to limit the size of each simple data [40].

Thermal Imaging

The Thermal Imaging (TI) approach includes feeding the dismantled e-waste fractions into the sorting system *via* a conveyor belt, supplying them into a hot chamber maintained at a temperature of 55±3 °C and transporting them from the hot chamber to the adjacent inspection zone to capture their thermograms using a thermal imaging camera, and their further classification into metallic fractions, such as copper and aluminum, and non-metallic fractions, including glass, printed circuit boards, and plastic. For the selected materials, their thermal radiation patterns were studied under the 8–15 μm wavelength interval corresponding to the long-wave infrared range (LWIR). A method was reported that involved setting thresholds for mean and standard deviations, as well as image sharpness, of thermogram intensities for the e-waste recyclable groups. The overall classification success rate of its implementation approached 84–96% [41].

Induction, Magnetic, and Eddy Current Sorting

The electric conductivity-based technique (induction sorting), along with magnetic sorting and eddy current sorting, is intended for the separation of the valuable metal content from the stream of non-metallic waste. Inductive sensors create magnetic induction to identify metallic objects. The sensor coils generate a magnetic field, which induces an electric current within a passing metallic object (or another conductive object). Such sensors generate an electronic signal that activates an ejector mechanism, for instance, an array of compressed air nozzles. To separate the detected objects from the waste flow, air jets push them over a diverting screen. This method enables the separation of metallized foils (and other metallized 2D materials, *e.g.*, multilayer waste), as the induction sensor's sensitivity can be increased to detect minute metallization amounts [26].

The principles of the methods listed in this section are illustrated in Fig. (**3**).

Prompt- and Delayed Gamma Neutron Activation

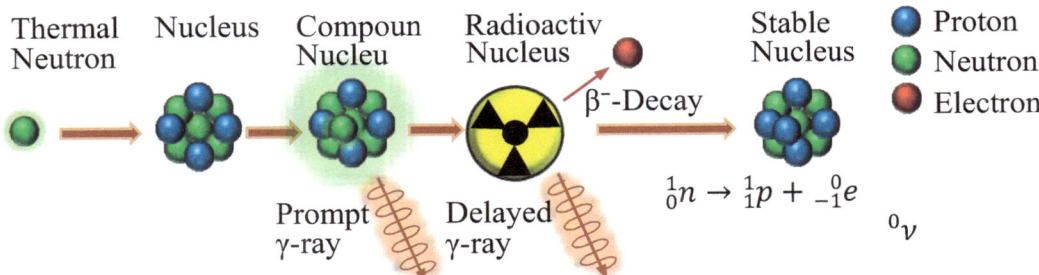

Measured effect: atomic radioluminescence.

Neutron Radiography	X-ray or γ-ray

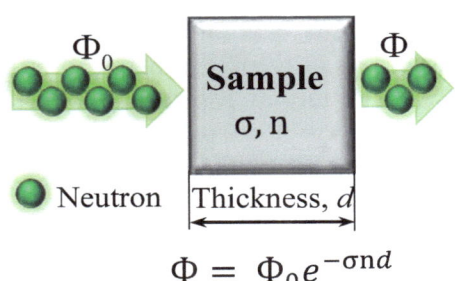

$$\Phi = \Phi_0 e^{-\sigma n d}$$

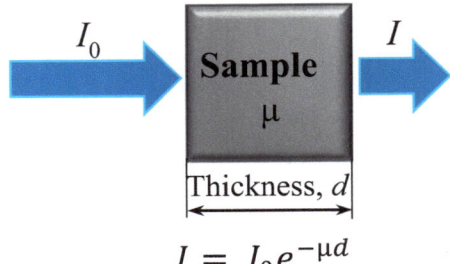

$$I = I_0 e^{-\mu d}$$

Φ – neutron flux, σ – neutron cross section, n – volumetric atom concentration.

Measured effect: transmitted neutron flux.

I – intensity, μ – linear attenuation coefficient.

Measured effect: transmitted X-rays or γ-rays.

(Fig. 3) contd.....

X-ray

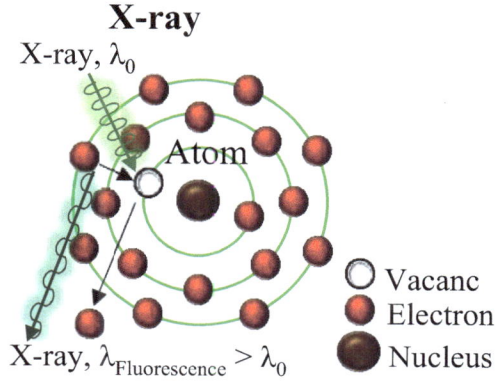

Measured effect: atomic fluorescence.

X-ray

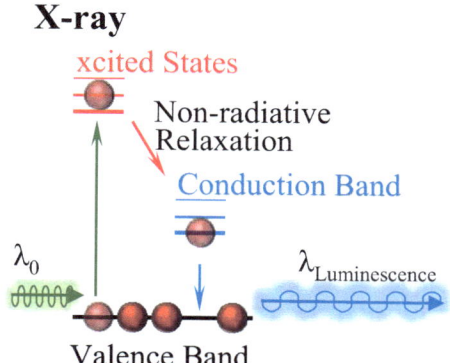

Measured effect: luminescence in visible range.

Laser-Plasma

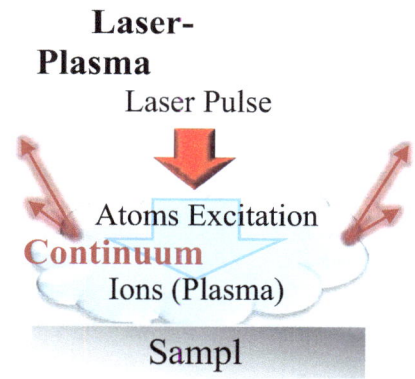

Measured effect: atomic fluorescence.

Laser-Fluorescence

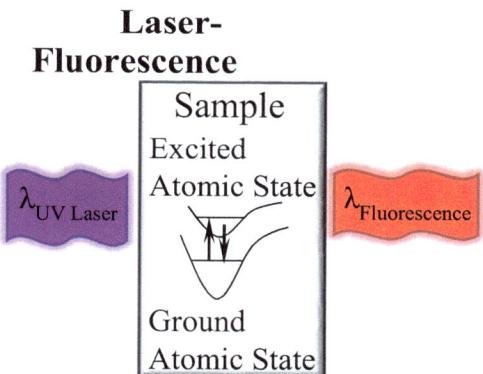

Measured effect: atomic fluorescence.

(Fig. 3) contd.....

Raman Spectroscopy

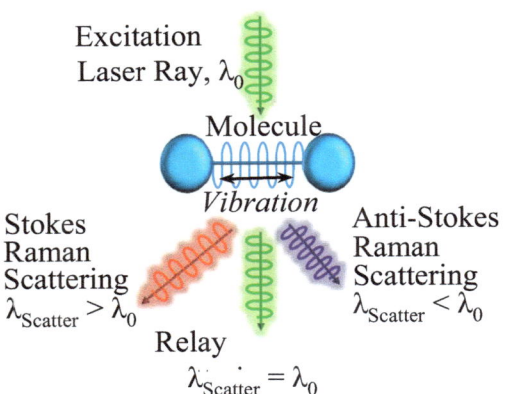

Excitation
Laser Ray, λ_0

Molecule

Vibration

Stokes
Raman
Scattering
$\lambda_{Scatter} > \lambda_0$

Anti-Stokes
Raman
Scattering
$\lambda_{Scatter} < \lambda_0$

Relay
$\lambda_{Scatter} = \lambda_0$

Measured effect: Raman scattering.

Visual Spectroscopy

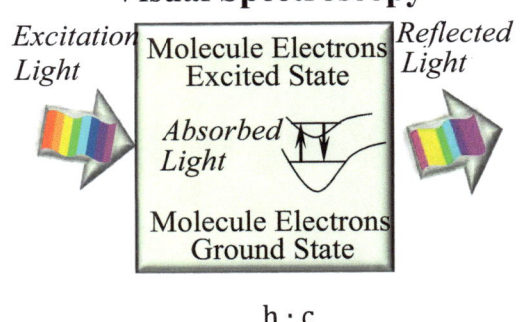

Excitation Light   Molecule Electrons Excited State   *Reflected Light*

Absorbed Light

Molecule Electrons Ground State

$$\lambda = \frac{h \cdot c}{\Delta E}$$

λ – wavelength, h – Plank's constant,

c – speed of light, ΔE – energy margin.

Measured effect: wavelengths of diffusely reflected light.

Infrared

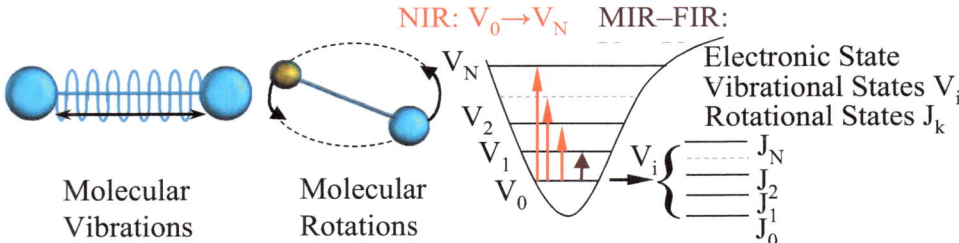

NIR: $V_0 \rightarrow V_N$ MIR–FIR:

V_N

V_2

V_1

V_0

V_i

Electronic State
Vibrational States V_i
Rotational States J_k

J_N

J_2
J_1
J_0

Molecular
Vibrations

Molecular
Rotations

Measured effect: wavelengths of absorbed radiation.

(Fig. 3) contd.....

THzTime-Domain Spectroscopy

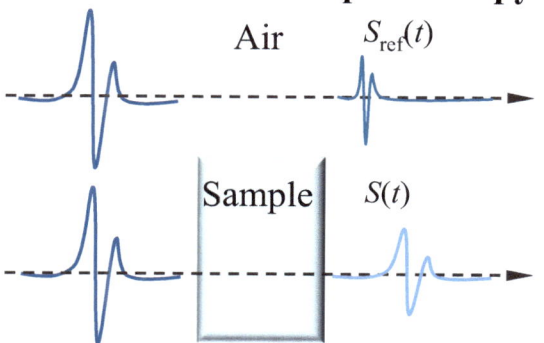

S_{ref} – reference waveform; t – time delay between emission and reception; S – signal transmitted by the sample.

Measured effect: transmitted terahertz radiation.

Microwave

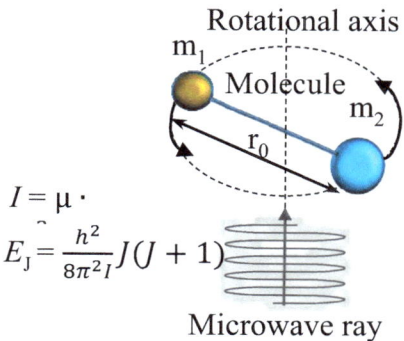

$$I = \mu \cdot$$
$$E_J = \frac{h^2}{8\pi^2 I} J(J+1)$$

r_0 – distance between mass centers, μ – reduced mass, I – moment of inertia; E_J – rotational energy; h – Plank's constant,

J – rotational quantum number.

Measured effect: transmitted microwave radiation.

Nuclear Magnetic Resonance

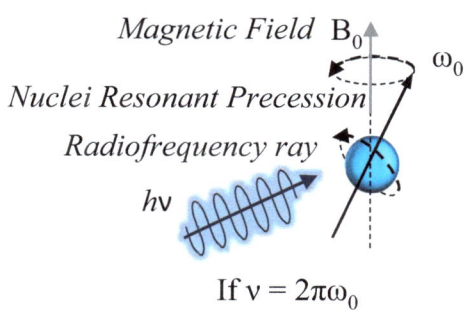

If $\nu = 2\pi\omega_0$

spin-up $S = \frac{1}{2} \rightarrow$ *spin-down* $S = -\frac{1}{2}$

Measured effect: transmitted terahertz radiation.

Acoustic Sorting

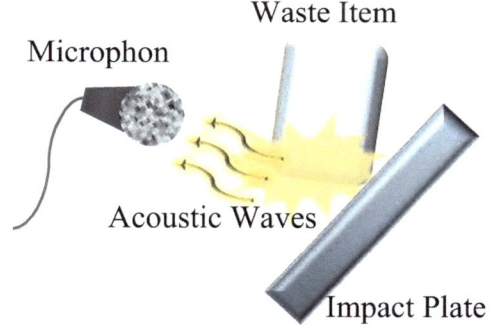

Measured effect: acoustic data derived from an impact.

(Fig. 3) contd.....

Thermal Imaging   ## Electric conductivity

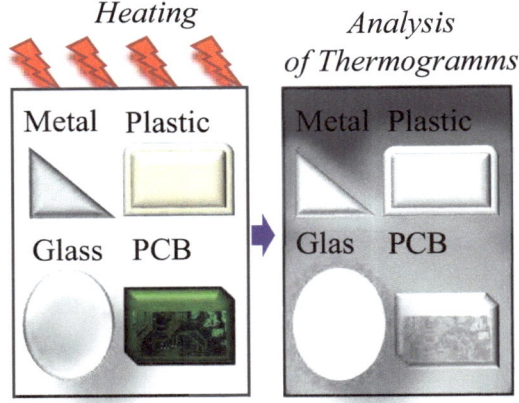
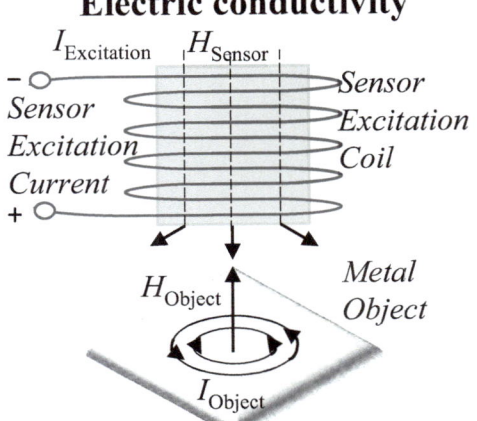

Measured effect: heat radiation.   *Measured effect: changes in*
                                                   electromagnetic field.

Fig. (3). Principles of the existing identification techniques for waste sorting.

Although a vast variety of identification methods for sorting waste components is known, not all of them have been introduced into industrial exploitation so far. According to the up-to-date data, the predominantly employed techniques include methods for analyzing the quantitative and qualitative elemental compositions (XRF and LIBS) and molecular structure (NIR and Raman spectroscopy). Sensor-based sorting for waste recycling primarily relies on IR-spectroscopy techniques (mainly NIR), assisted by VIS-spectroscopy and XRF, while LIBS is used less commonly [2, 7, 25, 42].

CONCLUDING REMARKS

The extraction of recyclable components from mixed waste flows involves the implementation of material identification methods. These methods are based on the analysis of signals emitted by samples affected by electrical and magnetic fields, electromagnetic radiation, and physical impacts. Metal fractions are effectively detected by induction, magnetic, and eddy current-based techniques. Non-metal materials are commonly analyzed *via* XRF to obtain their quantitative and qualitative elemental compositions, and NIR, combined with VIS spectroscopy, is used to establish their molecular structure. Recent studies reflect an increased interest in the application of LIBS, Raman spectroscopy, and XRT spectroscopy to waste component categorization. Emerging innovations in the domain of materials recognition encompass the use of terahertz spectroscopy, microwave imaging, NMR spectroscopy, acoustic impact analysis, and thermal imaging for waste characterization, as well as machine learning and artificial intelligence tools for data processing. The future prospects of waste component classification are expected to be based on advanced hybrid materials identification technologies, empowered by ML and AI tools.

REFERENCES

[1]   Y.H. Lin, W.L. Mao, and H.I.K. Fathurrahman, "Development of intelligent Municipal Solid waste Sorter for recyclables", *Waste Manag.,* vol. 174, pp. 597-604, 2024.
      [http://dx.doi.org/10.1016/j.wasman.2023.12.040] [PMID: 38145587]

[2]   C. Lubongo, M.A.A. Bin Daej, and P. Alexandridis, "Recent developments in technology for sorting plastic for recycling: The emergence of artificial intelligence and the rise of the robots", *Recycling,* vol. 9, no. 4, p. 59, 2024.
      [http://dx.doi.org/10.3390/recycling9040059]

[3]   A. Matuszewska, M. Owczuk, and K. Biernat, "Current trends in waste plastics' liquefaction into fuel fraction: A review", *Energies,* vol. 15, no. 8, p. 2719, 2022.
      [http://dx.doi.org/10.3390/en15082719]

[4]   "Plastics—The facts 2021—report plastics europe.," The association of plastics manufacturers in europe. , available online: https://plasticseurope.org/pl/knowledge-hub/plastics-the-facts-2021/ (accessed on 13 January 2022).

[5]   T. Xiong, W. Ye, and X. Xu, "Combination of dual-energy x-ray transmission and variable gas-ejection for the in-line automatic sorting of many types of scrap in one measurement", *Appl. Sci. (Basel),* vol. 11, no. 10, p. 4349, 2021.
      [http://dx.doi.org/10.3390/app11104349]

[6]   Jyoti Rawat, Manisha Nanda, Sanjay Kumar, Nishesh Sharma, Rohit Sharma, Harish Chandra Joshi, Mikhail S Vlaskin, Afzal Hussain, Vinod Kumar, "Integrating wastewater treatment to bio-stimulant and biochar generation for plant growth promotion using microalgae," Process Biochemistry, 187-194 2024.
      [http://dx.doi.org/10.1016/j.procbio.2024.06.031]

[8]   S.C. Chelgani, and A. Asimi Neisiani, "Sensor-Based separation", In: *Dry Mineral Processing.,* S.C. Chelgani, A. Asimi Neisiani, Eds., Springer International Publishing: Cham, 2022, pp. 125-148.
      [http://dx.doi.org/10.1007/978-3-030-93750-8_5]

[9]   A. Havenith, "QUANTOM−Non-destructive scanning of waste packages for material characterization", *EPJ Web Conf.,* vol. 225, p. 06013, 2020.
      [http://dx.doi.org/10.1051/epjconf/202022506013]

[10] J. Roult, "Prompt gamma neutron activation analysis with gamma-gamma coincidences for recycling waste characterization", *2022 IEEE Nuclear Science Symposium and Medical Imaging Conference (NSS/MIC),* pp. 1-8, 2022.
      [http://dx.doi.org/10.1109/NSS/MIC44845.2022.10398919]

[11] N.A. Buczkó, M. Papp, B. Maróti, Z. Kis, and L. Szentmiklósi, "Classification of electronic waste components through x-ray and neutron-based imaging techniques", *Materials (Basel),* vol. 17, no. 19, p. 4707, 2024.
      [http://dx.doi.org/10.3390/ma17194707] [PMID: 39410278]

[12] T. Ueda, S. Koyanaka, and T. Oki, "In-line sorting system with battery detection capabilities in e-waste using combination of X-ray transmission scanning and deep learning", *Resour. Conserv. Recycling,* vol. 201, p. 107345, 2024.
      [http://dx.doi.org/10.1016/j.resconrec.2023.107345]

[13]   W. Sterkens, D. Diaz-Romero, T. Goedemé, W. Dewulf, and J.R. Peeters, "Detection and recognition of batteries on X-Ray images of waste electrical and electronic equipment using deep learning", *Resour. Conserv. Recycling,* vol. 168, p. 105246, 2021.
[http://dx.doi.org/10.1016/j.resconrec.2020.105246]

[14]   H.I.D.I. Muri, and D.R. Hjelme, "Sensor technology options for municipal solid waste characterization for optimal operation of waste-to-energy plants", *Energies,* vol. 15, no. 3, p. 1105, 2022.
[http://dx.doi.org/10.3390/en15031105]

[15]   Z. Lang, *Validation and investigation of x-ray luminescence and x-ray transmission response for the recovery of diamonds using sensor based sorting.* University of Saskatchewan Saskatoon, 2024.

[16]   N. Siraj, B. El-Zahab, S. Hamdan, T.E. Karam, L.H. Haber, M. Li, S.O. Fakayode, S. Das, B. Valle, R.M. Strongin, G. Patonay, H.O. Sintim, G.A. Baker, A. Powe, M. Lowry, J.O. Karolin, C.D. Geddes, and I.M. Warner, "Fluorescence, phosphorescence, and chemiluminescence", *Anal. Chem.,* vol. 88, no. 1, pp. 170-202, 2016.
[http://dx.doi.org/10.1021/acs.analchem.5b04109] [PMID: 26575092]

[17]   U.K. Adarsh, and V.K. Unnikrishnan, "A multiparameter sensing spectroscopic system for remote plastic sorting with a single laser shot: a first information report", *Opt. Laser Technol.,* vol. 177, p. 111194, 2024.
[http://dx.doi.org/10.1016/j.optlastec.2024.111194]

[18]   A. Chauhan, Harish Chandra Joshi, "Sources for biofuels production from biomass," Trends in Mathematics, Part F3197, pp. 1 – 64.
[http://dx.doi.org/10.1007/978-981-99-7250-0_1]

[19]   G. Bonifazi, G. Capobianco, P. Cucuzza, S. Serranti, and V. Spizzichino, "Black plastic waste classification by laser-induced fluorescence technique combined with machine learning approaches", *Waste Biomass Valoriz.,* vol. 15, no. 3, pp. 1641-1652, 2024.
[http://dx.doi.org/10.1007/s12649-023-02146-z]

[20]   E.R.K. Neo, Z. Yeo, J.S.C. Low, V. Goodship, and K. Debattista, "A review on chemometric techniques with infrared, Raman and laser-induced breakdown spectroscopy for sorting plastic waste in the recycling industry", *Resour. Conserv. Recycling,* vol. 180, p. 106217, 2022.
[http://dx.doi.org/10.1016/j.resconrec.2022.106217]

[21]   M.J. Baker, C.S. Hughes, and K.A. Hollywood, *Biophotonics: Vibrational spectroscopic diagnostics.* Morgan & Claypool Publishers, 2016.
[http://dx.doi.org/10.1088/978-1-6817-4071-3]

[22]   E.R.K. Neo, J.S.C. Low, V. Goodship, and K. Debattista, "Deep learning for chemometric analysis of plastic spectral data from infrared and Raman databases", *Resour. Conserv. Recycling,* vol. 188, p. 106718, 2023.
[http://dx.doi.org/10.1016/j.resconrec.2022.106718]

[23]   U.K. Adarsh, E. Bhoje Gowd, A. Bankapur, V.B. Kartha, S. Chidangil, and V.K. Unnikrishnan, "Development of an inter-confirmatory plastic characterization system using spectroscopic techniques for waste management", *Waste Manag.,* vol. 150, pp. 339-351, 2022.
[http://dx.doi.org/10.1016/j.wasman.2022.07.025] [PMID: 35907331]

[24]   S. Marín-Cortés, M. Fernández-Álvarez, E. Enríquez, and J.F. Fernández, "Experimental characterization data on aggregates from construction and demolition wastes for the assistance in sorting and recycling practices", *Constr. Build. Mater.,* vol. 435, p. 136798, 2024.
[http://dx.doi.org/10.1016/j.conbuildmat.2024.136798]

[25]   C. Lubongo, and P. Alexandridis, "Assessment of performance and challenges in use of commercial automated sorting technology for plastic waste", *Recycling,* vol. 7, no. 2, p. 11, 2022.
[http://dx.doi.org/10.3390/recycling7020011]

[26]   K. Friedrich, G. Koinig, R. Pomberger, and D. Vollprecht, "Qualitative analysis of post-consumer and

post-industrial waste *via* near-infrared, visual and induction identification with experimental sensor-based sorting setup", *MethodsX,* vol. 9, p. 101686, 2022.
[http://dx.doi.org/10.1016/j.mex.2022.101686] [PMID: 35478596]

[27] J. Zheng, *Multispectral and hyperspectral infrared imaging for plastic waste sorting and recycling (SPIE Defense + Commercial Sensing).* SPIE, 2024.

[28] S. Serranti, P. Cucuzza, and G. Bonifazi, Hyperspectral imaging for VIS-SWIR classification of post-consumer plastic packaging products by polymer and color (SPIE Future Sensing Technologies). SPIE, 2020.

[29] C. Signoret, A.S. Caro-Bretelle, J.M. Lopez-Cuesta, P. Ienny, and D. Perrin, "Alterations of plastics spectra in MIR and the potential impacts on identification towards recycling", *Resour. Conserv. Recycling,* vol. 161, p. 104980, 2020.
[http://dx.doi.org/10.1016/j.resconrec.2020.104980]

[30] S. P. Gundupalli, S. Hait, and A. Thakur, "Thermal imaging-based classification of the E-waste stream."

[31] A. Sandagdorj, "Fusing Terahertz and Mid Infrared Technologies to Recycle E-waste Black Plastics," University of Waterloo, 2022.

[32] J. Yu, X. Liu, G. Manago, T. Tanabe, S. Osanai, and K. Okubo, "New terahertz wave sorting technology to improve plastic containers and packaging waste recycling in Japan", *Recycling,* vol. 7, no. 5, p. 66, 2022.
[http://dx.doi.org/10.3390/recycling7050066]

[33] N. Zhang, L. Z. X. Josie, and P. W. Ji, "Utilizing machine learning algorithms in conjunction with terahertz spectra for enhanced plastic sorting efficiency," in 2024 49th International Conference on Infrared, Millimeter, and Terahertz Waves (IRMMW-THz), pp. 1-2, 2024.
[http://dx.doi.org/10.1109/IRMMW-THz60956.2024.10697674]

[34] A. Chauhan, and H. Chandra Joshi, "Energetic efficiency of a biofuels production system mathematical modeling", *Trends in Mathematics,* no. Part F3197, pp. 337-357, 2024.
[http://dx.doi.org/10.1007/978-981-99-7250-0_8]

[35] B. Duan, E.R. Bobicki, and S.V. Hum, "Application of microwave imaging in sensor-based ore sorting", *Miner. Eng.,* vol. 202, p. 108303, 2023.
[http://dx.doi.org/10.1016/j.mineng.2023.108303]

[36] A. Chauhan, and H.C. Joshi, "Recent developments and applications in bioconversion and biorefineries", *Trends in Mathematics,* no. Part F3197, pp. 247-307, 2024.
[http://dx.doi.org/10.1007/978-981-99-7250-0_6]

[37] B. Walkley, and J.L. Provis, "Solid-state nuclear magnetic resonance spectroscopy of cements", *Mater. Today Adv.,* vol. 1, p. 100007, 2019.
[http://dx.doi.org/10.1016/j.mtadv.2019.100007]

[38] A. Giannattasio, V. Iuliano, G. Oliva, D. Giaquinto, C. Capacchione, M.T. Cuomo, S.W. Hasan, K.H. Choo, G.V. Korshin, D. Barceló, V. Belgiorno, A. Grassi, V. Naddeo, and A. Buonerba, "Micro(nano)plastics from synthetic oligomers persisting in Mediterranean seawater: Comprehensive NMR analysis, concerns and origins", *Environ. Int.,* vol. 190, p. 108839, 2024.
[http://dx.doi.org/10.1016/j.envint.2024.108839] [PMID: 38943925]

[39] J. Schmidt, M. Haave, J. Underhaug, and W. Wang, "Unlocking the potential of NMR spectroscopy for precise and efficient quantification of microplastics", *Microplast. Nanoplast.,* vol. 4, no. 1, p. 17, 2024.
[http://dx.doi.org/10.1186/s43591-024-00095-5]

[40] G. Lu, Y. Wang, H. Yang, and J. Zou, "One-dimensional convolutional neural networks for acoustic waste sorting," Journal of Cleaner Production, vol. 271, p. 122393, 2020/10/20/ 2020.
[http://dx.doi.org/10.1016/j.jclepro.2020.122393]

[41] S.P. Gundupalli, S. Hait, and A. Thakur, "Classification of metallic and non-metallic fractions of e-waste using thermal imaging-based technique", *Process Saf. Environ. Prot.,* vol. 118, pp. 32-39, 2018. [http://dx.doi.org/10.1016/j.psep.2018.06.022]

[42] Y. Zhao, and J. Li, "Sensor-based technologies in effective solid waste sorting: successful applications, sensor combination, and future directions", *Environ. Sci. Technol.,* vol. 56, no. 24, pp. 17531-17544, 2022.
[http://dx.doi.org/10.1021/acs.est.2c05874] [PMID: 36383409]

CHAPTER 7

Approach for Smart Municipal Solid Waste Management and a Sustainable Circular Economy

Mikhail S. Vlaskin[1,*], Vinod Kumar[2] and Harish Chandra Joshi[3]

[1] *Joint Institute for High Temperatures of the Russian Academy of Sciences, Moscow, Russia*

[2] *Algal Research and Bioenergy Lab, Department of Food Science and Technology, Graphic Era (Deemed to be) University, Dehradun, Uttarakhand, India*

[3] *Department of Chemistry, Graphic Era (Deemed to be) University, Dehradun 248002, Uttarakhand, India*

Abstract: The global increase in Municipal Solid Waste (MSW) presents a significant challenge to the sustainable development of humanity. The waste generation rate continues to increase with population growth, urbanization, and economic development. The presented chapter overlooks the main methods of MSW utilization in order to elucidate the current state of MSW management as a whole and to assess and compare the environmental impact of different MSW utilization pathways. Consideration includes references to regional aspects. The current state of MSW management worldwide and the assessment of environmental impact from different MSW utilization pathways are described in the presented work. Analysis is based on the collection of data on MSW and its component properties determined experimentally.

Keywords: Classification, Composting, Chemical composition, Generation rate, Heating value, Incineration, Landfill, Municipal solid waste, Management, Recycling, Utilization.

INTRODUCTION

Today, the waste utilization problem is one of the main challenges on the path to the sustainable development of the economy and society as a whole. Humanity forms waste through both its economic activities and household life. Municipal Solid Waste (MSW) is a type of waste generated primarily by the urban population. MSW requires special attention, such as sustainable management practices, due to its proximity to people. The primary methods for MSW utilization today are landfill disposal, composting, recycling, and incineration. A

---

* **Corresponding author Mikhail S. Vlaskin:** Joint Institute for High Temperatures of the Russian Academy of Sciences, Moscow, Russia; E-mail: vlaskin@inbox.ru

considerable portion of Municipal Solid Waste (MSW) is currently managed through landfill disposal, where the decomposition processes release landfill gases containing methane, a potent greenhouse gas contributing significantly to global warming. One of the key thermochemical methods utilized for MSW management is incineration. MSW incineration also leads to the emission of greenhouse gases into the atmosphere, but it is accompanied by useful energy generation and should reduce the consumption of primary energy resources.

The presented paper overlooks the main methods of MSW utilization in order to elucidate the current state of MSW management as a whole and to assess and compare the environmental impact of different MSW utilization pathways. Consideration includes references to regional aspects.

MSW management in different regions of the world varies from one another. It depends on many factors, including the economic development of a region, its degree of urbanization and industrialization, culture, and climate, among others. Modern methods of MSW utilization are primarily implemented in high-income countries, whereas in developing countries, only one or two main methods dominate the MSW management system [1, 2]. The conversion of MSW into valuable secondary materials or energy represents a central focus of global research and development initiatives in the field of waste management. Efforts are increasingly being directed towards diversifying waste-to-energy technologies through alternatives such as gasification, pyrolysis, and torrefaction [3 - 5]. Nevertheless, these methods encounter various challenges due to the heterogeneous nature of MSW, its low density, high Moisture Content (MC), and the presence of potentially hazardous substances. Therefore, knowing or predicting the properties of MSW is crucial for the effective application of any thermochemical or mechanical treatment method. One of the main objectives of this work is to study MSW characterization data to estimate the recycling level and energetic potential of MSW and to evaluate its environmental impact. This work includes the study of the main properties of MSW of different regions of the world, as well as separate components and ingredients that are usually presented in MSW.

MSW Definition

Solid waste streams are categorized based on their origins, methods of generation, production rates, compositions, and physical characteristics. According to the World Bank, there are eight primary sources of solid waste: residential, industrial, commercial, institutional, construction and demolition, municipal services, agricultural, and process-related sources. The specific types of waste produced by each of these sources are summarized in Table **1**. It is important to note that the term "municipal solid waste" is used variably in the literature to describe waste derived from diverse sources, which can lead to inconsistent reporting of MSW quantities for the same region. In practical application, MSW incorporates waste from multiple sources, depending on the region, often including residential, industrial, commercial, institutional, construction, municipal services, and process-related waste. Often, only residential waste is referred to as MSW, and in high-income countries, it accounts for 25-35% of the total solid waste [6].

Table 1. Major waste sources and waste compositions according to the World Bank classification [6].

Waste Sources	Waste Producers	Waste Composition
Residential	Single and multifamily dwellings	Food wastes, paper, cardboard, plastics, textiles, leather, yard wastes, wood, glass, metals, ashes, special wastes (*e.g.*, bulky items, consumer electronics, white goods, batteries, oil, tires), and household hazardous wastes.
Industrial	Light and heavy manufacturing, fabrication, construction sites, power, and chemical plants.	Housekeeping wastes, packaging, food wastes, construction and demolition materials, hazardous wastes, ashes, and special wastes.
Commercial	Stores, hotels, restaurants, markets, office buildings, etc.	Paper, cardboard, plastics, wood, food wastes, glass, metals, special wastes, hazardous wastes.
Institutional	Schools, hospitals, prisons, and government centers	Same as commercial.
Construction and demolition	New construction sites, road repairs, renovation projects, and demolition of buildings.	Wood, steel, concrete, dirt, etc.
Municipal services	Street cleaning, landscaping, parks, beaches, other recreational areas, water, and wastewater treatment plants.	Street sweepings; landscape and tree trimmings; general wastes from parks, beaches, and other recreational areas; sludge.
Process	Heavy and light manufacturing, refineries, chemical plants, power plants, mineral extraction, and processing.	Industrial process wastes.
Agriculture	Crops, orchards, vineyards, dairies, feedlots, farms.	Spoiled food wastes, agricultural wastes, and hazardous wastes (*e.g.*, pesticides).

MSW Generation Rate

The MSW problem consists of vast amounts of waste generated by households and the economic sector. Today total amount of waste generated in the world is about 4 billion tons, of which 1.6-2 billion tons is MSW [7].

The MSW generation rate in a specific region is influenced by its economy, urbanization level, industry development, culture, and climate. The increase in economic activity and urbanization usually leads to an increase in the MSW generation rate. Table **2** presents data on waste production rates for various regions of the world [8]. The lowest waste production rates corresponded to the countries located in. In Sub-Saharan Africa (0.65 kg/day/cap), the highest rates correspond to the countries of the Organization for Economic Cooperation and Development (OECD). The average waste generation rates for the other regions are from 0.95 to 1.2 kg/day/capita. Fig. (**1**) shows the MSW generation rates per capita and the values of Gross Domestic Product (GDP) (calculated for the midyear population per capita) reported by the World Bank for some OECD countries [9]. The increase in economic development usually leads to an increase in waste generation rates [10]. Countries with a GDP of more than $35,000 per capita have waste generation rates exceeding 1.5 kg/day/capita. Denmark has the highest MSW generation rate in Europe, which is about 2.1 kg/day/capita. In Eastern European countries with a GDP of $ 20,000 USD/cap or less, MSW generation rates are usually about 1.2 kg/day/cap or less. The problem of waste is more pronounced in cities. In 1900, 220 million urban people (13% of the world's population) produced 300,000 tons of waste per day, including construction waste, ashes, food scraps, and packaging [8]. In 2010, the urban population of the world (2,980 million people – 50% of the world's population) generated more than 3.5 million tons of solid waste per day. According to a recent study [11], in 2025, solid waste generation might be 6 million tons per day. By 2100, this value might be over 11 million tons per day, and the peak of waste generation rate might not be reached in the XXI century.

Table 2. Average rates of waste generation for different regions of the world [8].

Region Index	Urban Population, Million	Average Rates of Waste Generation	
		kg/day/man	Tons/day
AFR	260	0.65	169,119
EAP	777	0.95	738,958
ECA	227	1.1	254,389
LCR	399	1.1	437,545
MENA	162	1.1	173,545

(Table 2) cont.....

Region Index	Urban Population, Million	Average Rates of Waste Generation	
		kg/day/man	Tons/day
OECD	729	2.2	1,566,286
SAR	426	0.45	192,410
Total	2,980	1.2	3,532,252

AFR – Sub-Saharan Africa; EAP – East Asia and the Pacific; ECA – Europe and Central Asia; LCR – Latin America and the Caribbean; MENA – Middle East and North Africa; OECD – Organization for Economic Co-operation and Development; SAR – South Asia Region.

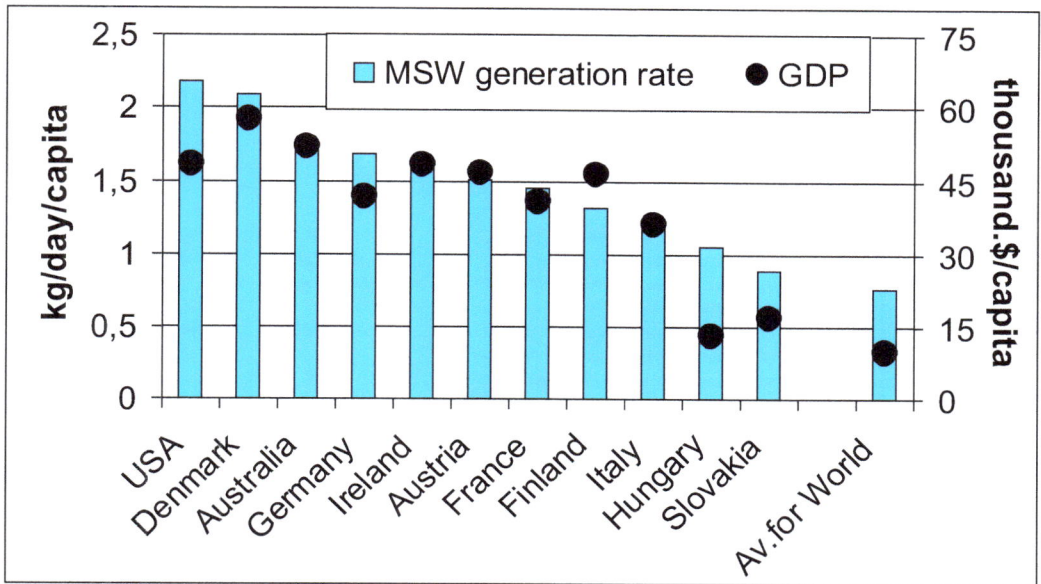

Fig. (1). MSW generation rate and GDP (calculated for midyear population) per capita for some OECD countries and the average for the world.

MSW properties

MSW utilization faces a number of difficulties related to heterogeneous composition, low density, high MC, presence of hazardous components, and others. The primary components of MSW include: food waste, paper, wood, plastics, ferrous and non-ferrous metals, glass, textiles, and rubber, among others. The specific composition of MSW is generally influenced by the geographical location and the season during which it is generated.

Fig. (**2**) (based on a wet-weight basis) [12] illustrates the proportional composition of MSW across different regions of the world. The proportions of food waste, paper, wood, and plastic range between 24% and 67%, 6% and 30%, 2% and 24%, and 3% and 14%, respectively. In higher-income countries, the percentage of food waste is generally lower, while packaging materials such as paper and glass are more prevalent. As economic conditions improve in a given region, the generation of plastic waste tends to rise due to its widespread use as a packaging material. However, restrictive measures regarding plastic packaging may subsequently lead to a reduction in the amount of waste generated.

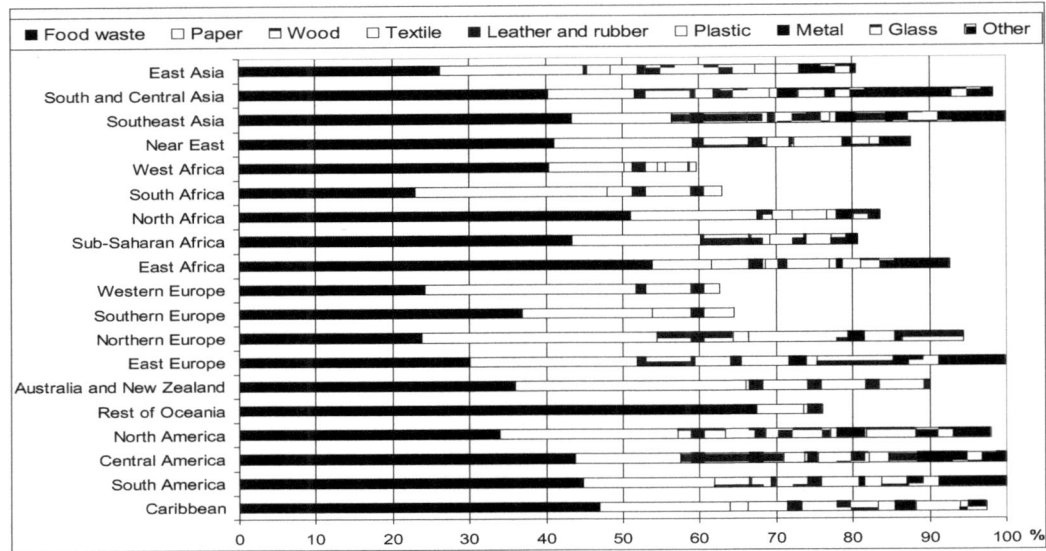

Fig. (2). Component composition of MSW for different parts of the world (wt.% % of MSW on wet basis).

Fig. (3) shows MSW component compositions for the USA and Germany. The total contents of food and garden waste in the USA and Germany are almost the same, equal to about 30%, with a rather high contribution from paper and plastic. The organic part, which includes food and garden wastes, wood, textiles, rubber, paper, plastic, and yard trimmings, totals nearly 80% of the MSW mass. The content of non-organic materials, such as metal scrap, glass, and other non-organic materials, is approximately 20%.

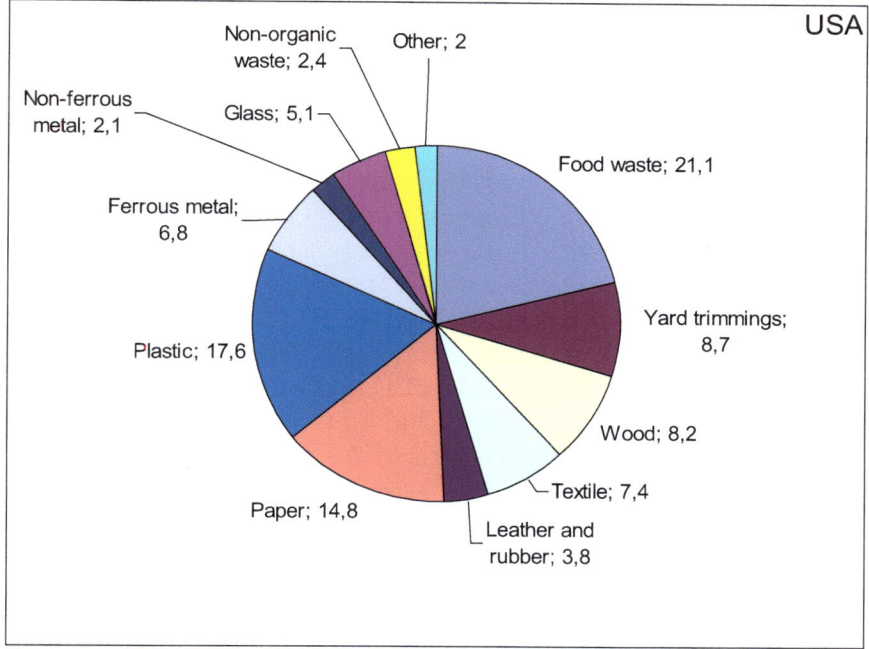

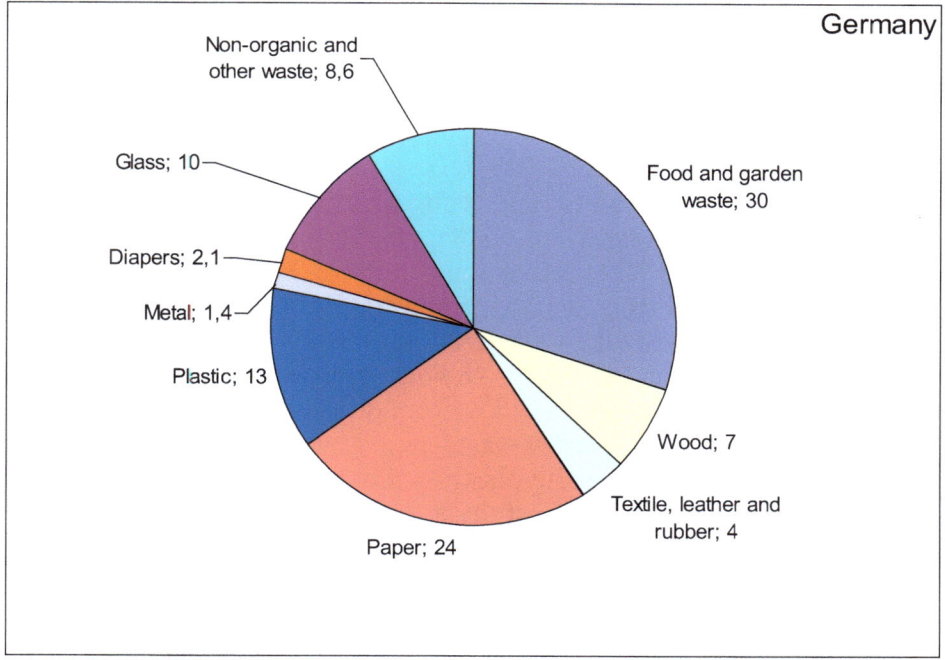

Fig. (3). Component composition (in %) for MSW produced in the USA and Germany [13 - 15].

Another major MSW property used to define the geometries of storage, transport, and recycling facilities is unit weight. The unit weight of MSW changes during its life cycle, generally increasing. MSW unit weight is affected by the initial unit weight, component composition, particle-size distribution, applied pressure, and compression period. The midyear density of MSW generated by households is 0.19-0.23 t/m³ [16], which corresponds to a unit weight of 1.9-2.3 kN/m³. The unit weight of MSW inside a refuse collection truck increases up to 3-5 kN/m³. At a landfill, MSW undergoes further compression by means of compaction rollers. MSW unit weight usually increases with the increase of soil and MC. MSW unit weights in the upper layers of landfills vary from 5 to 18 kN/m³ [17]. Typical unit weight depth distribution for a waste landfill is given in [18]. It shows that the unit weight for a surface layer of a landfill is approximately 6 kN/m³, and for a depth of 45 m, the unit weight achieves 13 kN/m³. According to a study [19], the capacity of a landfill with a burial depth of 60 m and an average MSW unit weight of 10 kN/m³ is 40% less than in the case of 15 kN/m³. The efficiency of energy and environmental performance in the thermochemical conversion of MSW largely depends on its chemical makeup. Analyzing MSW's chemical composition typically involves examining the elemental concentrations of carbon (C), hydrogen (H), nitrogen (N), oxygen (O), sulfur (S), and chlorine (Cl), as well as assessing fixed carbon, volatile matter, ash content, and moisture levels. Knowledge of the quantities of C, H, N, O, S, and Cl in MSW assists in forecasting the flue gas composition resulting from MSW incineration. Details regarding fixed carbon, volatile matter, ash, and moisture levels are crucial for optimizing processes such as MSW drying, thermochemical conversion, and ash disposal.

The concentrations of C, H, N, O, and S for specific materials typical in food, wood, paper, and plastic wastes are shown in Fig. (4) [20] (in this study, the materials were dried at 105 °C to remove moisture before chemical analysis). Fig. (5) illustrates the findings from the analysis of C, H, N, O, S, and Cl content in various MSW components taken from another study [21] (in this research, samples underwent drying at 103 °C for 48 hours before analysis). Fig. (6) depicts the C, H, N, O, S, and Cl content in MSW from different regions worldwide. The Carbon Content (CC) in food, garden waste, paper, wood, and textile is about 45-55% (on a dry ash-free basis), while plastic waste shows a CC of approximately 70-90%. Plastics exhibit the highest hydrogen content at around 10%. The hydrogen concentrations in garden waste, paper, wood, and textile are similar, about 6%. The nitrogen content exceeds 1% in food and garden wastes, plastics, and textiles. Sulfur content in MSW components, except for rubber, is typically less than 1%. Chlorine content in MSW usually ranges from 1-2%, with the highest levels found in plastic (1-2%), food, and garden wastes (slightly less than 1%). The Van Krevelen diagram for the materials listed in Figs. (4 - 6) is shown in Fig. (**7**). Most MSW components have H/C and O/C atomic ratios between 1.3 and 1.7, and 0.5 and 0.8, respectively. These figures are typical for biomass, such as *S. cerevisiae*, which has H/C and O/C atomic ratios of 1.613 and 0.557 [22]. The biomass region on the Van Krevelen diagram encompasses food and garden waste, wood, paper, and textiles, whereas plastic waste typically exhibits lower O/C atomic ratios, not exceeding 0.2. Because of the plastic content in MSW, the O/C atomic ratio of MSW is generally lower than that of biomass. Figs. (**8 - 10**) illustrate the ash, fixed carbon, and volatile contents for certain MSW components and MSW from various regions worldwide. It is observed that the ash content in food and garden wastes, paper, wood, and textiles generally does not exceed 12% (on a dry basis). The ash content in MSW ranges from 20% to 35% due to non-organic materials, including glass, metals, and other inorganic substances. Fig. (**11**) details the MC for different MSW components and MSW from various global regions. The highest MC is found in food and garden wastes (approximately 70% and 60%, respectively). Paper, plastic, and textiles have an MC of 20-30%. MC of MSW in different regions of the world also remains relatively high, ranging from 30 to 60%. Its composition and weather conditions influence the MC in MSW. Summarizing the data on MSW chemical composition, it is significant to note that MSW has a high energetic potential due to its substantial CC (up to 50-55%). The majority of MSW comprises organic materials, accounting for 60-90%, depending on the region. This raises the question of the viability of adopting waste-to-energy solutions either *via* direct incineration of MSW or through the generation of refuse-derived fuels accomplished by mechanical and/or thermochemical processes.

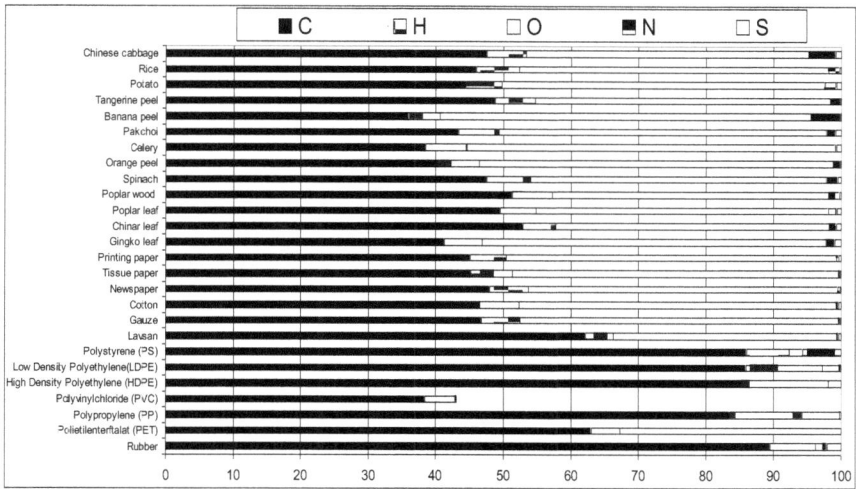

Fig. (4). The contents (in %) of C, H, N, O, S for different ingredients typical for food, wood, paper, and plastic waste (in dry ash-free basis) [20].

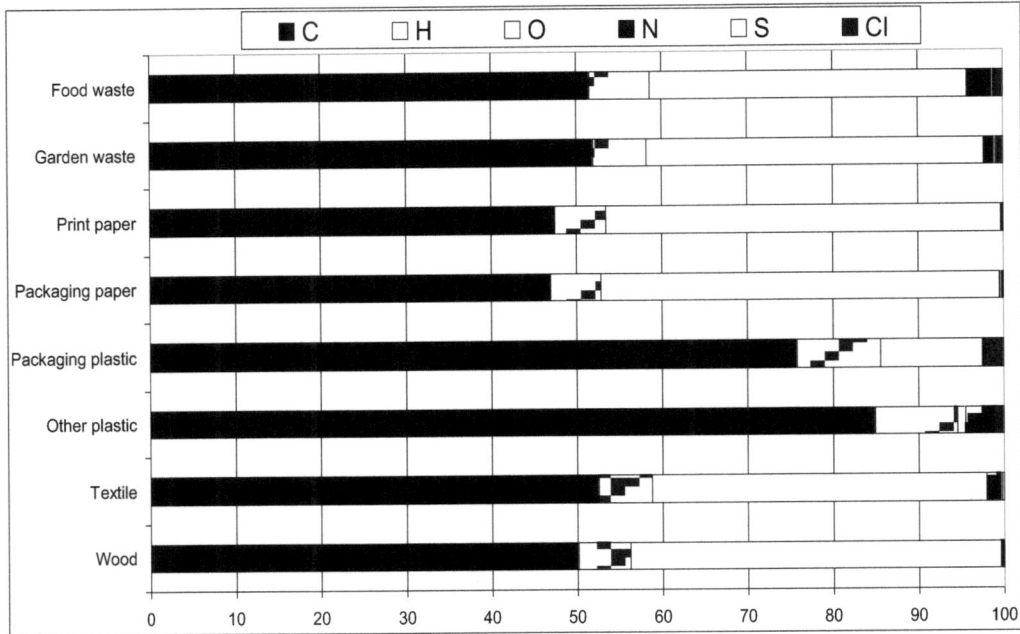

Fig. (5). The contents (%) of C, H, N, O, S, Cl for different MSW components (in dry ash-free basis) [21].

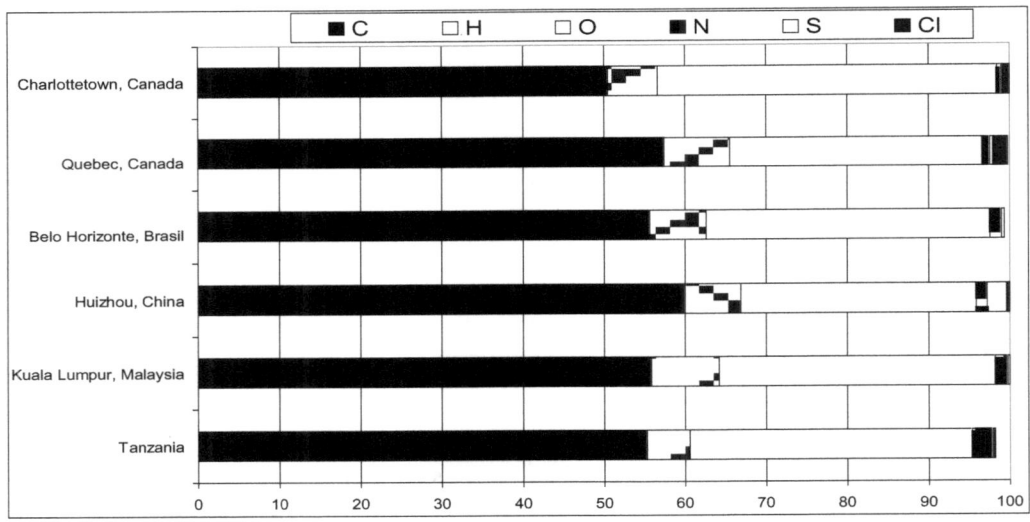

Fig. (6). The contents (%) of C, H, N, O, S, Cl for MSW from different regions of the world (in dry ash-free basis) [23 - 27].

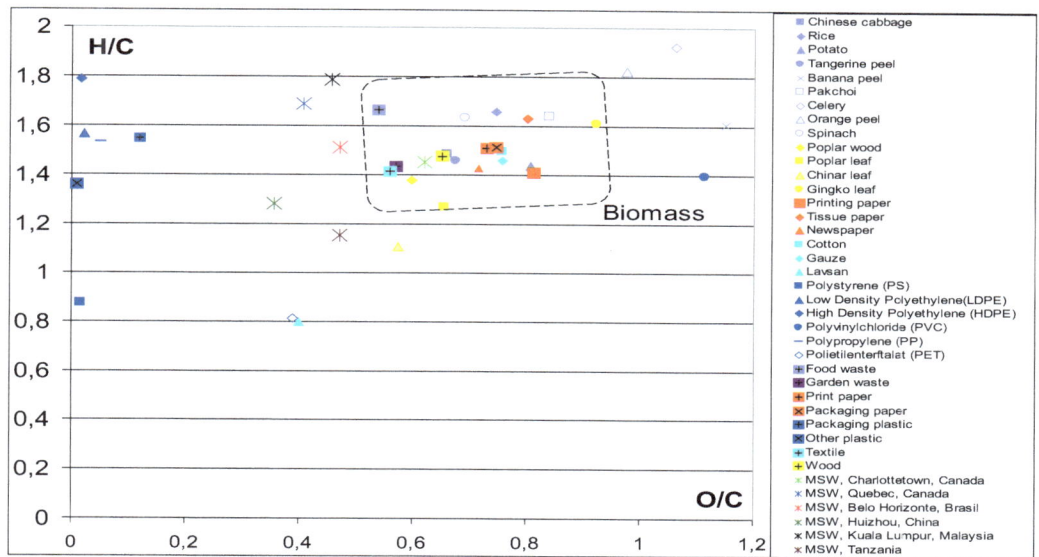

Fig. (7). H/C and O/C atomic ratios for different MSW components and ingredients and MSW of different regions (dry ash-free basis).

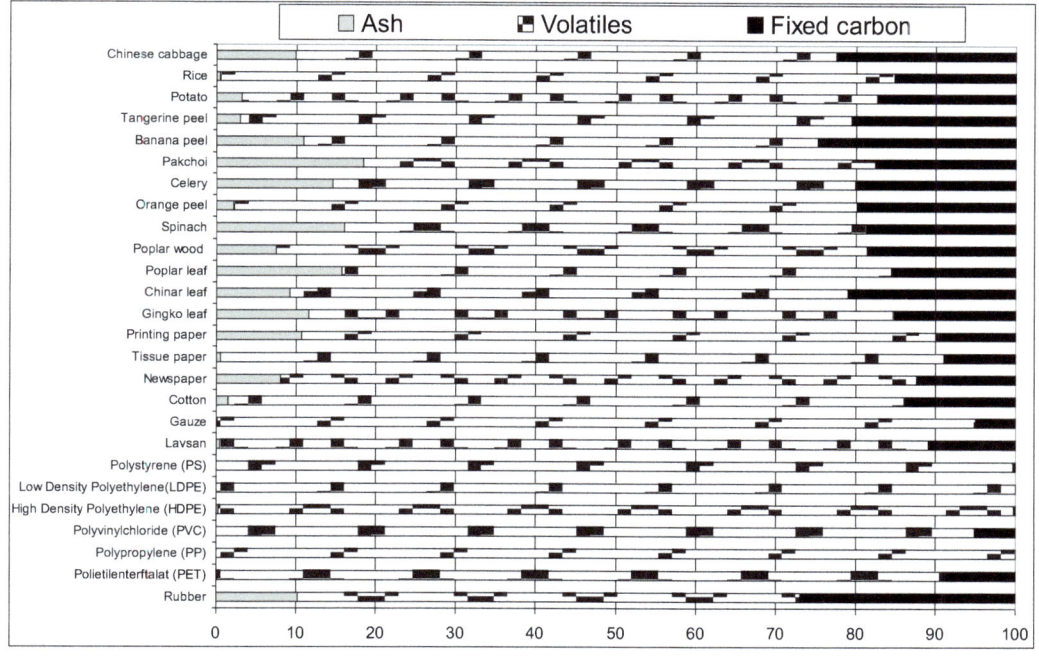

Fig. (8). Contents of ash, fixed carbon, and volatiles (%) in specific ingredients from food, wood, paper, and plastic waste (in dry basis) [20].

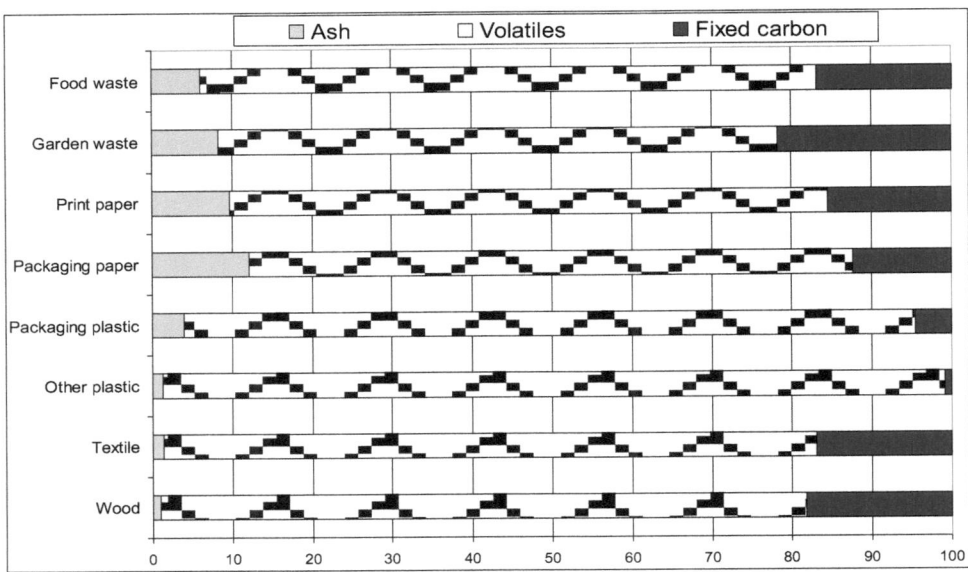

Fig. (9). Contents of ash, fixed carbon, and volatiles (%) in MSW components (in dry basis) [21].

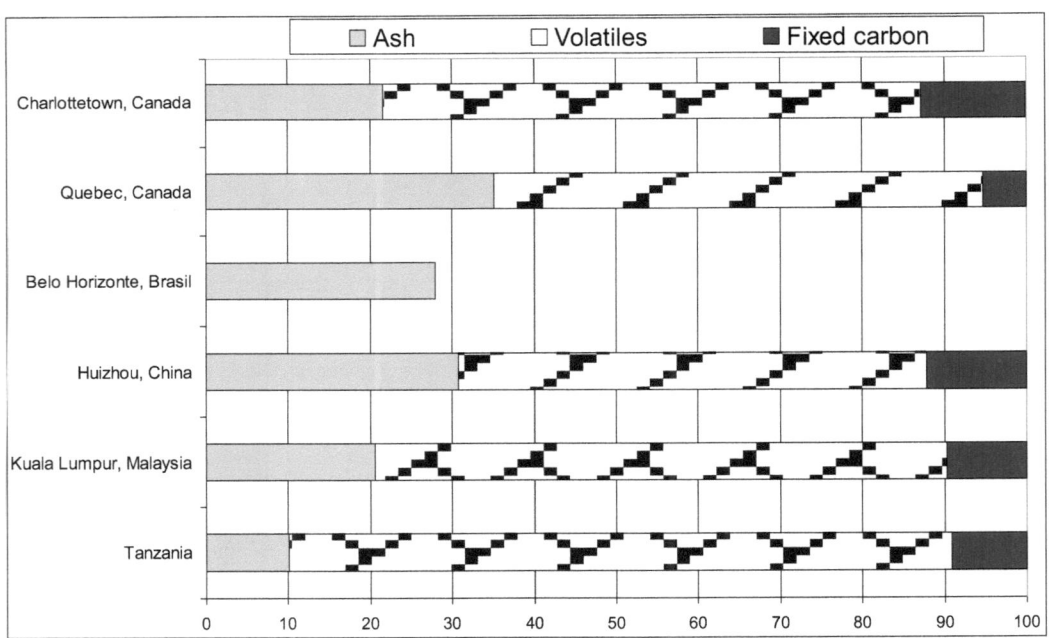

Fig. (10). Contents of ash, fixed carbon, and volatiles (%) in MSW of different regions of the world (in dry basis).

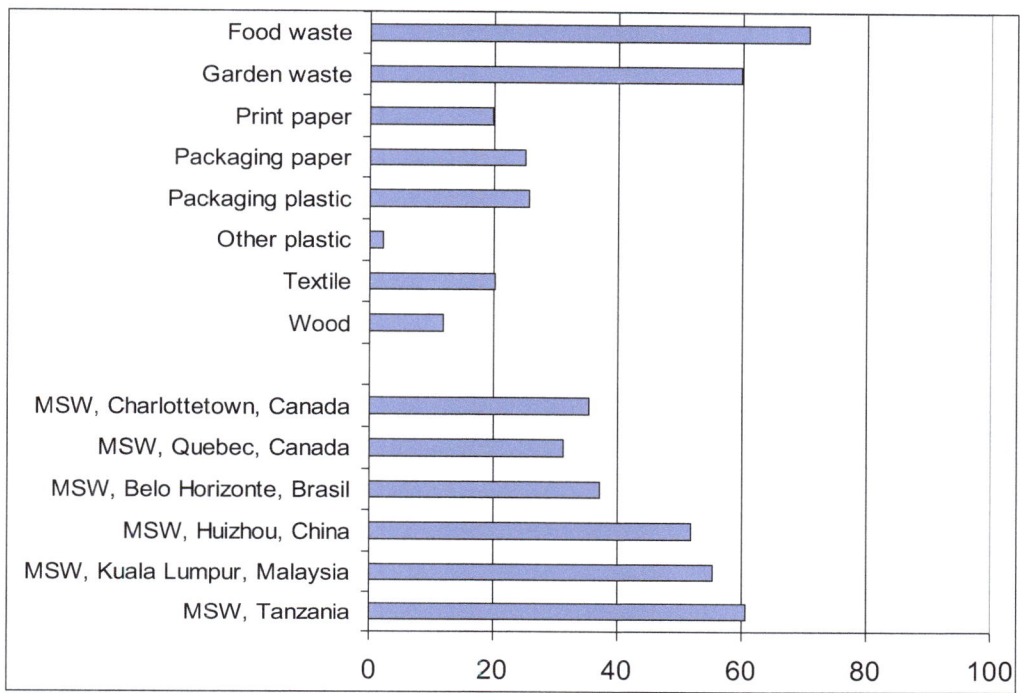

Fig. (11). Moisture contents (%) for different MSW components and MSW from different regions of the world (in wet basis).

MSW Utilization Methods

The primary methods for MSW utilization are disposal at landfills, composting, recycling, and incineration.

Nearly 70% of MSW produced in the world is disposed of in landfills, 11% is used for heat and electric energy generation through MSW thermochemical conversion, and 19% of MSW undergoes mechanical-biological treatment, including composting [7]. For instance, 53% of MSW produced in the USA is landfilled, 25% is recycled, another 13% is incinerated with energy generation, and 9% of MSW is composted [28]. Fig. (**12a**) shows the distribution of MSW utilization methods for countries in the European Union [29]. The current trend is to reduce landfilling and increase the share of incineration and recycling. It should be noted that in some European countries (Belgium, Denmark, Germany), the amount of MSW landfilled is already less than 1-2% (Fig. (**12b**).

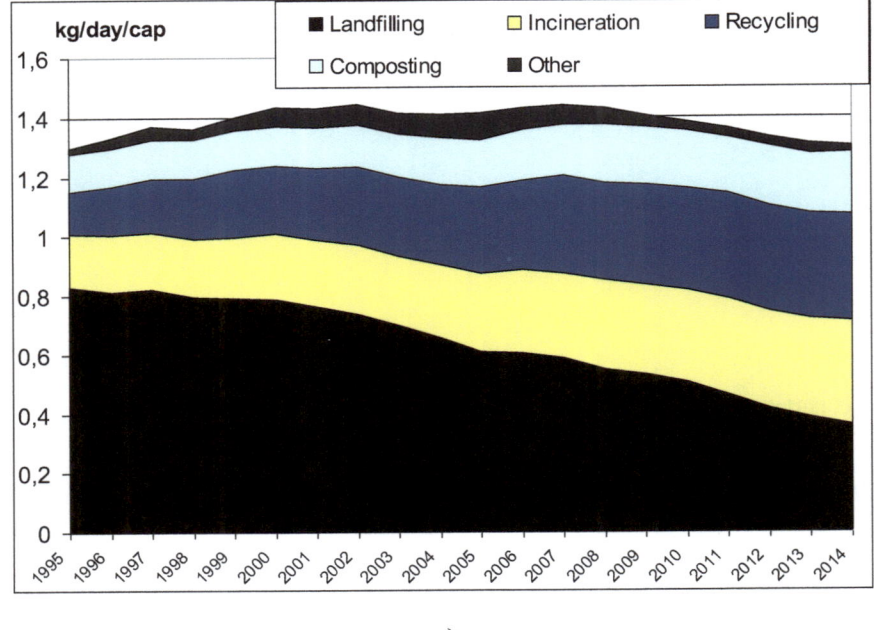

a)

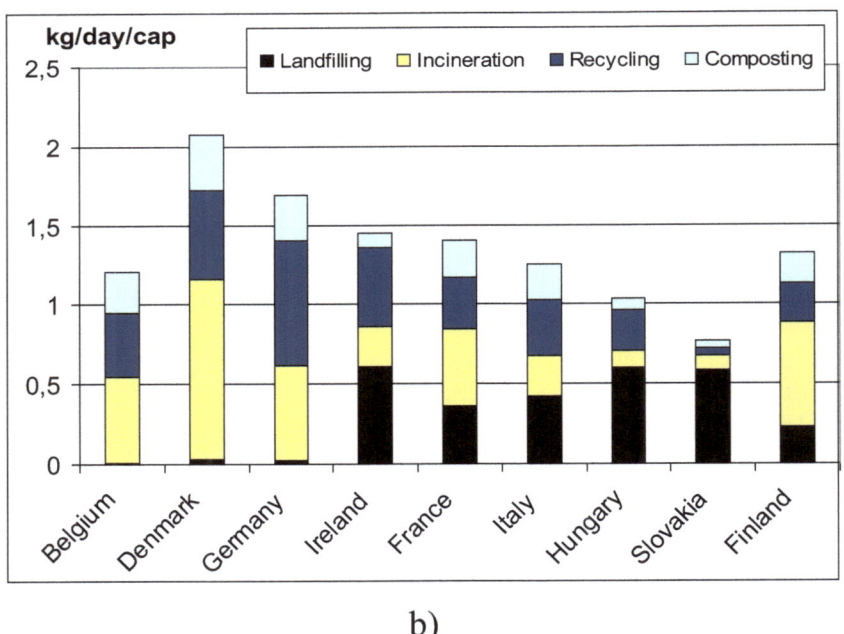

b)

Fig. (12). Distribution of MSW utilization methods for countries in the European Union in total (a) and for some countries in particular (b).

Landfilling

Landfilling is the most widely used method of MSW utilization globally. However, landfill creation and exploitation result in numerous harmful ecological impacts, including air and soil pollution. Landfill gas contains methane, carbon dioxide, carbon monoxide, and other greenhouse gases that escape into the atmosphere. If a waterproofing layer is not provided, pollution of nearby soil and water also occurs. Waste disposal methods have been permanently improved; however, new technical solutions are not employed everywhere. The current MSW disposal methods are based on different procedures for the collection and utilization of liquid and gaseous landfill products. A dump is a place of waste disposal that lacks environmental protection and waste compression facilities, and it does not have an exploitation plan. An MSW landfill is a specially designed earthwork facility used for waste collection and neutralization, accompanied by pollution abatement and control activities. Innovation procedures for MSW disposal at landfills employ more advanced systems for collection, utilization, and neutralization of gaseous and liquid products. At landfills, the organic matter of MSW begins to decompose due to the activity of aerobic bacteria. The residual part of the organic matter is then decomposed by anaerobic bacteria, resulting in the formation of basic compounds such as cellulose, sugars, and amino acids. These compounds, in turn, participate in fermentation, producing short-chained hydrocarbons that are consumed by methanogenic bacteria. Due to the activity of these bacteria, the fermentation products are transformed into stabilized organic compounds and biogas, which generally comprises methane and carbon dioxide. Biogas migrates to the upper layers of the landfill body, where some of it oxidizes, resulting in carbon dioxide production. At landfills equipped with biogas collection and utilization systems, gas is extracted from the layers below the aerobic layer. Biogas from landfill bodies consists of methane (35-65%), carbon dioxide (15-40%), nitrogen (5-40%), and oxygen (up to 5%) [30]. The rate of methane production primarily depends on the landfill's capacity and age. The average daily Methane Yield (MY) reached approximately 90 g per square meter of landfill. During MSW decomposition, the mass of solid organics decreases due to the formation of gaseous and liquid products. MY and CC in waste remained after anaerobic decomposition of various waste samples are shown in Table **3** [31, 32]. The greatest MY was observed in the case of food waste, office paper, cupboard boxes, and grass (more than 100 cm^3 of methane per 1 gram of dry sample). The highest decomposition degree was observed for food and garden wastes, and the lowest for paper and wood waste.

Table 3. Methane Yield (MY), Carbon Content (CC) in solid products of anaerobic decomposition, and decomposition degree for MSW and some MSW components. Decomposition degree is a ratio of the measured MY to MY corresponding to 100% conversion of cellulose or hemicelluloses (and proteins for food waste) contained in samples into methane and carbon dioxide.

Material	Initial CC in dry material, wt. %	CC in solid decomposition product, g per 1 g of dry material (standard deviation)	MY, cm³ per 1 g of dry material (standard deviation)	Decomposition degree, %
MSW	42	0.22 (0.01)	92 (4.1)	58.4
Grass	44.87	0.32 (0.02)	144.4 (15.5)	94.3
Leaves	49.4	0.54 (0.06)	30.6 (8.6)	28.3
Twigs	49.4	0.38 (0.02)	62.6 (13.3)	27.8
Food waste	50.8	0.08 (0.04)	300.7 (10.6)	84.1
Coated paper	34.3	0.34 (0.02)	84.4 (8.1)	39.2
Newspaper	49.2	0.42 (0.02)	74.3 (6.8)	31.1
Cupboard boxes	46.9	0.26 (0.01)	152.3 (6.7)	54.4
Office paper	40.3	0.05 (0.01)	217.3 (15)	54.6

Gaseous products of MSW decomposition contain 50% of the carbon initially presented in MSW [32, 33]. Approximately the same amount of MSW carbon (about 50%) is accumulated in a landfill body. Nearly 2% of carbon is contained in the liquid hydrocarbons removed with the filtrate. Using the results of [31], the evaluation of the greenhouse effect produced by landfill methane can be done. The greenhouse warming potential of methane produced during a century is 21 times higher than that of carbon dioxide for the same period [34]. Therefore, the contribution of carbon dioxide to the whole greenhouse effect produced by landfill gas can be neglected because carbon dioxide is generated at landfills simultaneously with methane, with mass ratio close to 1:1. In accordance with a study [31], 1 ton of dry MSW gives about 92 m³ of methane that is equal to 4,107 moles or 65.7 kg. Assuming that the entire amount of methane produced by MSW decomposition reaches the atmosphere, the greenhouse effect will be 1,365 kg of CO_2 equivalent per ton of dry MSW. Combustion of the same amount of MSW will result in the production of 1,537 kg of carbon dioxide per ton of dry MSW. We can see that MSW incineration leads to an insignificant increase in greenhouse gas emissions compared to MSW landfilling.

The average quantity of methane produced at landfills worldwide is estimated to be approximately 50 m³ or 35.7 kg per ton of MSW, with 10% of this amount being collected and utilized [35]. Given a global waste production rate of 2 billion tons per year, with 70% placed in landfills [7], and assuming that 10% of the generated methane is utilized at these sites [35], the resulting total amount of methane generated by landfills and emitted to the atmosphere is estimated to be 70 billion m³ or 50 million tons. This quantity of methane is equal to 1,050 billion tons of CO_2-equivalent. The total amount of CO_2 emitted by the power sector worldwide is approximately 32 billion tons of CO_2-equivalent [36]. Thus, the methane from landfills accounts for 3.3% of the world's greenhouse gas emissions, which agrees with [37], where this value is estimated at 4%.

Emission from landfills in terms of time scale is a useful characteristic as well. The first-order equation can represent the output of landfill methane:

$$A = L_0 \times R \times (e^{-kc} - e^{-kt}), \text{ where} \tag{1}$$

A is the amount of methane produced in a year (m³/year);

L_0 is the methane creation potential of 1 ton of MSW (m³/ton of MSW);

R is the average annual capacity of a landfill (tons of MSW/year);

k is the methane creation constant (g⁻¹), $k = \ln(2)/t_{50\%}$ ($t_{50\%}$ is the time required for the decomposition of 50% of waste);

c is the time since the closing of a landfill (years);

t is the time since the opening of a landfill (years).

The quantity of methane produced during a year depends on the time since the landfill was opened and the time since it was closed, the average capacity of the landfill, methane creation potential, and the MSW decomposition rate, as represented in equation (1) by the constant k. The time required for the decomposition of 50% of the waste is the MSW half-life, which depends on the MSW composition, climate, and technical conditions. Usually, it varies from 2 to 40 years. The low content of organic matter in MSW, combined with a cool and dry climate, results in an increased MSW half-life.

Fig. (**13**) shows the result of calculating annual methane emissions (in CO_2 equivalent) from a landfill with a capacity of 100,000 tons per year, hypothetically opened in 2000. Calculations were carried out for different operation periods and MSW half-lives ($t_{50\%}$). The methane creation potential (L0) was assumed to be 50 m³ per 1 ton of MSW. It can be seen from Fig. (**13**) that as the MSW half-life increases, the gas emission from the landfill becomes more prolonged. In the case of a half-life of 30 years (which is very typical for cool and dry climates), considerable methane emissions may be observed almost 100 years after landfill opening. It states that during this period, the territories around the landfill cannot be used for building due to the risk of methane accumulation in houses.

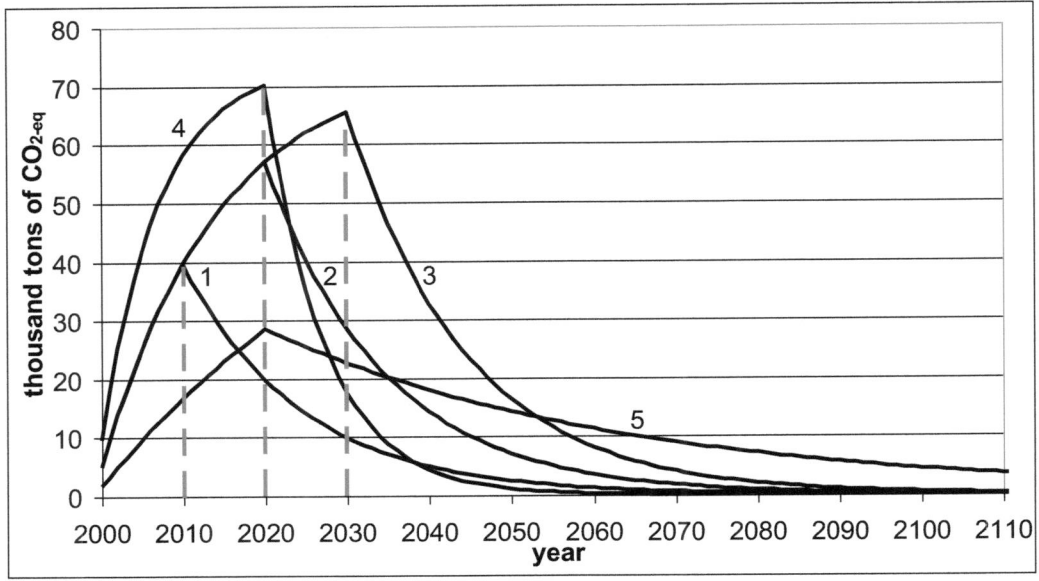

Fig. (13). Methane emissions (in CO_2 equivalent) from landfill with capacity of 100,000 tons per year hypothetically opened in 2000 for different landfill operation periods and MSW half-lifes of MSW. 1 – 10 years of operation, half-life period of 10 years; 2 – 20 years of operation, half-life period of 10 years; 3 – 30 years of operation, half-life period of 10 years; 4 – 20 years of operation, half-life period of 5 years; 5 – 20 years of operation, half-life period of 30 years. The methane creation potential is 50 m^3 per 1 ton of MSW.

Composting

Composting is a method for utilizing solid waste through the aerobic conversion of organic waste materials into a stable, solid product (compost) similar to humus. Composting is generally applied to waste with high organic content. This method can be used for the utilization of MSW and other types of organic waste, including agricultural waste. Compost can be made from a variety of materials, including twigs and leaves, grass, paper, food waste, peat, animal waste products, etc., and can be used then as fertilizer. Composting may last from several years (*e.g.*, in the case of compost heaps) or several days (for open-type or closed-type industrial facilities). The industrial composting process usually proceeds at elevated temperatures (up to 50-60 °C) with intensive stirring of the waste, providing higher uniformity of the produced compost and intensifying the process. Gaseous products of composting, such as CO_2 and CH_4, contain 45-60 wt.% and 0.1-5 wt.% of the initial carbon, respectively [38]. Approximately 40-50% of the carbon remains in the compost.

Assuming that half of the carbon transforms into CO_2 and the other half remains in the compost, we can evaluate the greenhouse effect from composting. Considering the same MSW (0.42 tons of carbon per 1 ton of dry MSW) as in the previous section, we get CO_2 emissions equal to $420 \times 0.5 \times (44/12) = 770$ kg of CO_2 per 1 ton of dry MSW. Greenhouse gas emissions in the case of composting are about two times lower than in the case of landfilling. However, in the case of composting, one of the main issues is where to dispose of the solid product in large quantities.

Recycling

MSW recycling involves the extraction of useful materials, including paper, glass, plastic, and metals, which can be returned to the market. Obtained secondary products can be subjected to further processing, thus reducing the consumption of the corresponding primary raw material. The extraction of secondary materials from MSW is carried out at special waste sorting plants *via* manual or automated waste sorting. Modern waste sorting plants aim not only for the highest extraction values but also for the production of high-quality secondary products with a high purity of the main component, high density, and other enhanced properties. The MSW sorting process typically includes the following stages: unloading MSW at a sorting site, shredding to achieve the desired particle size, waste sorting, and extraction of materials such as glass, polymeric materials, scrap paper, rechargeable batteries, aluminum, and ferromagnetic alloys. Extracted materials are usually pressed and bundled into briquettes. The residual part of MSW with a large content of organic matter goes through either compression with subsequent transportation to landfills or shredding to obtain the "Refuse-Derived Fuel" (RDF). RDF is supplied to incineration plants or biological treatment plants. MSW recycling as a method of MSW processing and utilization is widely used in high-income countries. The contents of some MSW components before and after utilization through recycling and composting in the USA are shown in Table **4** [13]. The contents of food waste, wood, textile, and plastic in the residual part of MSW considerably increase while the contents of paper and yard trimmings decrease. The highest utilization rates correspond to nonferrous metals (68%), paper (64.6%), and yard trimmings (57.7%). A high degree of yard trimmings is achieved mostly by composting. The total degree of utilization by recycling and composting for MSW in the USA is about 34.5%. For instance, in Germany, this value is close to 60% [5].

Table 4. MSW components before and after utilization by recycling and composting, and the corresponding utilization degrees in the USA [13].

Components	Share of Initial MSW, million tons	Share of Initial MSW, %	Share of MSW After Utilization, %	Utilization Degree, %
Food waste	36.43	14.5	21.1	4.8
Wood	15.82	6.3	8.2	15.2
Fabric	14.33	5.7	7.4	15.7
Leather and rubber	7.53	3	3.8	17.9
Paper	68.620	27.4	14.8	64.6
Plastic	31.75	12.7	17.6	8.8
Ferrous metal	16.8	6.7	6.8	33
Nonferrous metal	5.58	2.2	2.1	68
Glass	11.57	4.6	5.1	27.7
Yard trimmings	33.96	13.5	8.7	57.7
Nonorganic waste	3.9	1.6	2.4	0.05
Other	4.6	1.8	2	28.3
Total	250.89	100	100	34.5

Incineration

Incineration is a primary thermochemical process for MSW utilization worldwide. By MSW incineration, useful heat and electrical energy can be produced. MSW may be combusted at an incineration plant after preliminary preparations, such as shredding, homogenization, and drying, or without them.

The energy efficiency of MSW incineration is significantly influenced by the specific heat value of the MSW, which depends on its composition and is strongly affected by moisture and ash contents [39, 40]. The Tanner diagram is a widely used instrument to describe the self-sustained combustion of MSW [41]. The sides of the Tanner diagram represent three axes corresponding to the contents (on a wet basis) of moisture, organic matter (fixed carbon and volatile yield), and ash. For every point within the Tanner diagram, the sum of weight percentages of moisture, organic matter, and ash equals 100%. The area within the Tanner diagram, where self-sustained combustion of MSW (without the addition of any extra high-quality fuels) is possible, lies below a maximum moisture of 50%, below a maximum ash content of 60%, and above a minimum organic matter content of 25%. If the MSW composition does not meet these requirements, combustion is possible only with the addition of high-quality fuel with a higher heating value. These results are presented inside the Tanner diagram as a shaded area corresponding to self-sustained MSW combustion. In general, the results of the experimental study agree with the limits suggested by Tanner. The heat energy released by MSW combustion can be represented as the higher heating value (Qh) or the lower heating value (Ql). Both of these values can be normalized to the weight unit of either a wet or dry sample. In thermal engineering, the Lower Heating Value (LHV) per unit weight of a wet sample is typically used. This value includes the loss of energy that is spent on moisture evaporation.

$$Q_l^w = Q_l^d \times (1\ 0{,}01M)\ 2{,}43 \times 0{,}01M\ (MJ/kg),$$

where M – MC (in wet basis), in %; 2,43– latent heat of vaporization of water at a temperature of 25 °C, in MJ/kg.

Higher and lower heating values for dry sample Q_h^d and Q_l^d can be bounded in accordance with the procedure shown in a similar study [41]:

$$Q_l^d = Q_h^d\ 0{,}2122 \times H^d - 0{,}0008 \times (O^d + N^d\ (MJ/kg),$$

where H^d, O^d, and N^d are contents of hydrogen, oxygen, and nitrogen, respectively (in dry basis), in %.

Higher heating value for wet sample Q_h^w is bound with a higher heating value for dry sample Q_h^d. This is in accordance with a similar study [41]:

$$Q_h^w = Q_h^d \times (1 - 0{,}01M)\ (MJ/kg).$$

Specific heat values determined experimentally for some ingredients and MSW components are shown in Table **5**. According to these data, plastic, rubber, and textile have the highest heat values (40.6, 35.74, and 20.9 MJ/kg, respectively, on a dry basis). Dried food and garden waste also have relatively higher heating values (up to 20 MJ/kg). However, in the initial wet state, these kinds of waste have the lowest specific heat values.

Table 5. Specific heat values determined experimentally for some MSW components and ingredients and MSW from different regions of the world.

Material	Q_h^d	Q_h^w	Q_l^d	Q_l^w
	MJ/kg			
Food waste	20.3	6	18.8	3.8
Garden waste	18.7	7.5	17.5	5.5
Print paper	16.3	13.1	15.1	11.6
Packaging paper	15.5	11.6	14.4	10.2
Packaging plastic	25.9	19.3	23.9	17.1

(Table 5) cont.....

Material	Q_h^d	Q_h^w	Q_l^d	Q_l^w
	MJ/kg			
Other plastic	40.6	39.7	38.6	37.7
Textile	20.9	16.7	19.6	15.2
Wood	19.5	17.2	18.2	15.7
	-	-	-	-
MSW, Charlottetown, Canada	-	-	-	10.53
MSW, Quebec, Canada	-	-	-	9.79
MSW, Belo Horizonte, Brazil	-	-	-	7.98
MSW, Huizhou, China	15.04	-	-	-
MSW, Kuala Lumpur, Malaysia	-	-	-	9.11

It is interesting to compare the heating values of MSW components with fresh-cut wood. The higher heating value of wet wood from MSW is approximately 19.5 MJ/kg. Specific heat values for various species of wood (in a fresh-cut state with minimal moisture) are presented in Table **6** [42]. Fresh-cut wood has considerably higher MC and so a lower specific heat value.

Table 6. Minimum moisture content, density, and specific heat values for different species of wood (in fresh-cut state) [42].

Species of Wood	Wood Moisture, %	Density, kg/m3	Higher Heating Value, MJ/kg	Lower Heating Value, MJ/kg
Pine	46.3	775	10.91	8.83
Fir	46.3	680	10.91	9.02
Birch	40.3	869	12.13	10.31
Aspen	42.2	705	11.75	9.9

Table **5** demonstrates the specific heat capacities of MSW from various global regions. Owing to its reduced moisture levels (30–60%) and the inclusion of plastics, the lower heating value of MSW surpasses that of food and garden waste. The typical lower heating value of MSW ranges between 7 and 9 MJ/kg. For MSW incineration to proceed without the addition of external high-quality fuel, a minimum lower heating value of 5–6 MJ/kg is necessary [43]. Table **7** provides data on the lower heating values of MSW-based fuels utilized in several European waste-to-energy plants [44]. In modern incinerators, the lower heating value of fuel derived from MSW often exceeds 10 MJ/kg. This is generally achieved through preparatory procedures such as sorting, drying, shredding, and other processing steps. It is worth noting that the solid fuels obtained from MSW today have lower heating values than those of woody biomass, wood pellets, and even low-grade lignite coal.

Table 7. Lower heating values of solid fuels derived from MSW consumed by modern incineration plants in Europe [44].

Lower heating value, MJ/kg	Region
15.1	Naples (Italy)
14.5	Rudersdorf bei Berlin (Germany)
12	Heringen (Germany)
10.5	Arhus (Denmark)
12	Hameln (Germany)

There exist three primary methods for the incineration of MSW: the mobile grate system, rotary kiln furnace combustion, and fluidized-bed combustion. Among these, the mobile grate method is the most prevalent one. The energy produced from the burning of MSW can either be transformed into heat (or useful high-temperature steam), electricity, or employed for the cogeneration of heat and electricity. In Europe, approximately 40.8% of incineration facilities are used for the combined production of heat and electricity, roughly 18.4% solely generate electricity, and about 10.5% are used just for heat generation [45]. The conversion efficiency of heat to electricity at advanced MSW incineration facilities can exceed 30% [46], while the efficiency for solely heat production can reach up to 80% [47]. These efficiencies fall short compared to coal-fired power plants, due to factors such as lower steam temperature and pressure within the water-steam cycle, significant auxiliary power usage, elevated condenser pressure, and the comparatively smaller scale of waste incineration facilities [48]. Typically, a waste incineration plant utilizing mobile grate combustion produces approximately 546 kWh of electricity per ton of MSW with a lower heating value of 10.44 MJ/kg [46], which corresponds to an 18% conversion efficiency from heat to electricity, excluding auxiliary consumption. Considering an auxiliary power usage of 150 kWh per ton of MSW, the net electrical energy supplied to consumers is 396 kWh, reflecting a conversion efficiency of 13%.

Modern waste incineration facilities can reach maximum steam temperatures of 400 °C and pressures of 4 MPa [46]. These installations have the capability to produce 640 kWh of electricity per ton of MSW at a lower heating value of 10.44 MJ/kg, corresponding to an efficiency rate of 22% (excluding auxiliary power consumption). When accounting for an auxiliary power use of 120 kWh per ton, the available electrical energy to consumers is 520 kWh, resulting in an 18% conversion efficiency from heat to electrical power.

In Europe, there are 314 waste incineration facilities, of which 118 have capacities under 100,000 tons annually (37.6%), 124 range between 100,000 and 250,000 tons per annum (39.5%), and 72 plants exceed 250,000 tons per year (22.9%) [45]. For the majority of facilities (61.4%), the thermal output from MSW combustion (based on the lower heating value of MSW) falls below 35 MW [48]. Meanwhile, 36.3% of plants generate thermal outputs between 35 to 90 MW, and 2.3% exceed the 90 MW mark. The largest incineration facilities employing mobile grate combustion possess capacities approaching 120 MW.

CONCLUDING REMARKS

Large volumes of MSW generation prompt us to consider optimal ways of their disposal and utilization. Indeed, the total amount of MSW generated worldwide can be compared today with the world's consumption of primary hydrocarbon fuels [49].

The approach to smart municipal solid waste management and a sustainable circular economy must be based on knowledge of the composition and recycling methods. That is why, by this chapter, issues such as MSW properties and current MSW utilization methods have been considered.

Study of MSW physical and chemical properties showed that MSW possesses high energetic potential due to high CC (up to 50-55%). MSW properties depend on the region and season of its production; however, in most cases, organic components dominate the MSW composition, accounting for approximately 80%. The heating value of MSW is typically 7-9 MJ/kg. By performing additional preliminary operations such as sorting, drying, and shredding, this value can be increased by up to 10-15 MJ/kg. It exceeds the heating value of woody biomass, wood pellets, and low-grade brown coals.

Methods such as composting, recycling, and incineration should replace landfilling as MSW utilization methods. All secondary materials that can be sent for reuse must be extracted from MSW first.

The high calorific value of MSW encourages waste-to-energy implementation. It can be carried out through direct MSW incineration or through the production of refuse-derived fuels by mechanical or other thermochemical conversion methods (such as gasification, pyrolysis, or torrefaction). Waste-to-energy implementation may contribute to a decrease in the volume of waste landfilled and a decrease in the consumption of primary hydrocarbon fuels.

The implementation of modern thermochemical and mechanical methods for MSW utilization worldwide depends on several factors, including the economic development of a region, its degree of urbanization and industrialization, culture, and climate, among others.

Further research and development should focus on optimizing the waste management industry to increase the reuse of secondary materials and mitigate negative environmental impacts, such as soil and air pollution. Such optimization should be based on modern recycling technologies and artificial intelligence methods, taking into account the above-mentioned characteristics of waste and the locations where it is generated.

REFERENCES

[1]   A. Soni, S. K. Gupta, N. Rajamohan, and M. Yusuf, "Waste-to-energy technologies: a sustainable pathway for resource recovery and materials management," *Materials Advances*, vol. 6, no. 14, pp. 4598-4622, 2025.
        [http://dx.doi.org/10.1039/D5MA00449G]

[2]   M.S. Korai, R.B. Mahar, and M.A. Uqaili, "The feasibility of municipal solid waste for energy generation and its existing management practices in Pakistan", *Renew. Sustain. Energy Rev.,* vol. 72, pp. 338-353, 2017.
        [http://dx.doi.org/10.1016/j.rser.2017.01.051]

[3]   G.C. Fitzgerald, "5 - Pre-processing and treatment of Municipal Solid Waste (MSW) prior to incineration", In: *Waste to Energy Conversion Technology.* Woodhead Publishing, 2013, pp. 55-71.
        [http://dx.doi.org/10.1533/9780857096364.2.55]

[4]   N. Ungureanu, N.-V. Vlăduţ, S.-Ş. Biriş, M. Ionescu, and N.-E. Gheorghiţă, "Municipal solid waste gasification: Technologies, process parameters, and sustainable valorization of by-products in a circular economy," *Sustainability*, vol. 17, no. 15, p. 6704, 2025.
        [http://dx.doi.org/10.1533/9780857096364.2.146]

[5]   L. Matsakas, Q. Gao, S. Jansson, U. Rova, and P. Christakopoulos, "Green conversion of municipal solid wastes into fuels and chemicals", *Electron. J. Biotechnol.,* vol. 26, pp. 69-83, 2017.
        [http://dx.doi.org/10.1016/j.ejbt.2017.01.004]

[6]   What a Waste: Solid Waste Management in Asia (Urban Development Sector Unit East Asia and Pacific Region). Washington, D.C. 20433, USA.: The World Bank, 1999.

[7]   Globalization & waste management. Phase 1: Concepts and facts. ISWA, 2012, p. 48.

[8]   D. Hoornweg and P. Bhada-Tata, What a Waste: A Global Review of Solid Waste Management (Urban Development Series - Knowledge papers). The World Bank, 2012.

[9]   Available from: http://data.worldbank.org/indicator/NY.GDP.PCAP.CD?page=1

[10]   "Waste. Investing in energy and resource efficiency," United Nations Environment Programme2011.

[11]   D. Hoornweg, P. Bhada-Tata, and C. Kennedy, "Environment: Waste production must peak this century", *Nature,* vol. 502, no. 7473, pp. 615-617, 2013.
        [http://dx.doi.org/10.1038/502615a] [PMID: 24180015]

[12]   R. Pipatti, C. Sharma, and M. Yamada, *"Waste Generation, Composition and Management Data " IPCC Guidelines for National Greenhouse Gas Inventories.* vol. Vol. 5. Waste, 2006.

[13]   "Municipal Solid Waste Generation, Recycling, and Disposal in the United States. Tables and Figures for 2012", U.S. Environmental Protection Agency. Office of Resource Conservation and Recovery 2014.

[14]   S. Mühle, I. Balsam, and C.R. Cheeseman, "Comparison of carbon emissions associated with

municipal solid waste management in Germany and the UK", *Resour. Conserv. Recycling,* vol. 54, no. 11, pp. 793-801, 2010.
[http://dx.doi.org/10.1016/j.resconrec.2009.12.009]

[15]   Available
from:
http://www.bmub.bund.de/fileadmin/bmu-import/files/pdfs/allgemein/application/pdf/bericht_siedlung
sabfallentsorgung_2006.pdf

[16]   V.G. Sister, and A.N. Mirny, "Selection of waste treatment technologies taking into account their composition and properties", *Municipal Solid Waste,* no. 1, pp. 16-21, 2009.

[17]   D. Zekkos, J.D. Bray, E. Kavazanjian Jr, N. Matasovic, E.M. Rathje, M.F. Riemer, and K.H. Stokoe II, "Unit Weight of Municipal Solid Waste", *J. Geotech. Geoenviron. Eng.,* vol. 132, no. 10, pp. 1250-1261, 2006.
[http://dx.doi.org/10.1061/(ASCE)1090-0241(2006)132:10(1250)]

[18]   E. J. Kavazanjian, N. Matasovic, R. Bonaparte, and G. R. Schmertmann, "Evaluation of MSW properties for seismic analysis", Geoenvironment 2000, vol. 2, pp. 126-1141, 1995.

[19]   N. Matasović, and E. Kavazanjian Jr, "Cyclic characterization of OII landfill solid waste", *J. Geotech. Geoenviron. Eng.,* vol. 124, no. 3, pp. 197-210, 1998.
[http://dx.doi.org/10.1061/(ASCE)1090-0241(1998)124:3(197)]

[20]   H. Zhou, Y. Long, A. Meng, Q. Li, and Y. Zhang, "Classification of municipal solid waste components for thermal conversion in waste-to-energy research", *Fuel,* vol. 145, pp. 151-157, 2015.
[http://dx.doi.org/10.1016/j.fuel.2014.12.015]

[21]   S.S. Hla, and D. Roberts, "Characterisation of chemical composition and energy content of green waste and municipal solid waste from Greater Brisbane, Australia", *Waste Manag.,* vol. 41, pp. 12-19, 2015.
[http://dx.doi.org/10.1016/j.wasman.2015.03.039] [PMID: 25882791]

[22]   U. von Stockar, and J.S. Liu, "Does microbial life always feed on negative entropy? Thermodynamic analysis of microbial growth", *Biochim. Biophys. Acta Bioenerg.,* vol. 1412, no. 3, pp. 191-211, 1999.
[http://dx.doi.org/10.1016/S0005-2728(99)00065-1] [PMID: 10482783]

[23]   A.J. Chandler, *Municipal Solid Waste Incinerator Residues.* Elsevier, 1973, p. 973.

[24]   M.M.V. Leme, M.H. Rocha, E.E.S. Lora, O.J. Venturini, B.M. Lopes, and C.H. Ferreira, "Techno-economic analysis and environmental impact assessment of energy recovery from Municipal Solid Waste (MSW) in Brazil", *Resour. Conserv. Recycling,* vol. 87, pp. 8-20, 2014.
[http://dx.doi.org/10.1016/j.resconrec.2014.03.003]

[25]   S. Kathirvale, M.N. Muhd Yunus, K. Sopian, and A.H. Samsuddin, "Energy potential from municipal solid waste in Malaysia", *Renew. Energy,* vol. 29, no. 4, pp. 559-567, 2004.
[http://dx.doi.org/10.1016/j.renene.2003.09.003]

[26]   L. Zhu, L. Zhang, J. Fan, P. Jiang, and L. Li, "MSW to synthetic natural gas: System modeling and thermodynamics assessment", *Waste Manag.,* vol. 48, pp. 257-264, 2016.
[http://dx.doi.org/10.1016/j.wasman.2015.10.024] [PMID: 26525970]

[27]   A.M. Omari, "Characterization of municipal solid waste for energy recovery. A case study of Arusha, Tanzania", *Journal of Multidisciplinary Engineering Science and Technology,* vol. 2, no. 1, pp. 230-237, 2015.

[28]   "Advancing sustainable materials management: 2013 fact sheet. Assessing trends in material generation, recycling and disposal in the United States," 2015.

[29]   "Municipal waste generation and treatment, by type of treatment method," Available: http://ec.europa.eu/eurostat/tgm/refreshTableAction.do?tab=table&plugin=1&pcode=tsdpc240&langu age=en

[30]   Swedish Gas Centre: Basic Data on Biogas. 2012.

[31]   W.E. Eleazer, W.S. Odle Iii, Y.S. Wang, and M.A. Barlaz, "Biodegradability of municipal solid waste components in laboratory- scale landfills, Environmental Science and Technology", *Article,* vol. 31, no. 3, pp. 911-917, 1997.

[32]   M.A. Barlaz, "Carbon storage during biodegradation of municipal solid waste components in laboratory-scale landfills", *Global Biogeochem. Cycles,* vol. 12, no. 2, pp. 373-380, 1998.
[http://dx.doi.org/10.1029/98GB00350]

[33]   S. Manfredi, D. Tonini, T.H. Christensen, and H. Scharff, "Landfilling of waste: accounting of greenhouse gases and global warming contributions", *Waste Manag. Res.,* vol. 27, no. 8, pp. 825-836, 2009.
[http://dx.doi.org/10.1177/0734242X09348529] [PMID: 19808732]

[34]   "Climate Change 1995, The Science of Climate Change: Summary for Policymakers and Technical Summary of the Working Group" Intergovernmental Panel on Climate Change (IPCC)1995.

[35]   H. Scharff, H.Y. Soon, S. Rwabwehare Taremwa, D. Zegers, B. Dick, T. Villas Bôas Zanon, J. Shamrock, "The impact of landfill management approaches on methane emissions," (in eng), *Waste Manag Res*, vol. 42, no. 11, pp. 1052-1064, Nov 2024.
[http://dx.doi.org/10.1016/j.renene.2006.04.020]

[36]   2015, "CO_2 Emissions From Fuel Combustion Highlights. 2015 Edition", International Energy Agency.

[37]   "Good Practice Guidance and Uncertainty Management in National Greenhouse Gas Inventories", in "CH4 Emissions from Solid Waste Disposal " Intergovernmental Panel on Climate Change2001.

[38]   B.G. Hermann, L. Debeer, B. De Wilde, K. Blok, and M.K. Patel, "To compost or not to compost: Carbon and energy footprints of biodegradable materials' waste treatment", *Polym. Degrad. Stabil.,* vol. 96, no. 6, pp. 1159-1171, 2011.
[http://dx.doi.org/10.1016/j.polymdegradstab.2010.12.026]

[39]   I. Febijanto, S. Steven, N. Nadirah, H. Bahua, A. Shoiful, D. Dewanti, I.P. Kristyawan, K. Haris, M. Yuliani, M. Hanif, M. Robbani, N. Yusuf, Prihartanto, P. Alfatri, R. Pratama, W. Purwanta, Wiharja, R. Nugroho, S. Ramadhan, "Municipal Solid Waste (MSW) reduction through Incineration for electricity purposes and its environmental performance: A case study in Bantargebang, West Java, Indonesia", *Evergreen*, vol. 11, pp. 32-45, 03/30 2024.

[40]   T.R. Sarker, M.L. Khatun, D.Z. Ethen, M.R. Ali, M.S. Islam, S. Chowdhury, K.S. Rahman, N.S. Sayem, R.S. Akm, "Recent evolution in thermochemical transformation of municipal solid wastes to alternate fuels," *Heliyon*, vol. 10, no. 17, p. e37105, 2024/09/15/ 2024.
[http://dx.doi.org/10.1016/j.wasman.2013.09.023] [PMID: 24135625]

[41]   D. Komilis, K. Kissas, and A. Symeonidis, "Effect of organic matter and moisture on the calorific value of solid wastes: An update of the Tanner diagram", *Waste Manag.,* vol. 34, no. 2, pp. 249-255, 2014.
[http://dx.doi.org/10.1016/j.wasman.2013.09.023] [PMID: 24135625]

[42]   E.Y. Thomson, Y.A. Dolatsis, Yu.S. Khrol, and D.P. Turlais, "Calculation of the dependence of the heat of combustion of wood on humidity", *Fourth Russian National Conference on Heat Transfer,* 2006pp. 324-326 Moscow (in Russian)

[43]   Dezhen Chen, and T.H. Christensen, "Life-cycle assessment (EASEWASTE) of two municipal solid waste incineration technologies in China", *Waste Manag. Res.,* vol. 28, no. 6, pp. 508-519, 2010.
[http://dx.doi.org/10.1177/0734242X10361761] [PMID: 20375128]

[44]   A. Main, and T. Maghon, "Concepts and experiences for higher plant efficiency with modern advanced boiler and incineration technology", *18th Annual North American Waste-to-Energy Conference NAWTEC*
[http://dx.doi.org/10.1115/NAWTEC18-3541]

[45]   D. O. Reimann, "Results of Specific Data for Energy, R1 Plant Efficiency Factor and NCV of 314 European Waste-to-Energy (WtE) Plants," in "CEWEP Energy Efficiency Report (Status 2007-2010)", Bamberg, Germany 2012.

[46]   O. Gohlke, and J. Martin, "Drivers for innovation in waste-to-energy technology, *Waste Management and Research*", *Article,* vol. 25, no. 3, pp. 214-219, 2007.

[47]   T. Rand, J. Haukohl, and U. Marxen, *Municipal solid waste incineration: requirements for a successful project.* The World Bank: Washington, D.C, 2000, p. 118.

[48]   L. Lombardi, E. Carnevale, and A. Corti, "A review of technologies and performances of thermal treatment systems for energy recovery from waste", *Waste Manag.,* vol. 37, pp. 26-44, 2015. [http://dx.doi.org/10.1016/j.wasman.2014.11.010] [PMID: 25535103]

[49]   World Energy Outlook 2013. International Energy Agency, p. 690, 2013.

CHAPTER 8

Sustainable Carbon Dioxide Capture Technologies

Kirill G. Ryndin[1]**, Pankaj K. Chauhan**[2]**, Vinod Kumar**[3] **and Mikhail S. Vlaskin**[1,*]

[1] *Joint Institute for High Temperatures of the Russian Academy of Sciences, Moscow, Russia*

[2] *Faculty of Applied Sciences and Biotechnology, Shoolini University, Solan, Himachal Pradesh-173229, India*

[3] *Algal Research and Bioenergy Lab, Department of Food Science and Technology, Graphic Era (Deemed to Be University), Dehradun, Uttarakhand, India*

Abstract: This study presents a comparison of the primary methods for capturing CO_2, followed by an examination of CO_2 utilization techniques. Special attention is given to the bioutilization of CO_2, particularly the use of microalgae. The potential of microalgae for CO_2 capture and as a source of renewable energy is emphasized.

Keywords: Biofuel, Biomass, CO_2 capture, Greenhouse effect, Microalgae.

INTRODUCTION

Over the past two centuries, the world's population has grown from 1 billion to 8 billion, and the average life expectancy has nearly doubled. To provide energy for such a large population and support economic development, it is essential to have access to appropriate energy resources. For the past few decades, fossil fuels have been the primary source of energy. According to 2021 data, over 80% of all the energy consumed globally comes from fossil fuels, including coal, oil, and natural gas [1]. However, the extraction and use of these traditional energy sources are not environmentally sustainable, as they release gases into the atmosphere that contribute to the greenhouse effect. The greenhouse effect refers to the increase in temperature of the Earth's lower atmosphere compared to its thermal radiation balance, driven by a rise in the concentration of specific gases. Greenhouse gases include CO_2, CH_4, N_2O, NO, NO_2, O_3, water vapor, and certain chemical industry products, such as halogenated hydrocarbons and hydrofluorocarbons [1]. NO_2, O_3, water vapor, and certain chemical industry products, such as halogenated hydro-

* **Corresponding Author Mikhail S. Vlaskin:** Joint Institute for High Temperatures of the Russian Academy of Sciences, Moscow, Russia; E-mail: vlaskin@inbox.ru

Harish Chandra Joshi, Anand Chauhan, Mikhail Vlaskin & Maulin P. Shah (Eds.)

carbons and hydrofluorocarbons [1]. To account for the wide range of potential greenhouse gases, a simplified model is often used, measuring them in units of Gt CO_2, as CO_2 is the most prevalent. In 2020, CO_2 emissions totaled 35 gigatons (Gt), while overall greenhouse gas emissions were 60 gigatons (Gt) in CO_2 equivalents [2]. The increase in these concentrations in the atmosphere has led to a significant rise in global temperatures. Notably, the average temperature deviation during the peaks of 2016 and 2020 was +1.66 °C above the baseline established in 1862 [3].

Research on the greenhouse effect has led to the formation of various advisory groups, national political programs, and international meetings involving representatives from many countries. The to establish plans, criteria, and targets for reducing greenhouse gas emissions and transitioning toward a net-zero greenhouse gas policy. The first major intergovernmental agreement on this issue was the United Nations Framework Convention on Climate Change (UNFCCC), adopted in 1992. Since its inception, the agreement has been frequently updated as new data have become available, such as with the Kyoto Protocol in 1997, which serves as a supplementary agreement to the original UNFCCC. The most recent significant intergovernmental initiative on greenhouse gas reduction is the "Fit for 55" package, adopted by the European Union in 2022. This initiative aims to reduce greenhouse gas emissions by 55% compared to 1990 levels by 2030 [4]. However, current data indicate that these agreements, like others before them, have yet to result in significant reductions in greenhouse gas emissions. Notably, the only periods during which emissions decreased since 1992 were 2009, due to the global financial crisis, and 2020, due to the COVID-19 pandemic [3].

CO_2 Capture Methods

To address the challenges of reducing the greenhouse effect, several CO_2 capture technologies are being developed [2]. These technologies can be divided into 3 groups. Before considering them in more detail, it is worthwhile to examine their advantages and disadvantages, as presented in Table **1**. Due to the substantial investment required, pre-combustion and oxyfuel combustion technologies have undergone limited applied research and development. In contrast, post-combustion capture is widely used in the industry and is considered a relatively mature technology, offering high selectivity and efficient CO_2 capture [5]. The following sections will provide a brief overview of the various CO_2 capture methods, with a primary focus on those employed to capture CO_2 after the combustion of hydrocarbon fuels.

Methods of CO_2 Capture

Chemical Absorption

The chemical absorption method typically involves the use of solvents, such as weak alkaline solutions, to react with carbon dioxide, resulting in a compound that contains CO_2 in a bound form. The absorbent can be regenerated by desorption, which is achieved by altering the temperature or pressure of the reaction [6]. Commonly used absorbents include solutions of organic amines, ammonia, or sodium hydroxide. Among these, organic amines are considered the most widely used and effective for chemical absorption [6].

Table 1. Comparison of CO_2 capture methods [2, 5]

Capture Methods	Advantages	Disadvantages
Pre-combustion Capture	Small size of CO_2 capture equipment, and higher energy output.	Operational limitations and high investment costs.
Oxyfuel combustion	High CO_2 concentration in combustion products, and lower emissions of NO_X and SO_X.	High energy and investment costs.
Post-combustion capture	Low investment costs, rapid technology deployment, and flexible operations to reduce expenses. High technology deployment speed and lower investment requirements.	Large equipment size for CO_2 capture.

Adsorption by Solid-Phase Porous Materials

Adsorption with solid-phase porous materials relies on electrostatic or Van der Waals interactions between the adsorbent (solid-phase porous material) and the adsorbate (CO_2) [7]. This method is characterized by a high adsorption capacity and low cost of chemical processes [5, 8]. These advantages are achieved through the use of adsorbents with high capacity and selectivity for carbon dioxide, along with the ease of regenerating the adsorbent for repeated use [9, 10]. The most studied and widely used adsorbents include metal-organic frameworks (MOFs), Covalent Organic Frameworks (COFs), zeolites, mesoporous materials, and activated carbon [10, 11].

Membrane Separation

This method is based on gas separation due to the varying rates of gas permeation through membranes [12]. Organic membranes, primarily made of polymers, are widely used in the industry for gas separation (especially CO_2), but their selectivity is not optimal [13, 14]. In contrast, inorganic membranes offer improved selectivity, greater stability, and higher efficiency. However, their industrial application is limited by their high cost [14]. Overall, the membrane separation method is appealing due to its low technology and energy costs, simplicity of the process, and compact equipment size [15]. However, its gas separation efficiency remains relatively low, indicating that future research should focus on developing more advanced membrane materials [14].

Cryogenic Separation

Cryogenic separation is a process in which solid CO_2 is isolated from a gas mixture by desublimation at low temperatures (below -56.6 °C). This method is particularly attractive for producing highly pure liquid CO_2. While it yields high-purity CO_2, its effectiveness is primarily limited to gas mixtures with high CO_2 concentrations [16]. Additionally, one of the drawbacks of this method is its high energy consumption [17].

Hydrate Method

Hydrate formation is a process that occurs when temperature and pressure decrease, resulting in the formation of unstable compounds that contain water, known as hydrates. Gas separation in this process is based on the differing partial pressures of the gases involved [18, 19]. Under standard environmental conditions, it is reported that 160-180 m^3 of carbon dioxide can be captured within 1 m^3 of hydrate [20]. This method has several advantages, including simplicity, as well as the absence of pollution and corrosion [21, 22].

Microbiological Method

The microbiological method of CO_2 capture relies on the photosynthetic capabilities of microorganisms. During photosynthesis, carbon dioxide and water serve as the primary reactants, producing glucose and molecular oxygen as the end products. It is estimated that this method could potentially capture up to 7 billion tons of CO_2 per year [23], highlighting its promising potential. Additional advantages of this approach include its environmental friendliness and the absence of pollution.

Comparison of CO_2 Capture Technologies

Each technology has its advantages and disadvantages. A comparison and brief analysis of these technologies are summarized in Table **2**.

Table 2. Comparison of CO_2 capture technologies [10, 12, 14, 16, 24, 25]

CO_2 Capture Technology	Advantages	Disadvantages
Chemical Absorption	Mature technology, high capture efficiency and CO_2 selectivity.	Environmental damage from solvents, rapid corrosion, high energy demand, and expense.
Adsorption using solid-phase porous materials	Easy to use, low energy demand, and applicable for low CO_2 concentration.	Low capture speed and challenges with flue gases.
Membrane Separation	Low energy requirements, simplicity and low cost.	Fragility and short lifespan of membranes, low selectivity, and external conditions affecting membrane permeability.
Cryogenic Separation	High selectivity, suitable for high CO_2 concentration and pressure.	Operates at low temperatures.
Hydrate-Based Method	Eco-friendly, low energy demand and simple technology.	Requires high pressure and low temperatures.
Microbiological Method	Eco-friendly; simple process; low cost CO_2 captured supports biomass growth.	A wide variety of microorganisms requires further evaluation for method efficiency.

CO_2 Disposal Methods

The problem of reducing CO_2 emissions requires a solution not only for CO_2 capture but also for the disposal or beneficial use of CO_2 [26]. Therefore, below we briefly consider the main methods of CO_2 disposal and use that are currently in practice and have potential for increased application of CO_2 [27].

Chemical Utilization

The chemical method is a popular approach for CO_2 disposal [28]. Carbon dioxide is a commonly used compound in the chemical industry, where it is employed in the synthesis of various compounds. For example, it is used in the synthesis of alcohols, esters, organic acids, amines, and hydrocarbons [13].

Bioutilization

Bioutilization of CO_2 involves utilizing the process of photosynthesis by organisms that use CO_2 as a source of food. Such organisms are plants and observations that monitor CO_2 emissions. In the mode of photosynthesis, these organisms also release water to produce glucose and oxygen [26, 29].

Using Microalgae to Utilize Carbon Dioxide Emissions

Algae, particularly microalgae, are the primary producers of oxygen on Earth and play a crucial role in capturing and utilizing carbon dioxide [30]. Oxygen production in microalgae occurs through the process of photosynthesis [23]. A distinguishing feature of microalgae is their high growth rate, which is significantly higher than that of many plants by orders of magnitude [31]. To produce 1 ton of dry microalgae biomass, approximately 1.8 tons of carbon dioxide are required [30]. This CO_2 utilization method is environmentally friendly.

Additionally, microalgae are employed for wastewater treatment, biofixation, and the mitigation of greenhouse gases. The rate at which microalgae capture carbon dioxide is higher than that of terrestrial plants per unit area [31]. Potential applications of microalgae as renewable biomass are shown in Fig. (1).

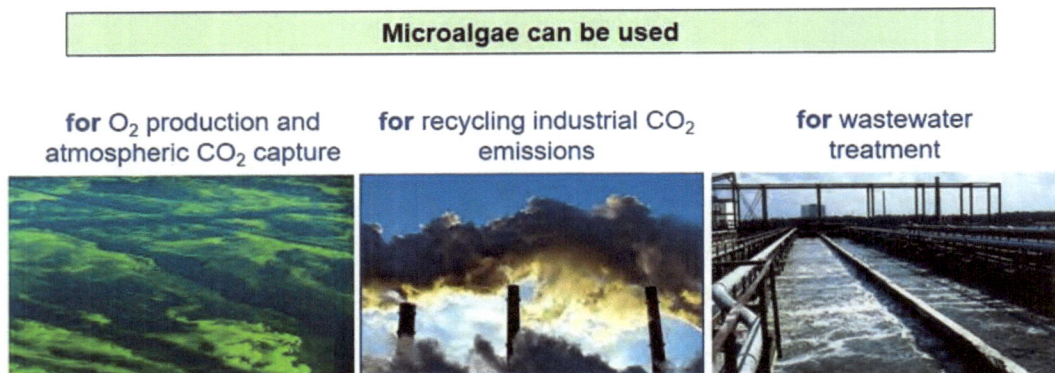

Microalgae can be used

for O_2 production and atmospheric CO_2 capture

for recycling industrial CO_2 emissions

for wastewater treatment

Fig. (1). Potential applications of microalgae as renewable biomass.

Production of Valuable Products from Microalgae Biomass

Microalgae are a promising direction for CO_2 capture, CO_2 processing, and wastewater processing. To assess the applicability of microalgae, it is worth assessing the microalgae market.

Assessments of the global microalgae market vary. For 2020, one study reports a market size of $ 3.4 billion USD [32], while another estimates it at $ 1.25 billion USD [33], and a third estimates the market for 2014 at € 2.4 billion [34]. However, these discrepancies should not be confusing. All studies agree on one key point: the microalgae market is expected to continue growing year over year [32 - 35]. The differences in market size estimates arise from the challenge of accurately assessing small, fragmented markets, particularly those with numerous small production facilities—such as the more than 400 production sites in the EU alone [36]—and the absence of dominant market players. Therefore, the market estimates in [32 - 34] should be considered approximations. Production volume can also be estimated. In 2021, the annual production of microalgae was approximately 25,000 tons [34]. This relatively small market is supported by government interest in developing the sector. As noted in [35], several companies are involved in microalgae production across various countries and regions. The commercialization of microalgae is further boosted by intergovernmental strategies promoting bioenergy development, particularly those focused on decarbonization policies [34]. These efforts align with the 17 Sustainable Development Goals (SDGs), which are set to be achieved by 2030 [37]. Microalgae are believed to contribute to several of these goals, including the eradication of hunger, improvements in health, access to clean water and sanitation, affordable and clean energy, and mitigation of climate change [37]. There is already a wide range of commercially successful microalgae species, including *Spirulina, Chlorella, Haematococcus, Dunaliella, Botryococcus, Phaeodactylum, Porphyridium, Chaetoceros, Crypthecodinium, Isochrysis, Nannochloris, Nitzschia, Schizochytrium, Tetraselmis*, and *Skeletonema* [38].

According to [39], the cost of producing 1 kg of microalgae varies between $ 20 and $ 200, with prices ranging from $ 0.4 to $ 100 per kg, depending on the application. For example, microalgae used as food or food supplements for humans can cost up to $ 100 per kg, while as raw materials for biofuel production, the price is around $0.40 per kg. However, the high production and application costs hinder the broader adoption of microalgae. For context, the EU accounts for around 70% of global microalgae production, while regions such as the Middle East, Africa, and Latin America collectively account for only about 6% [32]. The potential applications of microalgae are vast. They are used in CO_2 and flue gas biofixation, wastewater treatment, biofertilizer production, as biochemicals, in the food industry, cosmetics, medicine, and as raw materials for biofuel production [32, 34, 37, 38, 40].

One promising area is the extraction of valuable compounds from microalgae (Fig. **2**), with the remaining biomass being used for biofuel production [32]. Screening microalgae strains to increase the yield of these valuable compounds is a relatively straightforward process. Furthermore, a study [40] demonstrated successful screening, resulting in a high yield of valuable compounds, bio-oil, and the strain's ability to adapt to cultivation in a flue gas environment. Other strategies to reduce the cost of microalgae production have also been explored [32, 41]. These strategies can be divided into two categories: those that reduce the cost of one or more stages of microalgae production, and those aimed at lowering the cost across all stages of production [41]. It is important to note that a significant increase in microalgae production is hindered by the challenge of maintaining homogeneous strains during long-term cultivation. Cultivating microalgae in open systems is less expensive than in closed systems, but the cultivation conditions differ. In this case, the microalgae must undergo an adaptation stage again, which may result in changes to their characteristics.

Additionally, the cost of the nutrient medium plays a significant role in production costs, and using flue gases in large-scale microalgae production is complicated by the logistical challenges of delivering and supplying the gases. Furthermore, when scaling up production, specific regulatory requirements may change. For instance, the method used for harvesting microalgae may need to be adjusted.

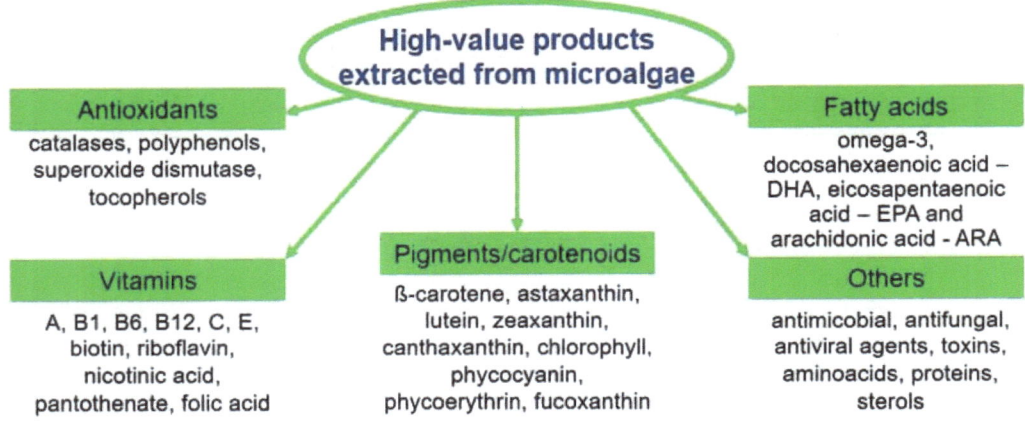

Fig. (2). The most common valuable products obtained from algae biomass.

Methods for obtaining Biofuels from Microalgae

The accumulated mass of microalgae can be used in various fields. One of these fields is the use of microalgae as a fuel source. Methods for obtaining biofuels from microalgae can be classified into two main groups: biochemical and thermochemical methods. These methods are illustrated in Fig. (3). Biochemical methods include transesterification, anaerobic digestion, and fermentation, which produce biodiesel, biogas, and bioethanol, respectively. Thermochemical processes are divided into pyrolysis, liquefaction, and gasification. Pyrolysis and hydrothermal liquefaction produce bio-oil, while hydrothermal carbonization generates biochar, and gasification produces synthesis gas.

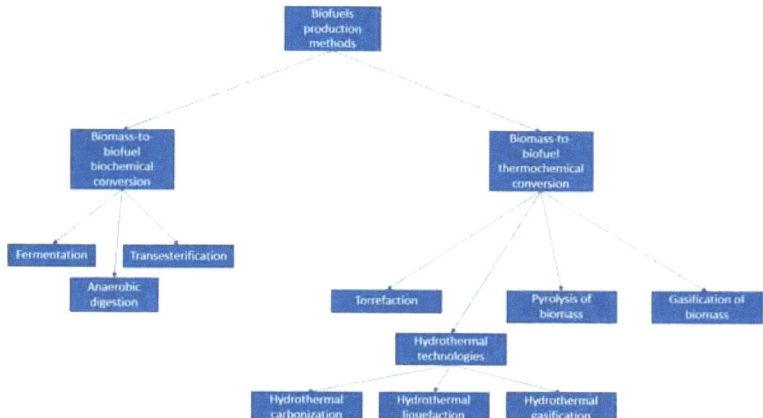

Fig. (3). Methods for producing biofuels from microalgae.

Third Generation Biofuels

Due to their high concentration of lipids, microalgae are promising raw materials for biofuel production [30]. Photosynthesis is the primary process for autotrophic microalgae. Photosynthesis is the primary process for autotrophic microalgae. CO_2, water are the starting materials in the photosynthesis reaction. Solar energy is the catalyst of the reaction. The products are glucose, oxygen, and water [42]. Currently, microalgae are cultivated for various purposes, including wastewater treatment, cosmetology, medicine, and the bioconservation of solar energy [42].

The primary barrier to the commercialization of third-generation biofuels is the high cost of biofuel production [43]. Microalgae contain a significant amount of lipids, have an oil yield an order of magnitude higher, are more productive, and have a higher growth rate compared to other traditional plant crops [42]. Additionally, microalgae are easier to cultivate, require less land, and do not compete with agricultural land for cultivation. However, the disadvantages include the relatively high production cost.

Table **3** presents the oil productivity values for various plants, including microalgae with different oil (lipid) contents [31]. From the comparison of the data presented in Table **3**, it can be concluded that for the production of the same amount of biofuel, microalgae require an order of magnitude smaller area of land for plant cultivation compared to other crops.

Currently, cultivating microalgae for subsequent biomass conversion into biofuels is considered a promising and relevant area of research and development. The advantages of microalgae over other plant crops include:

1. Cheaper transportation and collection of raw materials compared to traditional agricultural crops [44].
2. Microalgae require less water compared to other plant crops, reducing potential pressure on the food market. This eliminates competition with human food products. Growing microalgae requires much smaller land areas [30].
3. Microalgae can be grown in environments unsuitable for traditional agriculture (such as saltwater, reservoirs, wastewater treatment plant effluents, and non-arable land). Microalgae produce higher yields per unit area and are more environmentally friendly compared to other plant crops [45].
4. Microalgae contain 5-50% oil by dry weight and can achieve much higher productivity values [46].
5. Due to their small size, microalgae are more easily processed chemically. Additionally, microalgae are excellent systems for capturing solar energy, which they use to produce organic compounds [45].
6. Microalgae can help reduce atmospheric greenhouse gas levels, addressing one of the primary environmental concerns. For every ton of dry microalgae biomass, approximately 1.8 tons of carbon dioxide are consumed [30].

Table 3. Oil productivity values for various plants, including microalgae [31]

Plant	Oil content, mass %	Oil yield, l/ha/year	Required area (m^2) per 1 kg of biodiesel per year
Corn (*Zea mays L.*)	44	172	66
Hemp (*Cannabis sativa L.*)	33	363	31
Soybeans (*Glycine max L.*)	18	636	18
Jatropha (*Jatropha curcas L.*)	28	741	15
Ryžik (*Camelina sativa L.*)	42	915	12
False flax (*Brassica napus L.*)	41	974	12
Sunflower (*Helianthus annuus L.*)	40	1070	11
Castor oil plant (*Ricinus communis*)	48	1307	9
Palm (*Elaeis guineensis*)	36	5366	2
Microalgae	30	58700	0.2
Microalgae	50	97800	0.1
Microalgae	70	136900	0.1

CONCLUDING REMARKS

CO_2 capture methods can be conventionally divided into several types: methods in which chemical compounds play a key role, methods based on changes in thermodynamic parameters, and methods based on the use of photosynthesis.

Current environmental agendas limit the significant scaling of the first methods. The second method is constrained by energy costs and the unresolved issue of how to use large amounts of CO_2 effectively. These limitations can be addressed by methods that utilize photosynthesis. These methods are the most widespread, environmentally friendly, and inexpensive to operate, and are capable of relatively quickly utilizing large amounts of CO_2. Therefore, the scaling of this method could proceed more smoothly.

Microalgae are the primary representatives of this method. With their high growth rate and relatively fast adaptability, they can be widely used in commercial enterprises for the direct utilization of flue gases.

The further application of microalgae can only be estimated, as the market capacity for microalgae was relatively small (as of 2020, it was around $ 3.4 billion). A significant increase in the volume of microalgae on the market could lead to a reduction in the cost price and sale price of the main valuable products derived from microalgae. The residual biomass can be utilized for biofuel production, particularly bio-oil, through the HTC method, as this approach appears to be the most attractive considering its potential for significant production scaling. This would greatly contribute to resolving the important, unresolved issue of generating a large volume of practical data for a more accurate description of the economics of converting microalgae into valuable energy products. It would also simplify the achievement of the goals outlined in the "Fit for 55" program.

REFERENCES

[1]   H. van Asselt, "Governing fossil fuel production in the age of climate disruption: Towards an international law of 'leaving it in the ground'", *Earth System Governance,* vol. 9, p. 100118, 2021.
[http://dx.doi.org/10.1016/j.esg.2021.100118]

[2]   G. Rothenberg, "A realistic look at CO_2 emissions, climate change and the role of sustainable chemistry", *Sustainable Chemistry for Climate Action,* vol. 2, p. 100012, 2023.
[http://dx.doi.org/10.1016/j.scca.2023.100012]

[3]   P.R.M.R. Hannah Ritchie, https://ourworldindata.org/co2-and-other-greenhouse-gas-emissions

[4]   E. Commission, https://ec.europa.eu/commission/presscorner/detail/en/IP_21_3541

[5]   N.A. Rashidi, and S. Yusup, "An overview of activated carbons utilization for the post-combustion carbon dioxide capture", *J. CO_2 Util.,* vol. 13, pp. 1-16, 2016.
[http://dx.doi.org/10.1016/j.jcou.2015.11.002]

[6]   D. Su, Q. Zhang, and S. Xie, "Target segmentation in complex environment using fractal features",
[http://dx.doi.org/10.1109/COGINF.2006.365680]

[7]   A.I. Sarker, A. Aroonwilas, and A. Veawab, "Equilibrium and kinetic behaviour of CO_2 adsorption onto zeolites, carbon molecular sieve and activated carbons", *Energy Procedia,* vol. 114, pp. 2450-2459, 2017.
[http://dx.doi.org/10.1016/j.egypro.2017.03.1394]

[8]   J. Serafin, M. Ouzzine, O.F. Cruz Jr, J. Sreńscek-Nazzal, I. Campello Gómez, F.Z. Azar, C.A. Rey Mafull, D. Hotza, and C.R. Rambo, "Conversion of fruit waste-derived biomass to highly microporous activated carbon for enhanced CO_2 capture", *Waste Manag.*, vol. 136, pp. 273-282, 2021.
        [http://dx.doi.org/10.1016/j.wasman.2021.10.025] [PMID: 34737129]

[9]   T.L.P. Dantas, F.M.T. Luna, I.J. Silva Jr, D.C.S. de Azevedo, C.A. Grande, A.E. Rodrigues, and R.F.P.M. Moreira, "Carbon dioxide–nitrogen separation through adsorption on activated carbon in a fixed bed", *Chem. Eng. J.*, vol. 169, no. 1-3, pp. 11-19, 2011.
        [http://dx.doi.org/10.1016/j.cej.2010.08.026]

[10]   A. Pereira, A. F. P. Ferreira, A. E. Rodrigues, A. M. Ribeiro, and M. J. Regufe, "Additive manufacturing for adsorption-related applications—A review", *manufacturing for adsorption-related ,* 2022. vol. 4, no. 1, p. e10108
        [http://dx.doi.org/10.1002/amp2.10108]

[11]   W.M. Verdegaal, K. Wang, J.P. Sculley, M. Wriedt, and H.C. Zhou, "Evaluation of metal-organic frameworks and porous polymer networks for CO_2 -Capture Applications", *ChemSusChem,* vol. 9, no. 6, pp. 636-643, 2016.
        [http://dx.doi.org/10.1002/cssc.201501464] [PMID: 26840979]

[12]   J.L. Li, and B.H. Chen, "Review of CO_2 absorption using chemical solvents in hollow fiber membrane contactors", *Separ. Purif. Tech.,* vol. 41, no. 2, pp. 109-122, 2005.
        [http://dx.doi.org/10.1016/j.seppur.2004.09.008]

[13]   J. Godin, W. Liu, S. Ren, and C.C. Xu, "Advances in recovery and utilization of carbon dioxide: A brief review", *J. Environ. Chem. Eng.,* vol. 9, no. 4, p. 105644, 2021.
        [http://dx.doi.org/10.1016/j.jece.2021.105644]

[14]   R.L. Siegelman, E.J. Kim, and J.R. Long, "Porous materials for carbon dioxide separations", *Nat. Mater.,* vol. 20, no. 8, pp. 1060-1072, 2021.
        [http://dx.doi.org/10.1038/s41563-021-01054-8] [PMID: 34321657]

[15]   M. Kárászová, B. Zach, Z. Petrusová, V. Červenka, M. Bobák, M. Šyc, and P. Izák, "Post-combustion carbon capture by membrane separation, review", *Separ. Purif. Tech.,* vol. 238, p. 116448, 2020.
        [http://dx.doi.org/10.1016/j.seppur.2019.116448]

[16]   A.A. Olajire, "CO_2 capture and separation technologies for end-of-pipe applications – A review", *Energy,* vol. 35, no. 6, pp. 2610-2628, 2010.
        [http://dx.doi.org/10.1016/j.energy.2010.02.030]

[17]   J. Rawat, M. Nanda, S. Kumar, N. Sharma, R. Sharma, H.C. Joshi, M.S. Vlaskin, A. Hussain, and V. Kumar, "Integrating wastewater treatment to bio-stimulant & biochar generation for plant growth promotion using microalgae", *Process Biochem.,* vol. 145, pp. 187-194, 2024.
        [http://dx.doi.org/10.1016/j.procbio.2024.06.031]

[18]   J. Zheng, K. Bhatnagar, M. Khurana, P. Zhang, B.Y. Zhang, and P. Linga, "Semiclathrate based CO_2 capture from fuel gas mixture at ambient temperature: Effect of concentrations of tetra--butylammonium fluoride (TBAF) and kinetic additives", *Appl. Energy,* vol. 217, pp. 377-389, 2018.
        [http://dx.doi.org/10.1016/j.apenergy.2018.02.133]

[19]   J. Bai, X. Zhen, K. Yan, P. Li, S. Fang, and C. Chang, "The effect of additive molecular diameters on the hydrate-based CO_2 capture from simulated biogas", *Fuel,* vol. 278, p. 118370, 2020.
        [http://dx.doi.org/10.1016/j.fuel.2020.118370]

[20]   X. Wang, F. Zhang, and W. Lipiński, "Research progress and challenges in hydrate-based carbon dioxide capture applications", *Appl. Energy,* vol. 269, p. 114928, 2020.
        [http://dx.doi.org/10.1016/j.apenergy.2020.114928]

[21]   Y. Wang, D.L. Zhong, Z. Li, and J.B. Li, "Application of tetra-n-butyl ammonium bromide semi-clathrate hydrate for CO_2 capture from unconventional natural gases", *Energy,* vol. 197, p. 117209, 2020.

[http://dx.doi.org/10.1016/j.energy.2020.117209]

[22] J. Cai, Y. Zhang, C.G. Xu, Z.M. Xia, Z.Y. Chen, and X.S. Li, "Raman spectroscopic studies on carbon dioxide separation from fuel gas *via* clathrate hydrate in the presence of tetrahydrofuran", *Appl. Energy,* vol. 214, pp. 92-102, 2018.
[http://dx.doi.org/10.1016/j.apenergy.2018.01.055]

[23] A. Anand, S. Raghuvanshi, and S. Gupta, *Trends in Carbon Dioxide (CO_2) Fixation by Microbial Cultivations,* 2020.
[http://dx.doi.org/10.1007/s40518-020-00149-1]

[24] C. Song, M. Xie, Y. Qiu, Q. Liu, L. Sun, K. Wang, and Y. Kansha, "Integration of CO_2 absorption with biological transformation *via* using rich ammonia solution as a nutrient source for microalgae cultivation", *Energy,* vol. 179, pp. 618-627, 2019.
[http://dx.doi.org/10.1016/j.energy.2019.05.039]

[25] L. Wang, D. Chen, Y. Hu, Y. Ma, and J. Wang, "Towards enabling Cyberinfrastructure as a Service in Clouds", *Comput. Electr. Eng.,* vol. 39, no. 1, pp. 3-14, 2013.
[http://dx.doi.org/10.1016/j.compeleceng.2012.05.001]

[26] Z. Zhang, S-Y. Pan, H. Li, J. Cai, A.G. Olabi, E.J. Anthony, and V. Manovic, "Recent advances in carbon dioxide utilization", *Renew. Sustain. Energy Rev.,* vol. 125, p. 109799, 2020.
[http://dx.doi.org/10.1016/j.rser.2020.109799]

[27] W. Liu, L. Teng, S. Rohani, Z. Qin, B. Zhao, C.C. Xu, S. Ren, Q. Liu, and B. Liang, "CO_2 mineral carbonation using industrial solid wastes: A review of recent developments", *Chem. Eng. J.,* vol. 416, p. 129093, 2021.
[http://dx.doi.org/10.1016/j.cej.2021.129093]

[28] A Chauhan, and HC Joshi, "Lignocellulosic biomass for the conversion of bioethanol: production and optimization", *Trends Math,* pp. 187-214, 2024.
[http://dx.doi.org/10.1007/978-981-99-7250-0_1]

[29] X. Xu, G. J. O. Martin, and S. E. Kentish, "Enhanced CO_2 bio-utilization with a liquid–liquid membrane contactor in a bench-scale microalgae raceway pond", *Journal of CO_2 Utilization,* vol. 34, pp. 207-214, 2019.
[http://dx.doi.org/10.1016/j.jcou.2019.06.008]

[30] L. Brennan, and P. Owende, "Biofuels from microalgae—A review of technologies for production, processing, and extractions of biofuels and co-products", *Renew. Sustain. Energy Rev.,* vol. 14, no. 2, pp. 557-577, 2010.
[http://dx.doi.org/10.1016/j.rser.2009.10.009]

[31] "1A. Chauhan, Harish Chandra Joshi, Recent developments and applications in bioconversion and biorefineries", *Trends in Mathematics,* no. Part F3197, pp. 247-307, 2024.

[32] P. Loke Show, "Global market and economic analysis of microalgae technology: Status and perspectives", *Bioresour. Technol.,* vol. 357, p. 127329, 2022.
[http://dx.doi.org/10.1016/j.biortech.2022.127329] [PMID: 35589045]

[33] M. Onay, "Microalgae-based systems chapter 2 scope of the microalgae market: A demand and supply perspective", 2023.

[34] B. Vázquez-Romero, J.A. Perales, H. Pereira, M. Barbosa, and J. Ruiz, "Techno-economic assessment of microalgae production, harvesting and drying for food, feed, cosmetics, and agriculture", *Sci. Total Environ.,* vol. 837, p. 155742, 2022.
[http://dx.doi.org/10.1016/j.scitotenv.2022.155742] [PMID: 35526636]

[35] H. Vieira de Mendonça, P. Assemany, M. Abreu, E. Couto, A.M. Maciel, R.L. Duarte, M.G. Barbosa dos Santos, and A. Reis, "Microalgae in a global world: New solutions for old problems?", *Renew. Energy,* vol. 165, pp. 842-862, 2021.
[http://dx.doi.org/10.1016/j.renene.2020.11.014]

[36] D.B. Nguyen, T.T.Y. Doan, T.C.M. Phi, T.A. Ngo, L.D.H. Vu, and D.K. Dang, "Arthrospira production in Vietnam: Current status and prospects", *Bioresour. Technol. Rep.*, vol. 15, p. 100803, 2021.
[http://dx.doi.org/10.1016/j.biteb.2021.100803]

[37] A.G. Olabi, N. Shehata, E.T. Sayed, C. Rodriguez, R.C. Anyanwu, C. Russell, and M.A. Abdelkareem, "Role of microalgae in achieving sustainable development goals and circular economy", *Sci. Total Environ.*, vol. 854, p. 158689, 2023.
[http://dx.doi.org/10.1016/j.scitotenv.2022.158689] [PMID: 36108848]

[38] R. Sathasivam, R. Radhakrishnan, A. Hashem, and E.F. Abd Allah, "Microalgae metabolites: A rich source for food and medicine", *Saudi J. Biol. Sci.*, vol. 26, no. 4, pp. 709-722, 2019.
[http://dx.doi.org/10.1016/j.sjbs.2017.11.003] [PMID: 31048995]

[39] Y. Wang, S.H. Ho, C.L. Cheng, W.Q. Guo, D. Nagarajan, N.Q. Ren, D.J. Lee, and J.S. Chang, "Perspectives on the feasibility of using microalgae for industrial wastewater treatment", *Bioresour. Technol.*, vol. 222, pp. 485-497, 2016.
[http://dx.doi.org/10.1016/j.biortech.2016.09.106] [PMID: 27765375]

[40] J. Li, Y. Liu, J.J. Cheng, M. Mos, and M. Daroch, "Biological potential of microalgae in China for biorefinery-based production of biofuels and high value compounds", *N. Biotechnol.*, vol. 32, no. 6, pp. 588-596, 2015.
[http://dx.doi.org/10.1016/j.nbt.2015.02.001] [PMID: 25686716]

[41] G. Venkata Subhash, M. Rajvanshi, G. Raja Krishna Kumar, U. Shankar Sagaram, V. Prasad, S. Govindachary, and S. Dasgupta, "Challenges in microalgal biofuel production: A perspective on techno economic feasibility under biorefinery stratagem", *Bioresour. Technol.*, vol. 343, p. 126155, 2022.
[http://dx.doi.org/10.1016/j.biortech.2021.126155] [PMID: 34673195]

[42] N. Abdel-Raouf, A. A. Al-Homaidan, and I. B. M. Ibraheem, "Microalgae and wastewater treatment", *Saudi J Biol Sci*, vol. 19, no. 3, pp. 257-275, Jul 2012.
[http://dx.doi.org/10.1016/j.sjbs.2012.04.005]

[43] M. Balat, and H. Balat, "Progress in biodiesel processing", *Appl. Energy*, vol. 87, no. 6, pp. 1815-1835, 2010.
[http://dx.doi.org/10.1016/j.apenergy.2010.01.012]

[44] T.M. Mata, A.A. Martins, and N.S. Caetano, "Microalgae for biodiesel production and other applications: A review", *Renew. Sustain. Energy Rev.*, vol. 14, no. 1, pp. 217-232, 2010.
[http://dx.doi.org/10.1016/j.rser.2009.07.020]

[45] A. Chauhan, and H. Chandra Joshi, "Energetic efficiency of a biofuels production system mathematical modeling", *Trends in Mathematics*, no. Part F3197, pp. 337-357, 2024.
[http://dx.doi.org/10.1007/978-981-99-7250-0_8]

[46] M.K. Danquah, B. Gladman, N. Moheimani, and G.M. Forde, "Microalgal growth characteristics and subsequent influence on dewatering efficiency", *Chem. Eng. J.*, vol. 151, no. 1-3, pp. 73-78, 2009.
[http://dx.doi.org/10.1016/j.cej.2009.01.047]

CHAPTER 9

Ethanol 100: A New Approach for the Transportation Industry in India

Deepak Kumar[1], Neeraj Kumar[2,3] and Sujata Rathi[4,*]

[1] *Department of Physics, Graphic Era (Deemed to be) University, Dehradun, India*

[2] *Nanoscience Laboratory, Department of Physics, University of Trento, Via Sommarive 14, Povo (TN) 38123, Italy*

[3] *Department of Humanities and Applied Science, RIT Roorkee, Roorkee-247668, India*

[4] *Department of Botany, Multanimal Modi College, Modinagar-201204, India*

Abstract: The chapter "Ethanol 100: A New Approach for the Transportation Industry in India" delves into the potential of adopting 100% ethanol as the primary fuel source for India's transportation sector. It outlines the pressing issues of air pollution, fossil fuel dependency, and energy security that necessitate a shift toward alternative energy sources. The chapter highlights ethanol's advantages, including its role in reducing carbon emissions, enhancing energy independence, and supporting rural agricultural economies through increased demand for ethanol production. The discussion covers the technical, economic, and logistical aspects of implementing Ethanol 100 in India. It reviews the current infrastructure for ethanol production and distribution, vehicle compatibility, and necessary modifications to support a transition to a full ethanol system. Comparative case studies from other countries that have successfully integrated ethanol into their fuel mix offer valuable lessons and strategies. This chapter outlines the policy initiatives undertaken, particularly in India, to develop and promote various alternative fuels for road transportation, highlighting the progress made in comparison to global advancements. Innovations in ethanol production, distribution, and flex-fuel vehicle technologies make the adoption of E100 viable. With adequate investment in research, the required technologies can be developed to ensure the efficient and sustainable use of E100.

Keywords: Energy independence, Ethanol 100, Fossil fuel, Global advancements.

ENVIRONMENTAL AND ECONOMIC JUSTIFICATION FOR E100

The transportation subsector in the Indian economy is very crucial as it facilitates movement and trade throughout a large area of the country. But the sector also

* **Corresponding author Sujata Rathi**: Department of Botany, Multanimal Modi College, Modinagar-201204, India, E-mail: srathi84@gmail.com

Harish Chandra Joshi, Anand Chauhan, Mikhail Vlaskin & Maulin P. Shah (Eds.)

comes with challenges, primarily environmental sustainability, dependency on fossil fuels, or more broadly, energy security. These concerns make it clear that safer and better energy options are long overdue.

Overview of India's Transportation Sector Challenges

A large part of India's population is exposed to severe air pollution due to the Indian transport sector. This is mostly in urban areas that are relatively crowded. Pollution from vehicles leads to the degradation of the air with contaminants such as PM2.5 and PM10, NO_x, and VOCs. In major cities like Delhi alone, transport emissions account for as high as 30% of PM2.5 levels as per the findings of a study by Guttikunda and Goel (2019), which exerts a very negative impact on human health and the environment [1, 2]. The World Health Organization (WHO) has identified many Indian cities as the most polluted in the world, directly linking vehicle emissions to respiratory diseases, heart diseases, and premature death [3].

India's transportation energy dependence on fossil fuels, particularly gasoline and diesel, is a major problem for the country's carbon emissions. This dependency also poses economic risks due to the high availability of petroleum products, which account for about 80% of India's fuel needs [4]. The fluctuations in global oil prices further increase economic vulnerability, as seen in recent years. These fluctuations further create trade imbalances and affect currency stability. Moving towards alternative fuels can help reduce India's carbon footprint while reducing the risk of fluctuations in the global oil market [5].

Emerging demands for transportation services make energy security of utmost importance. Fossil fuels are finite resources, and their destruction poses a threat to energy security, along with the disruption of supply. With India's commitment to the Paris Agreement and focus on renewable energy, there is an urgent need to diversify energy sources in the transportation sector [6]. Switching to renewable and clean fuels, such as ethanol, biodiesel, and hydrogen, can increase energy security, reduce emissions, and support sustainable development goals.

Ethanol as an Alternative Fuel: Benefits and Global Context

Renewable biofuels, mostly derived from biomass sources, such as corn, sugar, and other crops, offer a promising alternative to fossil fuels. Their use as a transportation fuel is widespread in countries such as Brazil and the United States, where they have been observed to reduce greenhouse gas emissions and improve air quality. Ethanol emits less pollution than gasoline and diesel, particularly in terms of carbon monoxide, particulates, and greenhouse gases. Studies show that ethanol can reduce lifetime greenhouse gas emissions by 20–30% compared to natural gas [7]. Its use can also reduce smog production due to reduced nitrogen oxide and hydrocarbon emissions, which is especially beneficial in urban areas with high air pollution [8].

As a domestically produced fuel, ethanol can reduce India's dependence on imported fossil fuels, thereby contributing to energy security and economic stability. Ethanol production can also support the rural economy by creating employment in agriculture and biofuel processing. The Indian government has recognized the potential of ethanol and set an ambitious target to increase the blending of ethanol with gasoline to achieve a 20% blending target by 2025 [9]. The change is expected to save billions of dollars on oil imports and support energy independence. The enhanced production of ethanol can bring significant change in the rural economy with the increasing demand for agricultural products, such as sugarcane and maize. This can further create new job opportunities in rural areas, decrease the release of agriculture waste, and provide a new revenue stream for people residing in rural areas [10]. Importing a significant portion of oil makes the Indian economy vulnerable to the fluctuating global oil prices. The use of ethanol in the transportation sector is able to reduce dependency on imported oil. Scaling up ethanol production can act as double edge sword by providing both economic security and environmental sustainability. It can prevent India from an economic meltdown that could be threatened by the volatility of global oil markets.

Countries like Brazil have long supported ethanol as a primary fuel. Brazil's Proálcool program, launched in the 1970s, enabled the country to replace most of its fuel consumption with ethanol, reducing domestic emissions and creating a viable biofuel market [10]. Similarly, the United States has introduced ethanol into transportation fuels, mostly from corn-based ethanol, and this currently accounts for approximately 10% of the U.S. gasoline consumption [11]. These international events demonstrate the feasibility and quality of ethanol as a viable transportation fuel. By analyzing current policies, implementation challenges, and future prospects, this chapter aims to understand how ethanol can play a significant role in the transition to a cleaner and more sustainable transportation industry in India.

ETHANOL PRODUCTION AND FEEDSTOCKS

Ethanol can be produced through two primary processes: fermentation and chemical synthesis.

Fermentation

The most common method of producing ethanol is fermentation, in which sugar is converted into ethanol and carbon dioxide by yeast or bacteria. The fermentation process usually involves the following steps:

Sugar-rich materials (such as sugar or corn) are processed to extract sugars that are considered fermentable. The extracted sugar is fermented under anaerobic conditions to produce ethanol and carbon dioxide. The final product usually has an ethanol content of around 95% and can be further dehydrated to produce anhydrous ethanol (greater than 99% purity) for fuel use [8].

Chemical Synthesis

Ethanol can also be produced by chemical processes such as the hydration of ethylene extracted from petroleum. However, this method is less common in petroleum production compared to fermentation because it relies on fossil fuels and does not provide sustainable benefits [11].

Types of Feedstocks

India uses a variety of feedstocks to produce ethanol, focusing mostly on feedstocks that can be grown at home. The most common feedstocks include:

Sugarcane

Sugarcane is the main feedstock for ethanol production in India, accounting for a large portion of the country's ethanol production. The process involves extracting sucrose from orange juice and then fermenting it to produce ethanol. India has a well-developed sugarcane industry, and using this crop to produce ethanol can help reduce excess sugar while providing additional income to farmers [9].

Maize

Maize is another important food for ethanol production, especially in states like Karnataka and Maharashtra. Maize is processed to remove starch, which is then converted into fermentable sugars through enzymatic hydrolysis and then fermented [5]. Although the scale of maize ethanol production in India is smaller than sugar, it has the potential to grow as a food crop due to increased production and efficient processes that can be implemented.

Other Feedstocks

In addition to sugarcane and maize, other feedstocks such as potatoes and millets, as well as agricultural residues such as rice husk, rice straw, and bagasse, are also being utilized for ethanol production. The use of these feedstocks can help produce more ethanol, reduce waste, and provide economic benefits to farmers

[12]. Research and development into advanced biofuels, including cellulosic ethanol derived from lignocellulosic biomass, is ongoing and has the potential to diversify India's ethanol production landscape [13]. By utilizing more food resources, India can increase its ethanol production capacity, support rural development, and contribute to energy security and sustainability goals.

GLOBAL CASE STUDIES

India can adopt E100 as a transportation fuel by taking several countries into consideration that have successfully adopted E100. Several countries have successfully adopted E100.

Brazil

Brazil is a leading country of ethanol use in the transportation sector, with a robust network of flex-fuel vehicles that can run on E100. Nearly 80% of FFVs in Brazil are designed to run on both gasoline and ethanol. To meet the current need for ethanol in FFV production, large investments are made in the sugarcane industry and infrastructure to encourage widespread adoption. The Brazilian government offers encouragement for ethanol production and consumption to reduce fossil fuel consumption and emission of greenhouse gases [10]. Brazilian vehicles have verified trustworthiness and performance using E100, with widespread support from automakers for vehicle amendments and maintenance.

Sweden

Sweden has also incorporated ethanol into transportation fuels to fulfill its commitment to reduce greenhouse gas emissions and enhance the use of renewable energy resources. The automakers have been encouraged by different programs launched by the government to design vehicles compatible with E100 so that a wide range of models to be made available to consumers. All these initiatives have witnessed a significant reduction of up to 90% in greenhouse gas emissions as compared to the production of GHGs in fossil fuel use (International Energy Agency, 2021). The Swedish government is also focusing on sustainable agricultural practices and waste-to-energy solutions to boost its biofuel industry.

United States

In the United States, several states, especially those leading in maize production, have adopted programs vehicular for the use of E100. Vehicles have been modified to run on E100, particularly fleet vehicles used by government agencies. Studies from the U.S. have shown that these vehicles are performing well after modification in terms of fuel use and control to comply with E100, as reported by the Department of Energy. The U.S. is also exploring new incentives and techniques to boost the compatibility of vehicle fuel systems with ethanol-based fuels [7].

ALTERNATIVE FUELS COMPARISON

The potential of various biofuels, such as ethanol, methanol, biodiesel, CNG, electricity, and hydrogen, has been recognized in the future economic growth of India. The vision of the "Auto Fuel Policy, 2003" of the Government of India (GoI) had suggested among others: increment in the use of CNG/LPG in the cities with high numbers of vehicles; to increase the production of electricity, hydrogen and fuel cell-based vehicles; and to develop technologies for the production of locally available resources-based biofuels and compatible vehicles [14]. Recent developments focus on the utilization of renewable energy resources in India's growth. The utilization of transport assets is optimized by continuous developments in shared mobility solutions. The following sections cover the various policy initiatives adopted specifically in India for the production and promotion of different alternative fuels for road transportation, as well as prioritizing global achievements.

Ethanol

In India, the use of ethanol fuel began in 2003. Some of the key characteristics of an ideal fuel must be that it is inexpensive, available in large quantities, and non-polluting. This fuel has a C/H ratio of 0.333, a molecular weight of 46 (kg/kmol), and a lower calorific value of 26.9 MJ/kg in comparison to gasoline, which has a calorific value of 44-46000 kJ/kg. Certainly, gasoline is better as it is less hot than the other. Ethanol facilitates the use in high-compression engines owing to its high octane number and greater autoignition temperature, thereby reducing fuel wastage and enhancing engine output, which increases efficiency. Also, the use of ethanol as a fuel aids in the replacement of three of the most threatening compounds, including carbon monoxide, hydrocarbons, and nitrogen oxides. Ethanol is a readily available form of colorless liquid with a sweet taste, with a melting point of -114.1°C, a boiling point of 78.5°C, a density of 0.789 grams/ml at 20°C, and an octane number of 113. Ethanol is the most efficient fuel [15], but this is both its biggest advantage and biggest disadvantage, as it also uses ethanol in itself. Since you will be using more ethanol than gasoline, an appropriate gasoline should be used to get the desired outcome. Almost 98% of ethanol in India is produced from sugar molasses [16]. Ethanol can also be synthesized from materials that contain starch and cellulose, such as potatoes, algae, corn, wood waste, and agricultural as well as forestry waste products [17]. Ethanol is, however, used in different blends like E0R, which is 0% ethanol and 100% gasoline, E6.25P which is 93.75% premium gasoline and 6.25% ethanol, E10P which is 90% premium gasoline and 10% ethanol, as well as E20P which is 80% premium ethanol and 20% ethanol [15, 18, 19].

Methanol

Methanol comes from many sources, including cellulose, natural gas, biomass, and coal. To make synthetic fuel, carbon monoxide is passed through the air. Syngas is fed to the reactor along with the catalyst, which now produces methanol and water vapor. It can also be generated by heating fossils that contain methanol, such as coal and wood. The major producers of methanol in India are Gujrat Narmada Valley Fertilizers and Chemicals Limited, Rashtriya Chemicals and Fertilizers, Deepak Fertilizers, Assam Petrochemicals, and National Fertilizers Limited. Methanol has a C/H ratio of 0.25 and a molecular weight of 32.04 kg/km. Methanol has a lesser calorific value (15,910 kJ/kg) compared to gasoline (44,000 kJ/kg). The disadvantage of this oil is that its heat of vaporization is as high as 1102 kJ/kg, compared to 350-400 kJ/kg for gasoline, which is a big problem and requires more energy to convert the liquid state of methanol into gas. The density of methanol is 790 kg/m^3, the boiling point is 65.1°C, and the freezing point is -97.6°C. Because of its higher octane number (110), methanol has a higher ratio and, as a result, increases fuel efficiency. The flame temperature is lower than that of gasoline, which results in better combustion and reduced carbon dioxide and nitrogen oxide emissions. Methanol as a fuel will also significantly reduce carbon dioxide emissions. Methanol poses a smaller threat than gasoline in the event of an accident, as its evaporation rate is lower. Despite the rewards of methanol as a fuel, methanol has an energy disadvantage, as the energy essential to generate methanol is greater than the energy released when methanol is burned. In addition, the low cetane number can cause ignition delays and knock problems. Methanol's low fuel consumption is also a problem because it requires more fuel to travel longer distances. The corrosive and hygroscopic properties of methanol will eventually attract or absorb moisture from the air, further exacerbating storage problems. The best way to use methanol in overcoming this problem is to use it in a blended form.

E10-efficient cars are also suitable for the M3 blend, but since efficiency is required for more methanol products, equipment and special measures in the car and engine need to be changed. To resolve the transportation and storage problem, trucks need to be equipped with advanced equipment to control thermal expansion during transportation, short-term tilting during changeover, and temporary tilting. Transportation by train is a harmless way to transport ethanol as long as it is in an upright tanker.

Compressed Natural Gas

The fuel can be used as compressed natural gas, commonly known as CNG. As reported by the U.S. Department of Energy's Alternative Fuels Information Center, automobiles using CNG have an 11% reduction in greenhouse gas emissions. It is mostly made from petrol, which is composed of methane, and then compressed to 1% of its volume at normal atmospheric pressure. This fuel must be packed tightly into a container that is 2900-3600 psi. Storing and disposing of these containers is quite costly, so this is best handled by public transportation or government companies. Maintenance costs for natural gas-powered vehicles are lower than our gasoline-powered models because CNG does not dilute or pollute the lubricating oil. CNG fuel systems rarely absorb water or vaporize, and even though they leak like gasoline, they mix simply and regularly with air without causing any harm. CNG is the least likely to burn in hot or harsh temperatures due to its low flammability and autoignition temperature (540°C) [19]. CNG emits only 16.3 kg of CO_2 over 100 km, while a petrol car emits 22 kg of CO_2 for the same distance, so CNG emissions are much less CO_2. Despite all these benefits, CNG requires more storage than natural gas. Tests conducted in 2004 for CNG and diesel heavy vehicles under the Euro 6 standard showed that CNG produced a similar sound and CO_2 emissions, with higher fuel consumption but lower NOx emissions. There are currently over 3 million CNG vehicles in India, and the number continues to increase, accounting for about 3% of India's fuel consumption. As of 2020, the cumulative number of CNG vehicles is around 3.38 billion, with 2,207 CNG stations distributed across more than 20 states and 4 central districts in India. The government plans to set up 6,000 CNG stations by 2025 and plans to build a target of 10,000 CNG filling stations by the end of 2030 [20].

Liquefied Natural Gas

Liquefied Petroleum Gas or LPG is generally produced by mixing butane C_4H_{10} and propane C_3H_8 in equal proportions. Liquefied petroleum gas is a colourless and odourless gas. The density of butane at 15°C is 0.584 kg/dm³, and propane is 0.508 kg/dm³. While the vapor pressure of butane at 37.8°C is 2.6 bar and the vapor pressure of propane is 12.1 bar, the vapor pressure of butane at 37.8°C is 2.6 bar. C. The caloric value of butane is 10.920 MJ/kg⁻¹, and its energy value is 45 MJ/kg⁻¹. The caloric value of propane is 11.070 KJ/kg⁻¹, and its energy value is 46 MJ/kg⁻¹ [19]. The mixture of propane and butane is liquefied by cooling or compression. The mixture of propane and butane liquefies 260 times less than natural gas. LPG is very cheap as a fuel compared to gasoline and diesel. Due to its high-octane number, LPG will exhibit a relatively high compression ratio as an internal combustion engine fuel. LPG is inexpensive to use in internal combustion engines because it does not leave any residue after flowing into the ignition compartment in its gaseous form, so maintenance costs are very low. The usage of LPG in gasoline vehicles reduces CO emissions by 60%, HC emissions by 20%, and NO*x* emissions by 30%. In addition, LPG vehicles do not emit toxic substances, such as sulfur dioxide and aldehydes. Mission, Vision, and Policy 2025 was acquired by the Government of India in 2014. As of March 15, 2021, around 1,150 LPG filling stations have been established in over 564 cities [21]. Despite economic development and a friendly environment, the growth of LPG in India is still limited due to a lack of regulation, production, and marketing.

Biodiesel

Biodiesel is the ethyl ester or methyl ester of fatty acid, also recognized as FAME [22], which is a fatty acid methyl ester. It can be made from the most readily renewable materials, such as acid oil, non-edible vegetable oil, animal oil, or animal fat. Table **1** provides the overall characteristics of biodiesel from various plants. Vegetable oil is usually obtained from crops, such as sunflower, soybean, cotton, hazelnut, corn, and canola. The oil is extracted from the crops and then treated with various suitable alcohols such as methanol and ethanol, and catalysts such as sodium or potassium hydroxide to produce biodiesel and glycerol [23]. Diesel engines are generally expected to build on the path established to cope with the new emissions standards that are emerging at that time. In 2014, nearly 900 pharmacies in 350 cities started using low or high-biodiesel. Using biodiesel at low rates does not require changing the engine, but using biodiesel at high rates requires different types of modifications [24]. Compared to the use of biodiesel to produce diesel, biodiesel has 78% less carbon dioxide emissions, 15% less carbon dioxide emissions, 27% less hydrocarbon emissions, 50% less emissions, and 50% less particle emissions, resulting in a 22% reduction [25, 26]. Additionally, biodiesel emits no sulphur. Good lubrication or a good lubrication coefficient prevents deposits from accumulating in the engine, ultimately reducing engine wear. Biodiesel is safer to transport and store than diesel due to its high boiling point (>110°C). The high cetane content of biodiesel reduces the possibility of rollover and slows down fire. The energy content of biodiesel is 8% lower than that of diesel, which results in a 5% decrease in efficiency. The temperature of biodiesel is +3°C, which greatly affects its use in automobiles. This is because at high temperatures, the oil forms a gel-like structure, preventing it from flowing and causing blockages in the flow of the oil. The high viscosity of vegetable oil also causes serious problems in fuel injector operation due to the size of the oil droplets. The Indian government announced the National Biodiesel Mission in 2003 [27] and set a target of a 20% biodiesel blending rate by 2011-2012, suggesting that only non-biodiesel can be used in biodiesel production to increase the rate.

Table 1. Characteristics of diesel and biodiesel obtained from plants [28].

S. No.	Plant	Fuel Density (Kg/dm³)	Viscosity (mm²/S)		Cetane No	Flash Point (°C)	Energy Content (kJ/kg)
			27°C	75°C			
1	Corn	0.915	46	10.5	37.6	270-295	37,830
2	Sunflower	0.878	10	7.5	45-52	85	40,560
3	Cotton	0.874	11	7.2	45-52	70	40,580
4	Soyabean	0.872	11	4.3	37	69	39,760
5	Diesel	0.815	4.3	1.5	47	58	43,350

Hydrogen

Hydrogen is a colourless, odourless, tasteless, transparent, and the lightest element in nature. At 0°C, one litre of hydrogen weighs about 0.898 grams, so because of its lightness, it is not abundant in nature and is relatively rare. It can only be produced from a few sources, such as water, air, natural gas, and coal. It is a more suitable alternative fuel for Otto engines than diesel engines due to its high octane number and autoignition temperature (847-867 K at 1 atm pressure). It does not contain carbon content and therefore does not emit harmful substances, such as CO, CO_2, HC. The higher and lower calorific value of hydrogen is 119.93 kJ/g [29]. Using hydrogen as fuel in the engine will increase the compression ratio and lean mixture formation by 25%. Hydrogen can be a very easy fuel to use as an alternative fuel since its energy content is very low. Despite these benefits, hydrogen is unlikely to be used as an alternative fuel. The usage of hydrogen in gasoline engines resulted in both backfiring and preignition events. Ignition problems in hydrogen-fuelled engines arise from their low ignition energy. The emission of NO*x* is significantly larger when the combustion temperature of hydrogen fuel is high. One of the main problems with using hydrogen is that it is very difficult to liquefy, as it becomes liquid at about 20k and 2 bar. Components provide high-speed power worldwide for a variety of applications, including lighting, power, cooling, communications, and computing.

Electricity

Electricity is a very clean and efficient resource beyond the design stage. Motors used for rotation in many daily applications can operate at over 90% efficiency when powered by electricity. The properties of electricity make its energy very important for movement [30]. Vehicles powered by electricity are generally referred to as Electric Vehicles (EVs). The electricity used to run the generator can be obtained from the transmission line or a single or connected battery, or it can be generated onboard using FC [31, 32]. Electric locomotives, subway trains, and trains are powered directly by the grid, while other electric vehicles use batteries. The history of the electric car in Europe and the United States dates back to the 1880s and early 1920s, considered the first golden age of the culture in thousands of years. The emergence of vehicles based on the Internal Combustion Engine (ICE) and without the necessary electrical distribution and battery charging (including the time required for charging) is something that led to the end of the first golden age of the electric car [33]. Issues such as vehicle emissions, global climate change, and energy security have begun to attract the attention of policymakers and automakers, and interest in electric vehicles has been rekindled in the last decade [34]. While the transition to electric vehicles is still in its infancy in some countries, adoption is rapidly increasing in many of the world's largest automotive markets due to falling battery costs and the spread of EV charging infrastructure across multiple vehicle types, including cars, buses, taxis and ridesharing, Light Commercial Vehicles (LCVs), two/three-wheelers, and short-range vehicles (such as city delivery). As of the end of 2019, the number of electric cars in the world reached 7.2 million, while the number of light commercial vehicles and buses reached 370 thousand and 510 thousand, respectively [34].

INFRASTRUCTURE AND TECHNICAL FEASIBILITY

The first issue to be pointed out is the assessment of the current infrastructure for ethanol production, storage, and distribution, which is necessary to efficiently introduce E100 (100% ethanol) fuel in India.

Current Status of Ethanol Infrastructure in India

The Indian ethanol infrastructure, in general, is developing in line with the requirements of the Ethanol Blending Programme (EBP) aimed at bringing about 20% ethanol blended into petrol by the year 2025. This initiative has led to considerable growth in ethanol production facilities, as well as the expansion of the distribution network across the country, especially in states with strong agricultural dominance, such as Uttar Pradesh, Maharashtra, and Karnataka [9].

Ethanol Production Processes

Currently, most of the ethanol industries in India are located near sugar manufacturing companies, and they use molasses, a by-product of cane sugar production, as feedstock for ethanol production. There are over 300 distilleries in India that are engaged in the production of ethanol, and several other distilleries are also announced in different projects to increase this number [35].

Yet the ethanol production plants have to be modified in order to accommodate capacity expansion, which will facilitate E100 production as the filler blend percentage used is higher than normal, in this case, E100, which is ethanol. Moreover, E100 is produced from cellulosic ethanol through advanced biofuel facilities; hence, after technological improvements, more agricultural waste, like wheat and rice straw, can be used to enhance the sustainability of production to a higher extent [13].

Ethanol Market in India – Storage and Distribution

According to markets in the Defarge distributions industry, the current market for ethanol in India is for Ethanol blends as compared to pure ethanol, which seems to be the market for the future as well. This blending is carried out through mixing in earmarked ratios at the petrol pumps after the blending process. Ethanol is once procured from storage, which is typically located in tank farms, and blended with either sugar mills or the petroleum refining process. It is transported in such tankers to petrol pumps for sale purposes, and although blenders are used in specific regions across the roughly 80000 locations around the country as blending stations. There will be multiple dispensers at these petrol stations for passengers with flex-fuel vehicles. However, the majority of these stations are not yet ready to pump E100. Nevertheless, the tanks installed at these stations are mainly suitable for gasoline, which is a problem, especially because ethanol is a lot more hazardous and reactive than gasoline [6].

Necessary Alterations for Production, Storage, and Distribution

All these changes will require vast investments in human resources, as there will be vast changes required in the tanks and pumps due to the fact that ethanol and gasoline—although derived from similar compositions—contain different molecular compounds.

Production Enhancements

The current production capacity is mainly for E10 and E20. To be able to manufacture adequate E100 fuel, additional construction of ethanol production facilities is necessary, which emphasizes the development of first-generation (sugarcane-based) and second-generation (cellulosic) bioethanol technologies.

Depending on Feedstock Availability

The variability of feedstocks, such as sugarcane, corn, millets and agricultural residues, means that biomass processing plants may need to be upgraded or configured to effectively process various types of biomass [36]. The adoption of new technologies, such as enzymatic hydrolysis, will assist in converting syngas sourced from biomass into fuels. The feedstock fiber in question, which India is investing in, is intended to diversify its feedstock from sugarcane dependence Bharathiya Urja Sanrakshan Kendra, 2020.

Storage Facilities

Ethanol Transportation is done using Special Cans and Tanks: Tradition metallic tanks corrode with ethanol due to their hygroscopic nature, which draws water from the surroundings, making corrosion inevitable. Therefore, in order to avoid corrosion, metal tanks of some careful selection that are designed to prevent it, such as stainless steel or specially coated tanks, are needed. To reduce the moisture content in the stored items, other means like desiccant systems may also be necessary [11, 37]. E100 is prone to phase separation due to its sensitivity to water contamination. The withdrawal of other types of fuel storage facilities will minimize contamination and ensure the integrity and quality of ethanol fuel. In addition to ethanol being more volatile, it may be necessary to install vapor recovery systems in the storage facilities [8].

Distribution Upgrades

Pipelines and Tankers: The corrosive nature of ethanol can erode the steel pipelines used to distribute gasoline, which is a challenge for India since it will have to install new pipelines for ethanol or upgrade the old ones using corrosion-resistant materials and coatings. If pipelines are not an option, tankers that are designed to carry ethanol will be needed.

Upgrading Fuel Stations

It is a necessity that fuel dispensers and underground storage tanks at the fuelling stations should be compatible with E100. Dispensers that have difficulty dealing with E100 are common due to the nature of the substance, which contains ethanol that is more volatile and degrades rubber and plastic. Changes in storage pump options and seals, which can provide additional levels of ethanol safety standards, will be needed in that case, plus other safety changes to counter the elevated risk of vapor combustion [38].

Health and Safety Regulations

Since ethanol is considerably more flammable and volatile as compared to gasoline, more safety protocols and regulations will be required. This includes equipping facilities with fire suppression measures, and training personnel in safe infrastructure practices and the handling of pure ethanol during distribution and storage facilities processes.

Addressing these infrastructure upgrades will enable India to create a sturdy foundation for the use of E100 fuel. Besides, these upgrades will also enhance the energy infrastructure of India in terms of sustainability and resilience.

POLICY FRAMEWORK

The Government of India has planned a few policies (Fig. **1**) such as:

In 2003, 13 states and Union Territories mandated a 5% ethanol blend.

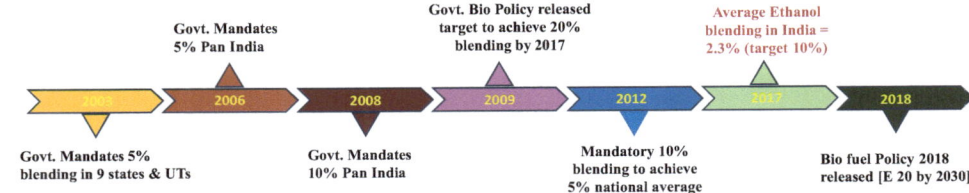

Fig. (1). Potential applications of microalgae as renewable biomass.

Pan India mandated 10% in 2008. In 2009, the country's biofuel policy promised to increase ethanol by 20%. In 2013, the required blend ratio was 10%, reaching the national average of 5%. In 2015, E85 (a blend of 85% ethanol and 15% gasoline) and E95 (a blend of 95% ethanol and 5% gasoline) were approved for use in vehicles.

In 2018, "Clean Air-Better Life" promised that Delhi would reach E10 (10% ethanol plus 90% gasoline) by 2022. The fuel is non-volatile, highly flammable, and has a distinctive sweet taste.

The adoption of ethanol as an alternative fuel in India has been promoted by a range of policies that are aimed at improving its production and utilization. The underlying sections critically review the policies, approaches, and future prospects for the promotion of ethanol as a suitable alternative fuel.

Review of Current Policies Promoting Ethanol Use in India

National Biofuel Policy (2018)

The policy was launched by the Government of India in 2018 with the aim of accomplishing the target of 20% blending of ethanol with petrol by 2025. The policy designs the framework for biofuel production and use, focusing on improving domestic production techniques and lowering fossil fuel consumption. The policy also emphasizes the use of various feedstocks to enhance the production of ethanol, incorporating both food and non-food sources (Ministry of Petroleum and Natural Gas, 2018).

Ethanol Blending Program

The program was launched in 2003 to enhance the blending of ethanol with petrol. With this initiative, public sector oil marketing companies are directed to avail ethanol for blending with petrol at fixed prices. This program ensures the price stability and a reliable market for ethanol producers. Undertaken steps in this program have become very significant in improving the production and blending of ethanol in recent years (Ministry of Petroleum and Natural Gas, 2021).

Financial Inducements and Subsidies

Several financial aids and subsidies have been provided by the government to induce the production of ethanol. It includes interesting schemes and research grants for ethanol-producing distilleries to improve infrastructure for storing ethanol and supplying it through pipelines. All these incentives aim to reduce the cost of ethanol production and increase the benefits for producers (NITI Aayog, 2021).

Regulatory Framework for Quality Standards

The quality standards have been established by the Bureau of Indian Standards (BIS) for ethanol that is compatible with the existing fuel system. These standards aim to boost the safe blending of ethanol for vehicle uses, with the aim to maintain the quality of fuel as well as the performance of the vehicle (Indian Standards Institution, 2020).

State-Level Initiatives

Ethanol production and use have been encouraged with the implementation of several policies introduced by several Indian states, such as Maharashtra and Uttar Pradesh. The biofuel policies include incentives such as taxes and subsidies for farmers and producers to boost the production of ethanol (Maharashtra Government, 2020).

CONCLUSION AND OUTLOOK

Implementing Ethanol 100 (E100) in India's transport sector can yield ample economic, environmental, and energy security benefits. The chapter has analyzed the viable opportunities that E100 fuel use raises, the benefits to the stakeholders, and the challenges and ways to resolve them in a sustainable manner. The following key points summarize the findings: The transport sector of India is an unavoidable menace to environmental pollution as it largely relies on fossil fuels. Furthermore, fossil fuel dependency poses a threat to energy security. This tension needs to be alleviated by adopting an alternative domestic fuel like ethanol. E100 presents lower emissions and reliance on fossil sources of energy, which would complement India's goal of energy diversification. However, there are several challenges with the adoption of E100 that need to be addressed and overcome through multifaceted strategies. These strategies incorporate key issues, such as infrastructure investment, consumer encouragement, interested policies to support producers, tax exemptions, *etc.* With all these measures, India will be able to develop a strong ethanol economy, making itself an energy secure country with less reliability on fossil fuels. India needs the collaborative efforts by engaging all government, industry, and public sectors.

Ethanol, most notably when produced from domestic sugarcane, is advantageous to the environment in that it lowers carbon emission levels and is a cleaner fuel. Moreover, the increasing global demand for clean energy makes ethanol more useful, especially when fuel diversity is being considered as an energy diversification strategy.

The introduction of the E100 would require the overhauling of existing fuel stations, storage facilities, and vehicle engines. However, international case studies from both Brazil and the United States suggest that these pilot projects are manageable, despite existing challenges. Advances in technology for the production and distribution of ethanol, as well as the development of flex-fuel vehicles, make the adoption of E100 a strong possibility. It is also reasonable to expect that with adequate funding for research, the requisite technology will be developed to facilitate increased or sustained usage of E100.

This chapter has highlighted that E100 has a good prospect to assist India in the achievement of its various targets in the transport sector, aimed at a cleaner environment and sustainable development. India indeed has the potential to manage the concerns associated with the adoption of E100 and make the country greener and more independent. Adopting E100 is not just a step towards environmental conservation, but also a means to encourage expansion, employment generation, and improve the rural economy. In conclusion, the successful use of ethanol as the main fuel in the transport system of India revolves around the central focus of strong policy support, technology transfer, and the commitment of all relevant sectors.

Future Outlook

Future Possibilities of Ethanol in India's Transportation Sector: India's transport sector is at a critical juncture, as the emphasis on the use of cleaner and renewable sources of energy increases. The ethanol fuel will also be one of the key elements in this shift, considering the pledges made by the government to lower emissions and energy self-sufficiency. In the near future, there are some long-range forecasts regarding the role of ethanol in the energy requirements in India in the coming years:

New Targets on Ethanol Blending

Considering the growing ethanol production capacity of India, the blending targets are expected to exceed the current set targets of 20%. This is because the long-term vision for particular segments of the vehicles is to target 100% ethanol, that is, E100. This transition is most likely to be gradual, starting from the blending target for the transportation of goods, followed by public transport. Some of the new market opportunities that arise for automakers due to the increased addition of ethanol to fuel are the development of flex-fuel vehicles and cars designed specifically for E100 fuel.

More Regional Feedstocks and Production Centres

Other than the development of these hubs, India is also said to create centres of production in other states rich in sugarcane, as Maharashtra and Uttar Pradesh. These states have the potential to serve nearby locations with ethanol hubs. These hubs may, in the future, also look into making second-generation ethanol from non-food sources, further increasing the diversity of feedstocks, as well as creating more sustainable options for the production of ethanol.

For a more stable production that will guarantee the supply of ethanol, incorporating agricultural residue, cellulosic biomass, and some other non-food sources into the feedstock will be an easier solution than solely relying on sugarcane for the production of ethanol.

REFERENCES

[1] J. Rawat, M. Nanda, S. Kumar, N. Sharma, R. Sharma, H.C. Joshi, M.S. Vlaskin, A. Hussain, and V. Kumar, "Integrating wastewater treatment to bio-stimulant & biochar generation for plant growth promotion using microalgae", *Process Biochem.*, vol. 145, pp. 187-194, 2024. [http://dx.doi.org/10.1016/j.procbio.2024.06.031]

[2] S.K. Guttikunda, and R. Goel, "Nature of air pollution, emission sources, and management in Indian cities", *Atmos. Environ.,* vol. 187, pp. 1352-2310, 2019.
[http://dx.doi.org/10.1016/j.atmosenv.2014.07.006]

[3] M.W. Tessum, S.C. Anenberg, Z.A. Chafe, D.K. Henze, G. Kleiman, I. Kheirbek, J.D. Marshall and C.W. Tessum, "Sources of ambient PM2.5 exposure in 96 global cities". *Atmospheric Environment,* vol. 286, p.119234 2022.
[http://dx.doi.org/10.1016/j.atmosenv.2022.119234]

[4] G. S. Kumar, "Anatomy of Indian energy policy: A critical review", Energy Sources, Part B: Economics, Planning, and Policy, vol. 12(11), pp. 976–985 2017.
[http://dx.doi.org/10.1080/15567249.2017.1336814]

[5] B. Mishra, S. Ghosh, K. Kanjilal, "Policies to reduce India's crude oil import dependence amidst clean energy transition, Energy Policy", vol. 183, pp.113804 2023.
[http://dx.doi.org/10.1007/978-981-99-7250-0_8]

[6] B. Dey, B. Roy, S. Datta and K.G. Singh, "Comprehensive overview and proposal of strategies for the ethanol sector in India". *Biomass Conversion and Biorefinery,* vol. 13(6), pp.4587-4618 2023.

[7] E.P. Borges Filho, and A. Dettmer, "Review of ethanol distillation process simulation: evolution, challenges, and perspectives" *Journal of Chemical Technology & Biotechnology,* 2025.

[8] S.M. Rosdi, M.F. Ghazali, and R. Mamat, "Evaluation of engine performance and emissions using blends of gasoline, ethanol, and fusel oil", *Case Studies in Chemical and Environmental Engineering,* vol. 11, pp.101065 2025.
[http://dx.doi.org/10.1016/j.rineng.2025.104273]

[9] H. Valera and A.K. Agarwal, "India's growing Ethanol Blending Program and implications of scalable and sustainable Methanol Blending Program for transport sector", *Transport Policy,* vol. 165, pp.179-193 2025.
[http://dx.doi.org/10.1016/j.tranpol.2025.01.007]

[10] J. Goldemberg, *The Brazilian biofuels industry.* vol. Vol. 6. Biotechnology for Biofuels, 2008.

[11] T.J. Lark, N.P. Hendricks, A. Smith, N. Pates, S.A. Spawn-Lee, M. Bougie, E.G. Booth, C.J. Kucharik and H.K. Gibbs, "Environmental outcomes of the US renewable fuel standard", *Proceedings of the National Academy of Sciences,* 119(9), p.e2101084119 2022.
[http://dx.doi.org/10.1073/pnas.2101084119]

[12] J. Goldemberg, S.T. Coelho, and P. Guardabassi, "The sustainability of ethanol production from sugarcane", *Energy Policy,* vol. 37, no. 6, pp. 2300-2311, 2019.
[http://dx.doi.org/10.1016/j.enpol.2008.02.028]

[13] S. Gangil, V.K. Bhargav, P. Diwan, M. Kumar, and P. Sahu, "Estimation of bioenergy potential based on collectable biomass: An important step for sustainable biomass utilization", *Sustainable Energy Technologies and Assessments,* vol. 82, pp.104535 2025.
[http://dx.doi.org/10.1016/j.seta.2025.104535]

[14] F.G. Barbosa, S. Sánchez-Muñoz, E. Mier-Alba, M.J. Castro-Alonso, R.T. Hilares, P.R.F. Marcelino, C.A. Prado, Campos, M.M. Cardoso, A.S. Santos, J.C. and Da Silva, S.S., "Challenges and perspectives on application of biofuels in the transport sector", *Bioethanol,* pp.463-499 2022.

[15] H. Yücesu, T. Topgül, and C. Cinar, Effect of ethanol–gasoline blends on engine performance and exhaust emissions in different compression ratios, https://www.sciencedirect.com/ science/article/pii/S1359

[16] F. Rosillo-Calle, and F.X. Johnson, *Sugarcane ethanol: Contributions to climate change mitigation and the environment,* 2019.https://www.routledge.com

[17] A. Chauhan, "Sources for biofuels production from biomass", *Trends Math,* pp. 1-64, 2024.
[http://dx.doi.org/10.1007/978-981-99-7250-0_1]

[18]   Y. Huang, Y. Zhou, and J. Liu, "Improving the performance of ethanol engines through advanced combustion strategies", *Energy,* vol. 203, p. 117921, 2020.

[19]   A. Verma, N.S. Dugala, and S. Singh, "Experimental investigations on the performance of SI engine with Ethanol-Premium gasoline blends", *Mater. Today Proc.,* vol. 48, pp. 1224-1231, 2022. [http://dx.doi.org/10.1016/j.matpr.2021.08.255]

[20]   A. Ugurlu, and S. Oztuna, "A comparative analysis study of alternative energy sources for automobiles", *Int. J. Hydrogen Energy,* vol. 40, no. 34, pp. 11178-11188, 2015. [http://dx.doi.org/10.1016/j.ijhydene.2015.02.115]

[21]   C. Bae, https://www.sciencedirect.com/science/article/pii/S1540

[22]   M.R. Nouni, P. Jha, R. Sarkhel, C. Banerjee, A.K. Tripathi, and J. Manna, "Alternative fuels for decarbonisation of road transport sector in India: Options, present status, opportunities, and challenges", *Fuel,* vol. 305, p. 121583, 2021. [http://dx.doi.org/10.1016/j.fuel.2021.121583]

[23]   N. Dugala, and G. Goindi, University ASJ of KS, https://www.sciencedirect.com/science/article/pii/S1016

[24]   T.Y. Khan, "A review of performance-enhancing innovative modifications in biodiesel engines", *Energies,* vol. 13(17), pp. 4395 2020.

[25]   S.S. Bhatti, S. Ali, and A. Ali, "Dual-fuel operation of a diesel engine using bioethanol and diesel", *J. Clean. Prod.,* vol. 225, pp. 819-826, 2019.

[26]   R. Gonzalez, and K. O'Donnell, "Genomic advancements in bioenergy crops: Potential implications for the ethanol industry", *Biotechnol. Adv.,* vol. 38, p. 107504, 2020.

[27]   A. Yadav, "N.S.dugala, "Production of methyl ester of mahua (Madhuca Biodiesel) for improving its cold flow characteristics"", *Indian J. Sci. Technol.,* vol. 11, no. 28, pp. 974-6846, 2018.

[28]   A. Saha, and N. Sharma Dugala, "An Incite on advanced alternative fuels Based on India", *IOP Conf. Ser. Earth Environ. Sci.,* vol. 1110, no. 1, p. 012008, 2023. [http://dx.doi.org/10.1088/1755-1315/1110/1/012008]

[29]   A. Chauhan, and H. Chandra Joshi, "Energetic efficiency of a biofuels production system mathematical modeling", *Trends in Mathematics,* no. Part F3197, pp. 337-357, 2024. [http://dx.doi.org/10.1007/978-981-99-7250-0_8]

[30]   M. Balat, "Potential importance of hydrogen as a future solution to environmental and transportation problems", *Int. J. Hydrogen Energy,* vol. 33, no. 15, pp. 4013-4029, 2008. [http://dx.doi.org/10.1016/j.ijhydene.2008.05.047]

[31]   G. Crabtree, Electricity.*Sustainable Energy Systems in Sustainability: A comprehensive foundation.,* T. Theis, J. Tomkin, Eds., U of I Open-Source Textbook Initiative, 2018, pp. 338-347.

[32]   A. Simpson, R. Salter, S. Dhar, and P. Newman, Electric vehicles.*Technologies for Climate Change Mitigation – Transport Sector.* UNEP, 2011.

[33]   P.R. Shukla, S. Dhar, M. Pathak, and K. Bhaskar, *Promoting low carbon transport in India – Electric vehicle scenarios and a roadmap for India.* UNEP DTU, 2014.

[34]   K.G. Høyer, "The history of alternative fuels in transportation: The case of electric and hybrid cars", *Util. Policy,* vol. 16, no. 2, pp. 63-71, 2008. [http://dx.doi.org/10.1016/j.jup.2007.11.001]

[35]   J. van Dyk, J.F. Görgens, and E. van Rensburg, "Ethanol production from whole sugarcane using solid-state fermentation", *BioEnergy Research,* vol. 18(1), pp.38 2025.

[36]   S. Rathi, M. Rani, A. Vashisth, N. Mittal, D. Kumar and A.K. Srivastava, "Occurrence of apomixis in *Eleusine coracana*" , *Flora,* 265, pp.151575 2020.

[37]   U.S. Department of Agriculture, *Economic Impact of the Ethanol Industry 2021*.https://www.usda.gov

[38]   M. Kovarthini, F. Josephine Lenta, S. Kannamudaiyar, "Economic and commercial impact of ethanol production-opportunities, challenges, and rural development for farmers in India", Telematique Vol. 24 No. 01 2025.

The Waste-to-Energy Nexus, 2026, 213-231

CHAPTER 10

Sustainable Future: Integrating Circular Economy Strategies into Renewable Energy for Effective Recycling and Resource Management

Shweta Vishnoi[1], Raj Kumar Goel[2], Sujata Rathi[3], Deepak Kumar[4],* and Anurag Tyagi[1]

[1] *Department of Physics, Noida Institute of Engineering & Technology, Greater Noida, Affiliated to Dr. A.P.J Abdul Kalam Technical University, Lucknow, U.P, India*

[2] *Department of Computer Application, North-Eastern Hill University, Tura Campus, 794002 Meghalaya, India*

[3] *Department of Botany, Multanimal Modi College, Modinagar-201204, India*

[4] *Department of Physics, Graphic Era (Deemed to be) University, Dehradun, India*

Abstract: The critical need to address climate change and implement circular economy principles is the driving force behind the background and motivation of this chapter. Numerous researchers have extensively published on circular economy strategies within the contexts of sustainability, recycling, resource management, and productivity. Simultaneously, debates on the impacts of climate change and mitigation strategies have intensified. The dwindling supply of limited fossil fuels and their harmful environmental impacts have prompted a global shift towards clean and renewable energy sources. This chapter also discusses the crucial role of circular economy principles in enhancing the sustainability of renewable energy systems, for example, wind turbines, solar panels, and Battery Energy Storage Systems (BESS). The primary focus is on how circular economy approaches can reduce costs, minimize environmental impacts, and promote a more sustainable energy future. We analyse the economic and environmental implications of these practices and offer recommendations to encourage stakeholders to adopt circular economy strategies. A comprehensive review of current practices, future trends, and approaches to integrating circular economy strategies into renewable energy systems will ultimately promote sustainable development and help achieve long-term environmental goals.

Keywords: BESS, Circular economy, Renewable energy systems, Sustainable development.

* **Corresponding author Deepak Kumar**: Department of Physics, Graphic Era (Deemed to be) University, Dehradun, India, E-mail: deepakphy@gmail.com

INTRODUCTION

Climate change is caused by the burning of fossil fuels for energy, deforestation, and industrial activities. Its negative effects include rising global average temperatures, extreme climate events, such as droughts and heavy rains, storms, melting of ice caps and glaciers, and impacts on agriculture. Due to the existence of these problems, countries around the world are considering how to mitigate the effects of climate change and have identified efforts to reduce emissions or adopt advanced technologies to limit greenhouse gas emissions [1 - 4].

Over time, the economies of countries have been growing, and the living standards of people have also been improving, resulting in increased consumption of goods, services, electronics, appliances, various consumer products, automation, and advanced manufacturing technologies, which have led to a significant increase in efficiency and productivity. As consumption increases, driving economic growth while reducing costs, it has also led to an increase in the volume of municipal solid waste, scrap metal, chemicals, and e-waste. The linear "take-make-leave" model, which has dominated industrial and economic practices, is no longer viable due to the lack of recyclable or reusable materials, environmental pollution, resource depletion, economic inefficiency, and the scarcity of rare earth metals and fossil fuels [5 - 7].

A multi-pronged approach is needed to effectively address environmental issues, achieve eco-efficiency that benefits both humans and the biosphere, and realize a self-sustaining future. One of the strategies is to incorporate the C2C (popularly called circular economy, (Fig. **1**) principles into the renewable energy sector, which can ensure significant changes and advancements in areas, such as bio-prospecting, bio-sorption, bio-leaching, bio-reduction, bio-flotation, microbial electrochemical technologies, recycling, and resource management to extract metals from e-waste [8].

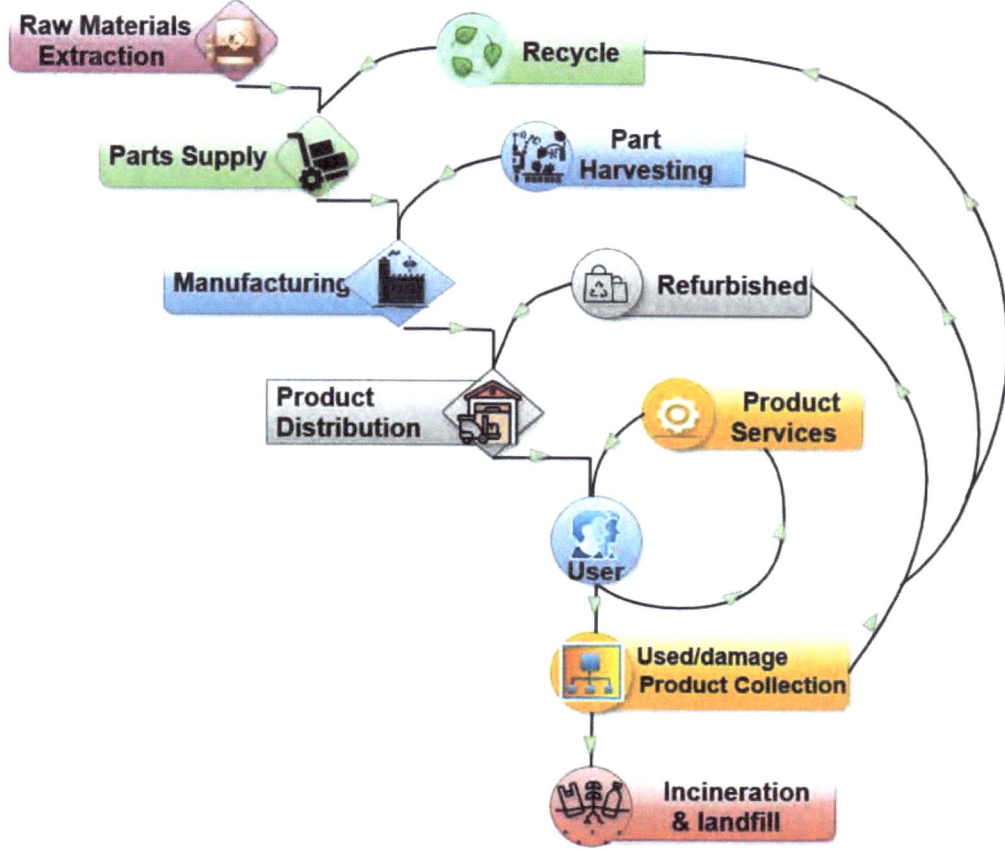

Fig. (1). Circular economy principles in the renewable energy sector [8]

The demand for Electric Vehicles (EVs) is increasing, and many countries are enacting legislation to promote the adoption of hybrid and fully electric vehicles. Global battery demand is expected to increase from 185 GWh in 2020 to over 2,000 GWh by 2030. One of the most popular choices is systems powered by lithium-ion batteries. As a result, the growing number of lithium-ion batteries will need an effective and sustainable end-of-life management program to manage their end-of-life, including recycling, reuse, landfill disposal, or incineration.

Recent research is focused on more sustainable approaches, such as remanufacturing, reusing, and reprocessing battery components for future use, as well as developing recycling methods for the sustainable disposal of large quantities of batteries in the near future. As the world moves towards renewable energy plants, it is important to integrate circular economy strategies to ensure that these technologies will contribute to a sustainable future and achieve SDG 7 [9, 10]. Given the pressing challenges posed by climate change and resource depletion, the need for sustainable development strategies is more critical than ever. One such strategy is the integration of circular economy principles with renewable energy systems, which we explore in detail below

LITERATURE REVIEW

The world is taking steps to adopt energy assets such as solar, wind, and hydropower, moving away from the worrying trend of dependence on fossil fuels. This is a suitable solution due to the growing climate crisis and the continuing shortage of these assets. The most notable is the integration of renewable electricity into the economy, which comes with numerous problems, especially in the context of resource management and waste disposal. Given the urgent need for sustainable systems, the circular economy provides a comprehensive framework that shifts the focus from linear consumption to regenerative cycles – this is particularly important in the renewable energy sector. Addressing these issues from a Circular Economy (CE) perspective enables the reduction of adverse environmental impacts and the improvement and redesign of products. This scientific review examines circular economy strategies at the level of individual components in renewable energy system development and their intersection: recycling, governance, and sustainability. This is how the organization operates and adheres to the 'take, make, dispose' model. The business cycle chart from Geissdoerfer *et al.* [11] suggests that business involves the progression of variable business factors considering reduction, reuse, and recycling. Its objective is to increase the product's lifespan, minimize its environmental impact, and reduce waste.

Blomsma and Brennan [12] stated that in recent years, CE has also been recognized as a credible approach to addressing resource and waste problems arising from the intensive use of renewable energy. This exercise poses a significant challenge in terms of resource management. This commentary demonstrates that renewable energy technologies, particularly PV panels, wind generators, and power storage batteries, rely on rare and specific non-renewable sources, such as lithium, cobalt, and Rare Earth Elements (REE). The excessive consumption, use, and final disposal of those materials raise sustainability concerns, including the sustainability of products, environmental impact, and waste management [13, 14]. For example, photovoltaic panels contain many materials that are difficult to recycle, such as silicon, cadmium, and aluminium. Additionally, wind turbines use many composite materials that can be cut on the cycle without recycling from the slinging part due to the integrated recycling technology within their lifecycle [15, 16]. There are also challenges associated with power storage batteries, and especially lithium-ion batteries. On the other hand, the demand for better cobalt resources for renewable technologies raises ethical issues related to the types of mining containing lithium or cobalt [17]. The circular financial system model is a crucial framework for addressing the environmental issues associated with production.

According to Kirchherr *et al.* [18], CE approaches focus on avoiding waste and contamination, increasing product utilization, and enhancing natural methods of forest improvement. Renewable energy satisfies these considerations at all stages of the product life cycle, from design to production to the final stage. Reuse and recycling are the primary goals in the design of any new technological invention under CE. For example, solar panels are manufactured with a section that can be easily removed during any demineralization process because the components, whose waste includes glass, aluminum, and silicon, are easier to recycle. The structural design of windmills also considers the efficient conversion of raw materials and waste into reusable resources, minimizing discard and promoting sustainability [19]. The study focuses on the production of more and more renewable vegetation, especially rechargeable batteries [20, 21].

Extended Producer Responsibility (EPR) is a legal concept that holds enterprises accountable for waste handling at the end of a product's lifecycle. Several studies have shown that EPR has the most significant impact on production, particularly in the application of solar panels and wind turbines, which motivates companies to adopt eco-design, eco-management, and recycling practices. Regarding the cognitive aspects, EPR can also support the introduction of recycling proposals and reverse logistics systems by enhancing the interconnectivity of renewable energy structures [22].

Renewable Records and Production; However, primary technology acts as an enabler in the realization of the renewable energy enterprise cycle. Great advances have been made in technologies, such as batteries and solar panels. For example, Tao and Yu. [23] demonstrated the potential of chemical energy and techniques to achieve higher efficiencies than conventional photovoltaic panels. Additionally, hydrometallurgical and pyrometallurgical methods have been explored to recover essential elements such as lithium and cobalt, which are used in batteries [24]. Furthermore, Schroeder *et al.* [25] highlighted policies that include recycling targets, financial incentives for developing circular economy infrastructures, and support for research and advancement of new technologies. A significant portion of ECU and round financial system motion plans exemplify policy initiatives that leverage technology with expertise in waste prevention, recycling, and sustainable business [26]. New research findings have emerged to analyze the economic outlook of renewable energy. For example, the Japanese solar industry has incorporated photovoltaic structures with higher recycling costs for materials such as aluminum and glass [27].

Similarly, Denmark has recently launched a novel wind turbine recycling project to recycle the overall material of turbine blades in the manufacturing industry, as the following findings suggest. Implementing the CESsIS assumption for renewable energy is useful in reducing the extraction of resources, reducing waste, and creating recycling jobs through phased processes. This process support is necessary to help us tackle the sustainability challenges related to control and waste production. CE provides an advanced framework that aims to address the overall environmental performance of renewable energy through the lifecycle and focuses on environmental design, renewable energy, and new technologies. However, successful implementation entails coverage support, cross-functional coordination, and nonstop innovation in the recycling method. For this reason, and with the increasing power consumption from renewable resources, a circular business model is important to realize the fate of sustainable electricity.

METHODOLOGICAL FRAMEWORK

To meet the growing demand for sustainable energy systems, this chapter proposes a structured framework to integrate circular economy principles into renewable energy development. The framework focuses on key strategies to improve resource efficiency, reduce waste, and enhance the resilience of energy systems. A core component is the deployment and optimization of Battery Energy Storage Systems (BESS), which are essential for balancing energy supply and demand on renewable energy grids. The following subsections outline the main elements of the framework, beginning with the energy management system of the BESS.

Energy Management System for BESS

According to Fortune Business Insights, countries around the world, including the United States, China, Australia, European nations, Japan, South Korea, and India, are adopting Battery Energy Storage Systems (BESS) to reduce greenhouse gas emissions, integrate more renewable energy into the grid, and achieve climate goals. As a result, the demand for BESS is increasing, given its critical role in storing and dispatching renewable energy (Fig. **2**), thereby facilitating the transition away from fossil fuels [28, 29]. Considering the importance of effective energy management, it is crucial to understand the structural and functional components that constitute a battery energy storage system.

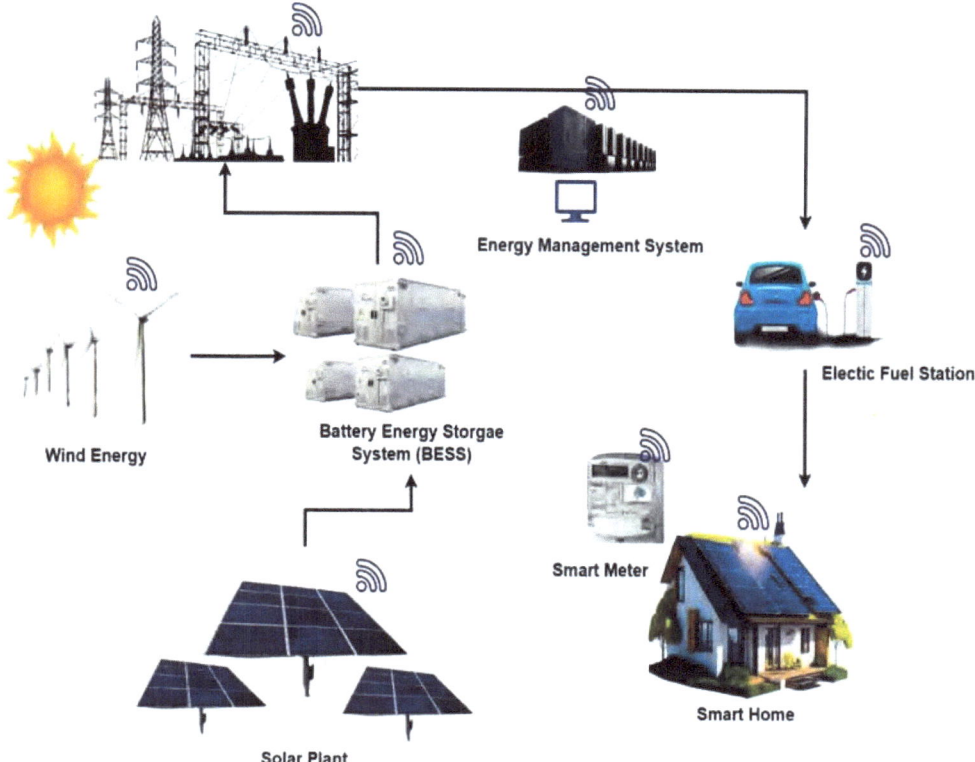

Fig. (2). BESS energy management framework for a sustainable future [28, 29]

Key Components of BESS

A complete BESS consists of several key components, such as batteries, charge controllers, inverters, and smart monitoring systems. These components work together to store electrical energy as chemical energy and manage the flow of electricity between the battery and the connected energy source (such as solar energy). The inverter converts the Direct Current (DC) stored in the battery into Alternating Current (AC), provides information on operating status, efficiency, and maintenance needs, and distributes power [30 - 32]. While hardware components form the backbone of BESS infrastructure, the integration of advanced digital technologies such as Artificial Intelligence (AI), the Internet of Things (IoT), and blockchain is revolutionizing how these systems are operated, monitored, and optimized.

Role of Emerging Digital Technologies

Artificial Intelligence (AI) and blockchain play a vital role in enhancing renewable energy systems by improving weather forecasting, energy management, and equipment maintenance. AI algorithms can predict energy production, identify potential losses, and predict system failures, making energy systems more reliable. This also helps overcome the limitations inherent in traditional power generation technologies. The incorporation of the Internet of Things (IoT) with AI further expands the capabilities of renewable energy. Equipped with billions of connected devices, IoT processes large amounts of data, facilitating better decision-making and optimization of energy systems. As a result, the synergy between AI and IoT is transforming the renewable energy sector, making it impossible to imagine a future without these technologies [33]. These technological advancements not only optimize current practices but also pave the way for more integrated, circular energy systems, encouraging broader, forward-thinking approaches to energy sustainability.

Combined with energy storage systems, these smart digital technologies significantly enhance real-time monitoring, predictive analysis, and system response capabilities. The integration of AI, IoT, and blockchain enables dynamic control of energy storage and distribution, making Battery Energy Storage Systems (BESS) smarter and more adaptable to fluctuations in supply and demand. This synergy is critical for improving the efficiency, resilience, and reliability of renewable energy infrastructure, especially in maintaining grid stability.

BESS for Grid Stability Enhancement

Controlled BESS maintains grid stability by performing frequency regulation and voltage support. They help stabilize grid frequency and maintain voltage levels by absorbing surplus power when supply exceeds demand (such as during high solar or wind energy generation) and releasing stored energy when demand exceeds supply. Additionally, BESS can provide emergency backup power, enhancing the resilience of energy systems during grid outages [34, 35].

Approximately 98% of the global BESS market relies on lithium-ion batteries due to their lightweight design, high energy density, low maintenance needs, large storage capacity, fast charging capability, and long lifespan. However, they are costly, sensitive to extreme temperatures, and overcharging can cause overheating. Other BESS options may include cheaper lead-acid batteries, which have lower energy density and high recycling rates, but are slow to charge, have a short life, and are bulky (see Table **1**). Nickel-cadmium batteries are known for their ruggedness, wide temperature operating range, and high discharge rates; however, they are not very efficient, suffer from memory effects, have low energy density, and are difficult to recycle [36, 37].

Table 1. Comparative evaluation of BESS technologies for sustainable energy systems.

Parameter	Lithium-ion	Lead-acid	Flow Battery	Sodium-sulfur (NaS)
Energy Density (Wh/kg)	150–250	30–50	20–50	150–240
Power Rating	kW–MW	kW–MW	MW	MW
Cycle Life (**Efficiency %**)	3000–10,000+ (90–95)	500–1000 (75–85)	10,000–20,000 (65–85)	2500–4500 (75–90)
Depth of Discharge % (Response Time)	80–100 (Milliseconds)	50–80 (Seconds)	100 (Seconds)	-100 (Seconds)
Operating Temp (°C)	15 - 45	0 - 40	10 - 40	300 - 350
Capital Cost ($/kWh)	300–700	100–300	400–800	300–500
Maintenance	Low	Moderate–High	High	High
Maturity	Commercially mature	Mature	Emerging	Limited commercial use
Use Cases	Grid, EV, Home	Backup power	Large-scale storage	Utility grid storage
Environmental Impact	Medium (recyclable)	High (toxic lead)	Low (recyclable)	High (high temp, reactive)

Circular Economy Strategies for End-of-Life BESS

Promoting a circular economy to enhance grid reliability and maximize renewable energy utilization requires effective end-of-life management of battery energy storage systems (see Fig. **3**). This approach aligns with Sustainable Development Goal (SDG) 7, which aims to ensure access to reliable, sustainable, affordable, and modern energy for all. While theoretical models lay the foundation, the practical application of circular principles requires a systematic approach, in particular through life cycle assessments and tailored strategies in renewable energy systems. Large numbers of batteries are being superannuated. If not reused or recycled, these countless superannuated batteries having volatile chemical elements could end up in landfills, causing environmental and economic harm. Thus, giving batteries a second life offers benefits such as lowering manufacturing costs, supporting grid applications in transportation, and minimizing waste from direct disposal. However, it also presents challenges related to battery collection, storage, treatment, and recycling [38]. Table **2** provides a comprehensive overview of circular economy strategies applicable to end-of-life (EoL) Battery Energy Storage Systems (BESS). These strategies cover the entire value chain—from design and monitoring to reuse, recycling, and policy intervention—providing actionable tools and technologies to support sustainable resource management and extend battery life.

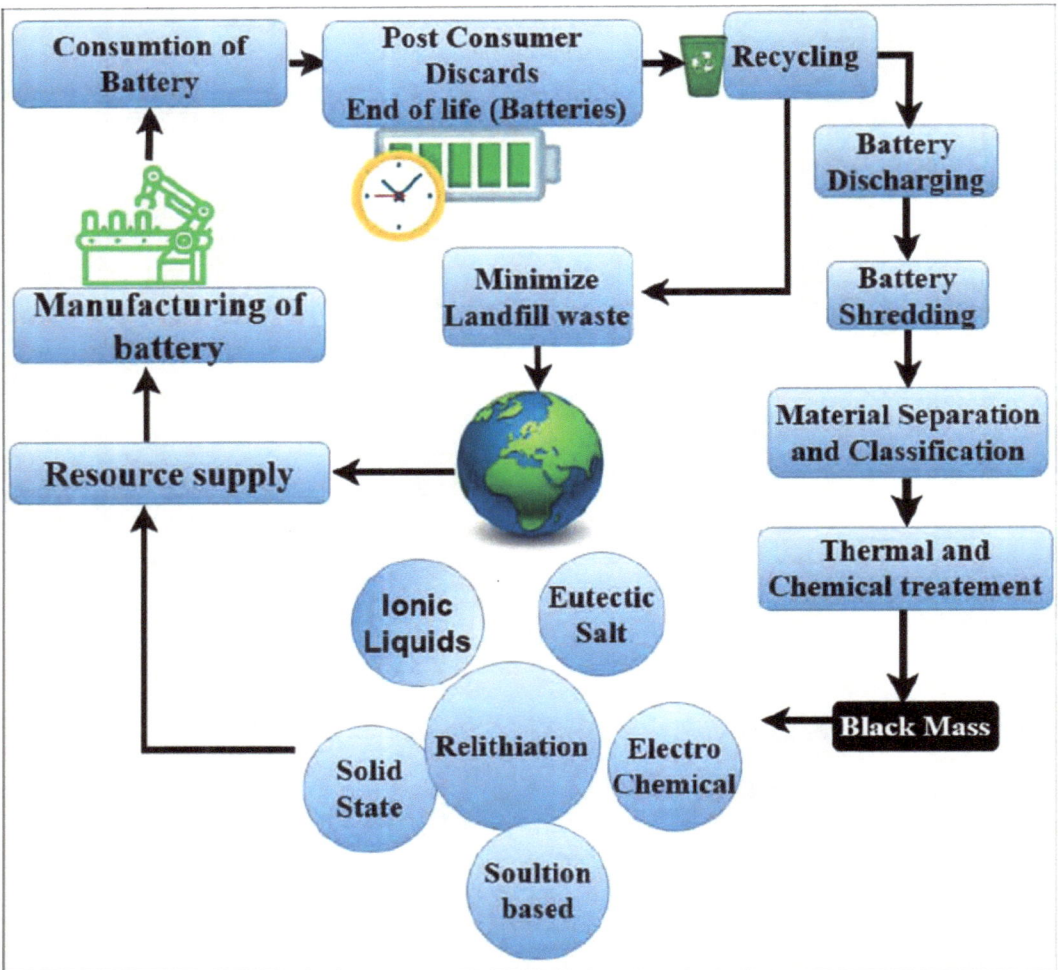

Fig. (3). Circular economy framework for effective recycling and resource management of BESS at end-of-life.

Table 2. Tools and Techniques Supporting Circular Economy for End-of-Life BESS

Strategy	Purpose	Key Actions/Methods	Tools/Technologies	Outcomes/Benefits	case study examples
Design for Disassembly	Enable easier recycling or refurbishing	Modular designs, standardized components	CAD, PLM (Product Lifecycle Management) tools	Simplifies reuse, remanufacture, recycling	Bosch – Modular battery design for easier disassembly

(Table 2) cont.....

Strategy	Purpose	Key Actions/Methods	Tools/Technologies	Outcomes/Benefits	case study examples
Life Cycle Assessment (LCA)	Quantify environmental impacts	Assess impacts from cradle-t--grave	SimaPro, OpenLCA, GaBi	Supports informed sustainability decisions	EU Horizon 2020 – LCA integration in BESS R&D
Battery Health Monitoring	Extend lifespan or enable second use	SoH (State of Health) diagnostics, predictive analytics	BMS (Battery Management Systems), AI/ML algorithms	Enables reuse, reduces early disposal	Tesla – Battery analytics in EVs and Powerwall systems
Reuse and Repurposing	Use BESS in less-demanding applications	Second-life deployment (e.g., in solar homes, EV charging)	Testing rigs, capacity grading software	Cost-effective, reduces raw material demand	Audi – Second-life EV batteries in grid storage (Germany)
Remanufacturing	Refurbish to "like-new" condition	Replace degraded cells/modules, test functionality	Disassembly tools, quality testing instruments	Extends product lifecycle	Nissan – Reusing Leaf batteries for stationary storage
Recycling & Material Recovery	Recover valuable materials	Mechanical shredding, hydrometallurgy, pyrometallurgy	Robotic disassembly, solvent extraction, smelting equipment	Reduces resource extraction, environmental load	Redwood Materials – Lithium/cobalt recovery (USA)
Digital Twin & Traceability	Improve circular tracking and resource use	Real-time tracking of battery history	Blockchain, Digital Twin platforms, QR/Barcode IDs	Enhances transparency, traceable recycling	Circulor – Blockchain tracking for battery supply chains
Policy & Extended Producer Responsibility (EPR)	Ensure compliance and accountability	Collection programs, take-back mandates	Regulatory platforms, producer registries	Shifts responsibility upstream	India EPR Framework – BESS take-back mandates
Eco-labeling & Certification	Promote sustainable choices	Certification based on recyclability, carbon impact	UL, ISO, IEC standards	Informs customers, drives green innovation	EU Battery Passport Initiative
Localized Recovery Infrastructure	Reduce logistics & emissions	Regional recycling centers, mobile labs	Modular recycling systems, mobile refurbishing units	Promotes local economy, reduces costs/emissions	ENVACORE – Local BESS refurbishing in rural India

As part of its ambitious climate agenda, the EU Green Deal promotes the circular economy as a fundamental strategy for long-term sustainable development. One important initiative is the EU Battery Regulation (2023), which includes mandatory collection rates, recycling efficiency targets, and a minimum recycled content in new batteries. The regulation also introduces a digital battery passport, a blockchain-based system for tracking battery materials that ensures transparency from sourcing to end-of-life. These frameworks not only support circularity but also promote innovation in digital compliance and lifecycle monitoring.

BESS remanufacturing restores used battery components or entire systems to working order, including replacing or repairing batteries, repairing electronics, and updating firmware. This process extends the life cycle of the battery system and reduces the environmental impact associated with new battery production. Recycling metals from electronic waste presents a promising opportunity and innovative business model for Small and Medium Enterprises (SMEs). To effectively manage e-waste and capitalize on the expanding metal recovery market, several auditing and assessment policies have been developed; some of them are discussed below:

Material Flow Analysis

Material Flow Analysis (MFA) is a widely used and essential method in industrial systems, aimed at quantifying the input and output flows of materials within an ecosystem. It measures how materials are used, reused, and ultimately disposed of, thereby supporting sustainable development in modern society. Accurate data collection is crucial for effective MFA and typically involves reviewing production records, conducting surveys, and directly measuring material flows. Various modeling techniques are employed to simulate and analyze these flows, ranging from simple spreadsheets to advanced software tools. STAN (Substance Flow Analysis) and system dynamics models are specialized tools used in MFA to visualize and analyze complex material flows [39, 40].

Extended Producer Responsibility (EPR)

Extended Producer Responsibility (EPR) is an environmental strategy that shifts the responsibility for managing the entire life cycle costs of products from governments and taxpayers to producers and consumers. EPR policies encourage manufacturers to recycle products, reduce the use of hazardous materials, and utilize recycled materials in production to minimize environmental impact. The core principle of EPR is that producers should be accountable not only for the products they sell but also for their environmental effects throughout the product life cycle. This includes ensuring that products are properly collected, recycled, or safely disposed of at the end of their use. EPR is expected to enhance the circularity of value chains affected by regulation [41, 42].

Life Cycle Assessment (LCA)

Life Cycle Assessment (LCA) is a vital tool for evaluating the environmental impacts of products and processes across all stages, from raw material extraction and manufacturing to transportation, use, recycling, and disposal. It aids in analyzing design choices and assessing potential environmental effects. Over the past decade, a key challenge in process systems engineering has been developing tools that effectively integrate these environmental considerations into design and analysis. LCA provides a comprehensive perspective on how each stage of a product or process affects the environment, enabling organizations and consumers to make more informed and sustainable decisions [43].

Advanced Recycling Fee (ARF)

The Advance Recycling Fee (ARF) policy is designed to encourage the recycling of electronic products, promote sustainable development, improve recycling practices, and enhance the management of end-of-life products. It does so by integrating recycling costs into the economy and reducing environmental impacts. Under this strategy, the government takes responsibility for product reprocessing and imposes a post-purchase recycling fee on consumers. This fee is applied when purchasing new or refurbished electronic products, and retailers transfer the collected fees to local governments to support the product take-back system. Notably, the ARF must be itemized on purchase receipts to inform consumers that the products they buy are subject to recycling requirements [44, 45].

DISCUSSION

Industrial batteries, such as Battery Energy Storage Systems (BESS), are transforming the modern energy landscape by storing surplus energy generated from renewable sources like solar and wind. BESS ensures that excess energy is not wasted and can be released during periods of high demand or low generation, thereby facilitating greater integration of green energy into the grid. In addition to supporting clean energy adoption, BESS enhances energy security and grid stability. By reducing reliance on fossil fuel-based generation during peak demand, BESS contributes to a more sustainable, efficient, and resilient energy system [46 - 48].

The initial list of barriers, drivers, and facilitators for BESS end-of-life management was highly supported by all stakeholders, with only some differences emerging. Driven by triple bottom line issues and a greater emphasis on research, academic institutions, and economic and environmental drivers, the main barriers identified included lack of profitability, inadequate regulation and incentives, and limited awareness of safe disposal options. Notably, only government respondents highlighted the deficiency of onshore reprocessing infrastructure as a significant barrier [49]. Effective end-of-life management of BESS is critical to minimizing environmental impact. This includes not only recycling materials but also finding secondary uses for batteries. For example, batteries that are no longer suitable for grid-scale storage may still have useful potential for less demanding applications, such as home energy storage systems or small renewable energy projects [50, 51].

Developing markets and systems for the use of secondary batteries will be key to enhancing the sustainability of BESS in the circular economy. Corporate social responsibility encourages companies to adopt proactive, self-regulatory measures to reduce environmental and health impacts throughout their value chains. Companies with a strong focus on environmental responsibility are motivated more by the goal of boosting market competitiveness and improving their green image than by the economic gains from recycling end-of-life products [52, 53].

Governments and entrepreneurs should use not only regulatory measures to promote recycling activities, but also support the development of new technologies through economic incentives and other approaches. In this direction, regulatory approaches can be implemented in which governments set stringent rules and standards that make recycling activities mandatory. Non-regulatory approaches include awareness programs, community participation, and volunteer efforts. Economic incentives, such as subsidies, tax breaks, or grants, can encourage the development of new recycling skills. Research companies should be encouraged to collaborate with the government and industry to develop more effective and efficient recycling technologies. Partnerships and collaboration between various stakeholders can implement more effective and sustainable measures. Such types of efforts can not only solve environmental problems but also create economic growth and new employment opportunities [54, 55].

CONCLUSION

In light of the discussed strategies, tools, and cases, circular economy integration is not just a theoretical ideal, but also a practical way to achieve long-term sustainable development. Integrating circular economy ethics into the lifecycle management of solar panels, wind turbines, and battery storage systems can ensure a sustainable energy future. These principles not only address ecological and environmental concerns but also stimulate innovation and improve the overall efficiency of renewable energy solutions. To encourage the adoption of circular economy practices, stakeholder engagement can be enhanced through workshops, seminars, and case studies. Awareness should be raised through the use of informative materials and clear communication campaigns that highlight the environmental and economic benefits of circular products. Integrating circular principles into renewable energy is not only a pathway but also a necessity to achieve a sustainable planet.

In addition, governments can play a crucial role in promoting the development of sustainable practices by providing incentives, such as tax exemptions and subsidies, for the adoption of circular practices, as well as by implementing regulations that mandate or encourage recycling and responsible product management. Here, BESS and grid interaction have been observed, highlighting the important roles played by BESS in grid stability and reliability, renewable energy integration, energy arbitrage, backup power, peak shaving, and demand response.

These concepts are no longer just theoretical; numerous initiatives worldwide are actively working to integrate circular approaches into renewable energy systems. Some notable examples illustrate this shift. Tesla, Inc., Samsung SDI, LG Energy Solution, Envision AESC, Northvolt, Fortum, Ecobat, and Umicore are among the leading companies integrating circular economy strategies into Battery Energy Storage System (BESS) business models. Their efforts help reduce the environmental impact of battery production and disposal, promote sustainable practices, and enhance the overall efficiency of energy storage systems.

Looking ahead, companies like Stem, Inc., which specializes in energy storage, are employing remanufacturing technologies to extend the lifespan of their battery systems. Additionally, Redwood Materials, founded by former Tesla CTO JB Straubel, is revolutionizing the circular economy of battery materials by repurposing used Electric Vehicle (EV) batteries into sustainable energy storage solutions.

REFERENCES

[1] M. Yang, L. Chen, J. Wang, G. Msigwa, A.I. Osman, S. Fawzy, D.W. Rooney, and P.S. Yap, "Circular economy strategies for combating climate change and other environmental issues", *Environ. Chem. Lett.,* vol. 21, no. 1, pp. 55-80, 2023.
[http://dx.doi.org/10.1007/s10311-022-01499-6]

[2] A. Kumar, *Global warming, climate change and greenhouse gas mitigation*, 2018.
[http://dx.doi.org/10.1007/978-81-322-3763-1_1]

[3] S. Vishnoi, and R.K. Goel, "Climate smart agriculture for sustainable productivity and healthy landscapes", *Environ. Sci. Policy,* vol. 151, p. 103600, 2024.
[http://dx.doi.org/10.1016/j.envsci.2023.103600]

[4] J. Rawat, M. Nanda, S. Kumar, N. Sharma, R. Sharma, H.C. Joshi, M.S. Vlaskin, A. Hussain, and V. Kumar, "Integrating wastewater treatment to bio-stimulant & biochar generation for plant growth promotion using microalgae", *Process Biochem.,* vol. 145, pp. 187-194, 2024.
[http://dx.doi.org/10.1016/j.procbio.2024.06.031]

[5] F.O. Ongondo, I.D. Williams, and T.J. Cherrett, "How are WEEE doing? A global review of the

management of electrical and electronic wastes", *Waste Manag.*, vol. 31, no. 4, pp. 714-730, 2011.
[http://dx.doi.org/10.1016/j.wasman.2010.10.023] [PMID: 21146974]

[6]   S. Rashid, and S.H. Malik, "Transition from a linear to a circular economy", In: *Renewable Energy in Circular Economy*. Springer International Publishing: Cham, 2023, pp. 1-20.
[http://dx.doi.org/10.1007/978-3-031-42220-1_1]

[7]   A. Chauhan, and H.C. Joshi, "Recent developments and applications in bioconversion and biorefineries", *Trends in Mathematics,* no. Part F3197, pp. 247-307, 2024.
[http://dx.doi.org/10.1007/978-981-99-7250-0_6]

[8]   R.K. Goel, C.S. Yadav, and S. Vishnoi, "Self-sustainable smart cities: Socio-spatial society using participative bottom-up and cognitive top-down approach", *Cities,* vol. 118, p. 103370, 2021.
[http://dx.doi.org/10.1016/j.cities.2021.103370]

[9]   T. Nogueira, E. Sousa, and G.R. Alves, "Electric vehicles growth until 2030: Impact on the distribution network power", *Energy Rep.,* vol. 8, pp. 145-152, 2022.
[http://dx.doi.org/10.1016/j.egyr.2022.01.106]

[10]   J. Arambarri, J. Hayden, M. Elkurdy, B. Meyers, Z.S. Abu Hamatteh, B. Abbassi, and W. Omar, "Lithium ion car batteries: Present analysis and future predictions", *Environ. Eng. Res.,* vol. 24, no. 4, pp. 699-710, 2019.
[http://dx.doi.org/10.4491/eer.2018.383]

[11]   M. Geissdoerfer, P. Savaget, N.M.P. Bocken, and E.J. Hultink, "The Circular Economy – A new sustainability paradigm?", *J. Clean. Prod.,* vol. 143, pp. 757-768, 2017.
[http://dx.doi.org/10.1016/j.jclepro.2016.12.048]

[12]   F. Blomsma, and G. Brennan, "The emergence of circular economy: A new framing around prolonging resource productivity", *J. Ind. Ecol.,* vol. 21, no. 3, pp. 603-614, 2017.
[http://dx.doi.org/10.1111/jiec.12603]

[13]   X. Zeng, J. Li, and N. Singh, "Recycling of spent lithium-ion battery: A critical review", *Crit. Rev. Environ. Sci. Technol.,* vol. 44, no. 11, pp. 1129-1165, 2018.

[14]   E. Dominish, N. Florin, and S. Teske, *Responsible minerals sourcing for renewable energy.* UN Environment Programme, 2019.

[15]   P. Liu, and C.Y. Barlow, "Wind turbine blade waste in 2050", *Waste Manag.,* vol. 62, pp. 229-240, 2017.
[http://dx.doi.org/10.1016/j.wasman.2017.02.007] [PMID: 28215972]

[16]   A. Chauhan, "Harish Chandra Joshi, Energetic efficiency of a biofuels production system mathematical modeling", *Trends in Mathematics,* no. Part F3197, pp. 333-357, 2024.

[17]   L. Gaines, "The future of automotive lithium-ion battery recycling: Charting a sustainable course", *Sustainable Materials and Technologies,* vol. 1-2, pp. 2-7, 2014.
[http://dx.doi.org/10.1016/j.susmat.2014.10.001]

[18]   J. Kirchherr, D. Reike, and M. Hekkert, "Conceptualizing the circular economy: An analysis of 114 definitions", *Resour. Conserv. Recycling,* vol. 127, pp. 221-232, 2017.
[http://dx.doi.org/10.1016/j.resconrec.2017.09.005]

[19]   T. Domenech, and B. Bahn-Walkowiak, "Transition towards a resource-efficient circular economy in Europe: Policy lessons from the EU and the member states", *Ecol. Econ.,* vol. 155, pp. 7-19, 2019.
[http://dx.doi.org/10.1016/j.ecolecon.2017.11.001]

[20]   X. Zhao, G. Dragan, and N.M. Schmidt, "Circular economy in energy storage: Research pathways to a sustainable future", *J. Clean. Prod.,* vol. 275, p. 123048, 2020.

[21]   H.C. Joshi, Reetika, Waseem, Bhawana, Nishesh Sharma, "A review on carbonaceous materials for fuel cell technologies:An advanced approach", *Vietnam J. Chem.,* pp. 1-10, 2024.
[http://dx.doi.org/10.1002/vjch.202300407]

[22] M. Haupt, C. Vadenbo, and S. Hellweg, "Do we have the right performance indicators for the circular economy? Insight into the Swiss waste management system", *J. Ind. Ecol.,* vol. 21, no. 3, pp. 615-627, 2017.
[http://dx.doi.org/10.1111/jiec.12506]

[23] M. Tao, and Z. Yu, "A review of recycling processes for photovoltaic modules", *Sol. Energy Mater. Sol. Cells,* vol. 141, pp. 108-115, 2015.
[http://dx.doi.org/10.1016/j.solmat.2015.05.005]

[24] J.F. Peters, M. Baumann, B. Zimmermann, J. Braun, and M. Weil, "The environmental impact of Li-Ion batteries and the role of key parameters – A review", *Renew. Sustain. Energy Rev.,* vol. 67, pp. 491-506, 2017.
[http://dx.doi.org/10.1016/j.rser.2016.08.039]

[25] P. Schroeder, K. Anggraeni, and U. Weber, "The relevance of circular economy practices to the sustainable development goals", *J. Ind. Ecol.,* vol. 23, no. 1, pp. 77-95, 2019.
[http://dx.doi.org/10.1111/jiec.12732]

[26] "Circular Economy Action Plan: For a cleaner and more competitive", *Europe,* 2020.

[27] D. Sica, O. Malandrino, S. Supino, M. Testa, and M.C. Lucchetti, "Management of end-of-life photovoltaic panels as a step towards a circular economy", *Renew. Sustain. Energy Rev.,* vol. 82, pp. 2934-2945, 2018.
[http://dx.doi.org/10.1016/j.rser.2017.10.039]

[28] C. Market, *Fortune Business Insights.* Cosmeceuticals Market, 2023.

[29] A. Chauhan, and H.C. Joshi, "Lignocellulosic biomass for the conversion of bioethanol: production and optimization", *Trends in Mathematics,* no. Part F3197, pp. 187-214, 2024.
[http://dx.doi.org/10.1007/978-981-99-7250-0_4]

[30] S. Atcitty, J. Neely, D. Ingersoll, A. Akhil, and K. Waldrip, Battery energy storage system.*Power Electronics for Renewable and Distributed Energy Systems: A Sourcebook of Topologies.* Control and Integration, 2013, pp. 333-366.
[http://dx.doi.org/10.1007/978-1-4471-5104-3_9]

[31] C. Zhang, Y.L. Wei, P.F. Cao, and M.C. Lin, "Energy storage system: Current studies on batteries and power condition system", *Renew. Sustain. Energy Rev.,* vol. 82, pp. 3091-3106, 2018.
[http://dx.doi.org/10.1016/j.rser.2017.10.030]

[32] R.K. Goel, and S. Vishnoi, "Strengthening and sustaining health-related outcomes through digital health interventions", *Journal of Engineering Science and Technology Review,* vol. 12, no. 2, pp. 10-17, 2023.
[http://dx.doi.org/10.25103/jestr.162.02]

[33] T. Genish, and S. Boopathiraja, "Biomass renewable energy: introduction and application of AI and IoT", In: *AI-Powered IoT in the Energy Industry,* S. Vijayalakshmi, . S., B. Balusamy, R.K. Dhanaraj, Eds., Springer: Cham, Switzerland, 2023.
[http://dx.doi.org/10.1007/978-3-031-15044-9_8]

[34] E. Reihani, S. Sepasi, L.R. Roose, and M. Matsuura, "Energy management at the distribution grid using a Battery Energy Storage System (BESS)", *Int. J. Electr. Power Energy Syst.,* vol. 77, pp. 337-344, 2016.
[http://dx.doi.org/10.1016/j.ijepes.2015.11.035]

[35] B. Xu, A. Oudalov, J. Poland, A. Ulbig, and G. Andersson, "BESS control strategies for participating in grid frequency regulation",
[http://dx.doi.org/10.3182/20140824-6-ZA-1003.02148]

[36] G.J. May, A. Davidson, and B. Monahov, "Lead batteries for utility energy storage: A review", *J. Energy Storage,* vol. 15, pp. 145-157, 2018.
[http://dx.doi.org/10.1016/j.est.2017.11.008]

[37] J. McDowall, "Nickel-cadmium batteries for energy storage applications", *Fourteenth Annual Battery Conference on Applications and Advances. Proceedings of the Conference (Cat. No.99TH8371),* 1999pp. 303-308 Long Beach, CA, USA
[http://dx.doi.org/10.1109/BCAA.1999.796008]

[38] M. Shahjalal, P.K. Roy, T. Shams, A. Fly, J.I. Chowdhury, M.R. Ahmed, and K. Liu, "A review on second-life of Li-ion batteries: Prospects, challenges, and issues", *Energy,* vol. 241, p. 122881, 2022.
[http://dx.doi.org/10.1016/j.energy.2021.122881]

[39] C. Sendra, X. Gabarrell, and T. Vicent, "Material flow analysis adapted to an industrial area", *J. Clean. Prod.,* vol. 15, no. 17, pp. 1706-1715, 2007.
[http://dx.doi.org/10.1016/j.jclepro.2006.08.019]

[40] T.E. Graedel, "Material flow analysis from origin to evolution", *Environ. Sci. Technol.,* vol. 53, no. 21, pp. 12188-12196, 2019.
[http://dx.doi.org/10.1021/acs.est.9b03413] [PMID: 31549816]

[41] M. Compagnoni, "Is Extended Producer Responsibility living up to expectations? A systematic literature review focusing on electronic waste", *J. Clean. Prod.,* vol. 367, p. 133101, 2022.
[http://dx.doi.org/10.1016/j.jclepro.2022.133101]

[42] Y. Gupt, and S. Sahay, "Review of extended producer responsibility: A case study approach", *Waste Manag. Res.,* vol. 33, no. 7, pp. 595-611, 2015.
[http://dx.doi.org/10.1177/0734242X15592275] [PMID: 26185163]

[43] L. Jacquemin, P.Y. Pontalier, and C. Sablayrolles, "Life Cycle Assessment (LCA) applied to the process industry: a review", *Int. J. Life Cycle Assess.,* vol. 17, no. 8, pp. 1028-1041, 2012.
[http://dx.doi.org/10.1007/s11367-012-0432-9]

[44] I.H. Hong, Y.T. Lee, and P.Y. Chang, "Socially optimal and fund-balanced advanced recycling fees and subsidies in a competitive forward and reverse supply chain", *Resour. Conserv. Recycling,* vol. 82, pp. 75-85, 2014.
[http://dx.doi.org/10.1016/j.resconrec.2013.10.018]

[45] F. Shan, W. Xiao, and F. Yang, "Comparison of three E-Waste take-back policies", *Int. J. Prod. Econ.,* vol. 242, p. 108287, 2021.
[http://dx.doi.org/10.1016/j.ijpe.2021.108287]

[46] S. Amiri, *Comprehensive Review and Comparative Analysis of Hydropower Integration Strategies and Energy Storage Technology.* Comprehensive Review, 2022. Available at: https://hdl.handle.net/10589/214674

[47] C. Zhao, P.B. Andersen, C. Træholt, and S. Hashemi, "Grid-connected battery energy storage system: A review on application and integration", *Renew. Sustain. Energy Rev.,* vol. 182, p. 113400, 2023.
[http://dx.doi.org/10.1016/j.rser.2023.113400]

[48] A. Chauhan, "Sources for biofuels production from biomass", *Trends in Mathematics,* no. Part F3197, pp. 1-64, 2024.

[49] H.K. Salim, R.A. Stewart, O. Sahin, and M. Dudley, "End-of-life management of solar photovoltaic and battery energy storage systems: A stakeholder survey in Australia", *Resour. Conserv. Recycling,* vol. 150, p. 104444, 2019.
[http://dx.doi.org/10.1016/j.resconrec.2019.104444]

[50] R. Kijak, and E. Gashi, "Strategic planning for the Battery Energy Storage System (BESS) safety and reliability over the BESS life cycle", *2024 International Conference on Renewable Energies and Smart Technologies (REST),* 2024pp. 1-5
[http://dx.doi.org/10.1109/REST59987.2024.10645454]

[51] A.T. Nguyen, S. Chaitusaney, and A. Yokoyama, "Optimal strategies of siting, sizing, and scheduling of BESS: Voltage management solution for future LV network", *IEEJ Trans. Electr. Electron. Eng.,* vol. 14, no. 5, pp. 694-704, 2019.

[http://dx.doi.org/10.1002/tee.22856]

[52] M. Koese, C.F. Blanco, V.B. Vert, and M.G. Vijver, "A social life cycle assessment of vanadium redox flow and lithium-ion batteries for energy storage", *J. Ind. Ecol.,* vol. 27, no. 1, pp. 223-237, 2023.
[http://dx.doi.org/10.1111/jiec.13347]

[53] R.K. Goel, and S. Vishnoi, "Urbanization and sustainable development for inclusiveness using ICTs", *Telecomm. Policy,* vol. 46, no. 6, p. 102311, 2022.
[http://dx.doi.org/10.1016/j.telpol.2022.102311]

[54] H.K. Salim, R.A. Stewart, O. Sahin, and M. Dudley, "Drivers, barriers and enablers to end-of-life management of solar photovoltaic and battery energy storage systems: A systematic literature review", *J. Clean. Prod.,* vol. 211, pp. 537-554, 2019.
[http://dx.doi.org/10.1016/j.jclepro.2018.11.229]

[55] W. Lu, L. Du, and Y. Feng, "Decision making behaviours and management mechanisms for construction and demolition waste recycling based on public–private partnership", *Environ. Sci. Pollut. Res. Int.,* vol. 29, no. 54, pp. 82078-82097, 2022.
[http://dx.doi.org/10.1007/s11356-022-21221-x] [PMID: 35748991]

CHAPTER 11

Smart Bioethanol Production System From Algal Biomass Harvested From River Water

Harish Chandra Joshi[1]**, Shubham Kumar Singh**[2]**, Anand Chauhan**[2,*] **and Abhinav Goel**[2]

[1] *Department of Chemistry, Graphic Era (Deemed to be) University, Dehradun, Uttarakhand, India*

[2] *Department of Mathematics, Graphic Era (Deemed to be) University, Dehradun, Uttarakhand, India*

Abstract: Researchers emphasize that prioritizing sustainable and renewable energy sources is essential for mitigating the rapid progression of global warming driven by excessive fossil fuel consumption. Bioethanol presents a promising alternative to fossil fuels, as it can be used in engines without modification, helping to reduce carbon emissions and air pollution. This study investigates the potential of algal biomass for bioethanol production and outlines a model for the complete conversion process. To achieve this, the research developed a smart bioethanol production system with variable production rates that effectively reduced energy consumption. This system effectively reduced contaminants in bioethanol by utilizing an automated inspection framework. The impure bioethanol undergoes a purification process to enhance its quality. The developed mathematical model accounts for both upstream and downstream processes, providing flexibility and adaptability for bioethanol production. This study aims to reduce the total cost of an algal-based bioethanol production system, thereby achieving the optimal quantity of bioethanol. The model's findings highlight that the main economic challenges arise from the high costs of harvesting and extracting algal biomass, particularly from riverbanks. This research contributes to the body of knowledge on sustainable energy production and underscores the need for targeted policy measures to transition from conventional fossil fuels to renewable biofuels.

Keywords: Algal biomass, Autonomation, Bioethanol, Smart production.

* **Corresponding author Anand Chauhan**: Department of Mathematics, Graphic Era (Deemed to be) University, Dehradun, Uttarakhand, India; E-mail: dranandchauhan83@gmail.com

INTRODUCTION

When it comes to the generation of second-generation bioethanol, it is noted that such feedstock could be potentially produced from lignocellulosic materials and is beneficial in several ways . Several feedstock pre-treatments have presented various yield and product inhibitory processing challenges. However, the pre-treatment of lignocellulosic biomass is a critical step in bioethanol production due to the inherently recalcitrant nature of these materials. The most prevalent carbonous fuel is biomass, which is ultimately derived from both microorganisms and plants. It is also recognized as lignocellulose-based biomass, which has now become a common biomass to facilitate the better production of bioethanol. The most common constituents of plant tissue cells are the cellulosic components, which are composed of cell structural carbohydrates, including cellulose, hemicellulose, and other unbranched structures, such as lignin. The contents of these components may vary significantly depending on the species, variety, and climatic conditions. In addition, several pre-treatment methods, including physical, chemical, biophysical, and biological [1, 2], have been developed to enhance enzyme access to cellulosic fibers. Ethanol is produced through a series of key steps, including the hydrolysis of cellulose and hemicellulose into fermentable reducing sugars, the fermentation of these sugars into ethanol, the separation of lignin residues, and the recovery and purification of ethanol to meet fuel-grade specifications [3]. Glucose and xylose are fermented in separate reactors in the traditional bioethanol production process, yielding a substantial amount of ethanol. This procedure significantly increases the yield and productivity of ethanol and can be carried out in a bioreactor using hydrolysate fractions rich in glucose and xylose [4 - 7].

Lignocellulosic biomass has a complex structure that includes both fermentable and non-fermentable sugars. With a compositional analysis of 33–47%, cellulose is the most abundant component of lignocellulosic biomass (LCB). Another abundant component is hemicellulose, which accounts for 19–27% of the LCB. The surface area that is available for enzyme activation, preventing degradation, and forming a barrier against external entry [8, 9].

The appropriate industrial use of biomass requires a well-planned and controlled production infrastructure. By making deliberate decisions about the production structure, modes of transportation, environmental impacts, social benefits, and node locations, it is possible to industrialize a bioethanol production system [10, 11]. Several researchers have examined the traditional process of producing bioethanol. Since labour and production problems are frequently caused by the traditional techniques used in bioethanol production, an innovative manufacturing system is vital. As traditional manufacturing methods often generate defective goods, intelligent production systems should strive to reduce their production. Consequently, implementing an energy-efficient bioethanol production system that reduces bioethanol pollutants is essential [12]. The smart production system uses automation techniques to identify any contaminants in the bioethanol. The impure bioethanol undergoes a purification process after each in-house production of bioethanol. After purification and asmart inspection process, the purified form of bioethanol is transferred to the bioethanol marketplaces.

Compared to typical ethanol systems, they have lower yields, require less land-use change, and emit fewer CO_2 emissions (like corn or sugarcane). Higher biomass yields, the ability to utilize non-arable land, and the potential for net CO_2 capture make microalgae-based systems more sustainable overall (Table **1**).

Table 1. Comparative study of traditional ethanol with algal-based ethanol

Parameters	Algal-based ethanol	Traditional ethanol
Yield	5000-10000 gallons/acre/ year	400-500 gallons/acre/year
C-absorptivity	Captures	Emit during processing
Feedstock	Macro/micro, Sunlight, CO_2 and nutrients	Food Crops
Land use	Can also flourish on marginal	Need fertile land

Mathematical Model

This study develops a mathematical model to minimize the total production cost of algal-based bioethanol. The model focuses on optimizing the allocation of algal biomass to the biorefinery after its harvest from a freshwater river, as well as the distribution of purified bioethanol to market demand zones. A liquid extraction process, utilizing the Soxhlet method, is employed to extract bioethanol from the algal biomass. Key cost components considered in the model include the collection of algal biomass, interim storage, and loading the biomass into transport equipment. The model also determines the variable production cost of bioethanol, which includes expenses such as purification (the process of converting impure bioethanol into pure bioethanol), autonomous inspection, and inventory maintenance at the biorefinery. Transportation costs for moving algal biomass from riverbanks to the biorefinery, as well as the distribution of bioethanol to market demand zones, are also accounted for in the total cost minimization. The bioethanol inventory behaviour in the biorefinery is shown in Fig. (**1**).

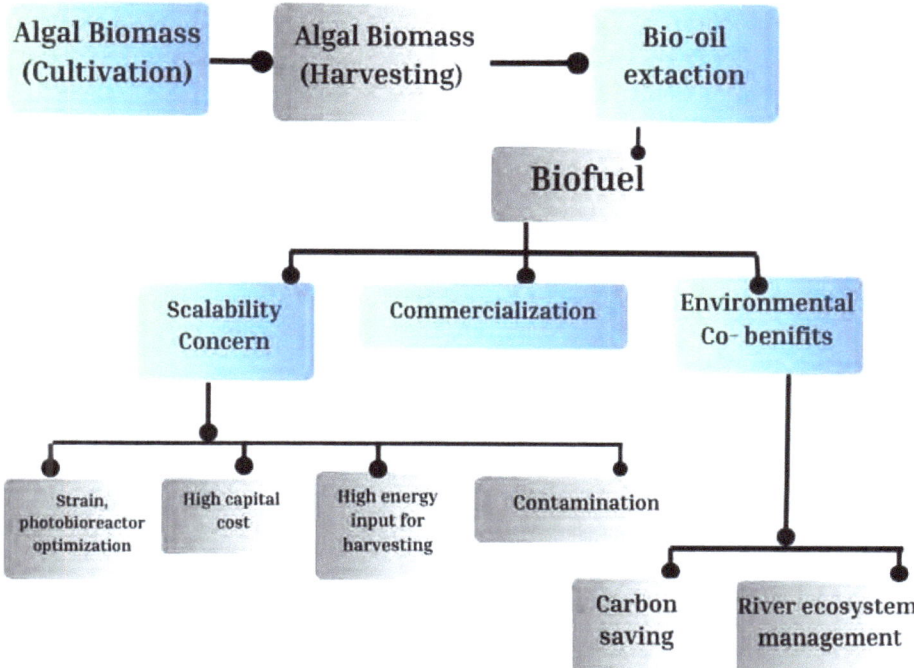

Fig. (1). Flow diagram for the production, environmental benefits, and scalability concerns.

Assumptions

- In the proposed algal-based bioethanol smart production model, the bioethanol variable production rate is denoted as P_1. A variable production rate offers the advantage of enabling management to adjust output in response to fluctuations or disruptions in the production process.
- Assumedly, the bioethanol production system has the random ability to generate $\lambda\%$ of contaminants in bioethanol at a rate of x, with x= λP_1, and if the condition (P1-x-D)>0 is satisfied, where D represents the bioethanol demand, then production processes can be maintained, ensuring no shortages occur.
- All the impure bioethanol is purified at the rate of P_2 at the end of each in-house production cycle with additional cost (Table **2**).
- The model incorporates smart autonomation inspection techniques. After the purification process, no impure bioethanol remains in the biorefinery, as confirmed by a thorough inspection. The first inspection of the entire bioethanol batch is performed using advanced smart inspection methods. In contrast, the second inspection, conducted after purification, is carried out by quality inspectors to ensure the bioethanol meets quality standards.

Table 2. List of notations for bioethanol smart production model [12].

r	locations for riverbanks (r = 1, 2, 3 ..., R)
d	bioethanol market demand zones (d = 1, 2, 3 ..., D)
Q_r	Net amount of algal biomass delivered from riverine banks to the site of the biorefinery.
Q	Produced a volume of bioethanol at the biorefinery
Q_d	Volume of bioethanol transferred from the biorefinery to the demand zone d
A_r	Available area for algal biomass at the riverbank r
π	The rate of conversion of algal biomass to bioethanol at the biorefinery
H_1	When in-house production is finalized, the highest inventory of bioethanol
H_2	When purification is completed, the highest inventory of bioethanol
C_c	The unit cost of collecting algal biomass at the riverbank r
C_s	The unit cost of temporary storage for algal biomass at the riverbank r
C_l	Unit cost of loading algal biomass into loaders at the riverbank r
C_{hr}	Harvesting cost per unit of algal biomass
C_{ex}	extraction cost per unit of algal biomass
ρ	Development cost per unit of bioethanol production
δ	Material cost per unit of bioethanol production
$α_1$	Scaling factor associated with the tool/die cost function for bioethanol production
$α_2$	Shape parameter associated with the tool/die cost function for bioethanol
ϒ	Cost of inspection for bioethanol
ψ	Inspection cost for impure bioethanol
INF	The fixed cost for a bioethanol quality inspector
CP	Per unit purification cost for impure bioethanol at the biorefinery
h_1	Inventory cost for per-unit bioethanol holding at the biorefinery
h_2	Holding cost per unit of bioethanol undergoing purification in a biorefinery
CCT_r	Transportation cost per unit of algal biomass from the riverbank r to the biorefinery
CCT_d	Transportation cost per volume of bioethanol from the biorefinery to the demand zone d

Notations

After completing in-house production, the biorefinery reached its maximum bioethanol inventory, as shown in Equation (1) and illustrated in Fig. (**1**).

$$H_1 = \left[\left((1 - E[\lambda]) * P_1\right) - D\right] * t_1 \tag{1}$$

Upon completion of the purification process, the biorefinery held its largest inventory of bioethanol, as indicated by Equation (2) and illustrated in Fig. (**1**).

$$H_2 = H_1 + (P_2 - D) * t_2 \tag{2}$$

Using assumptions,

$$t_2 = \frac{E[\lambda] \times Q}{P_2} \tag{3}$$

$$t_1 = \frac{Q}{P_1} \tag{4}$$

The total production cycle time

$$T = t_1 + t_2 + t_3 \tag{5}$$

Algal biomass cost

Equation (6) represents the total cost associated with collecting algal biomass, including temporary storage and loading onto transport equipment.

$$COST_{algal} = \sum_{r=1}^{R}(C_c + C_s + C_l) * Q_r \tag{6}$$

Liquid extraction cost

Algal biomass is harvested directly from a freshwater river, followed by liquid extraction using the Soxhlet extraction method.

$$COST_{lqe} = (C_{hr} + C_{ex}) * Q_r \tag{7}$$

Production cost

The biorefinery's bioethanol production cost is calculated by summing the tool and die costs, material costs, and development costs.

$$COST_{prod} = \left(\frac{\rho}{P_1} + \delta + \alpha_1 P_1^{\alpha_2}\right) * Q \tag{8}$$

Smart inspection cost

An autonomous inspection process is implemented to detect contaminants in the bioethanol. The machinery automatically manages the inspection and removes any detected impurities, ensuring high precision. This automation approach represents a smart, error-free inspection method [13]. However, a two-stage inspection is employed to further ensure that bioethanol is free of contaminants. Following the automated process, a quality inspector conducts a final analysis to confirm the purity of the bioethanol (Fig. **2**).

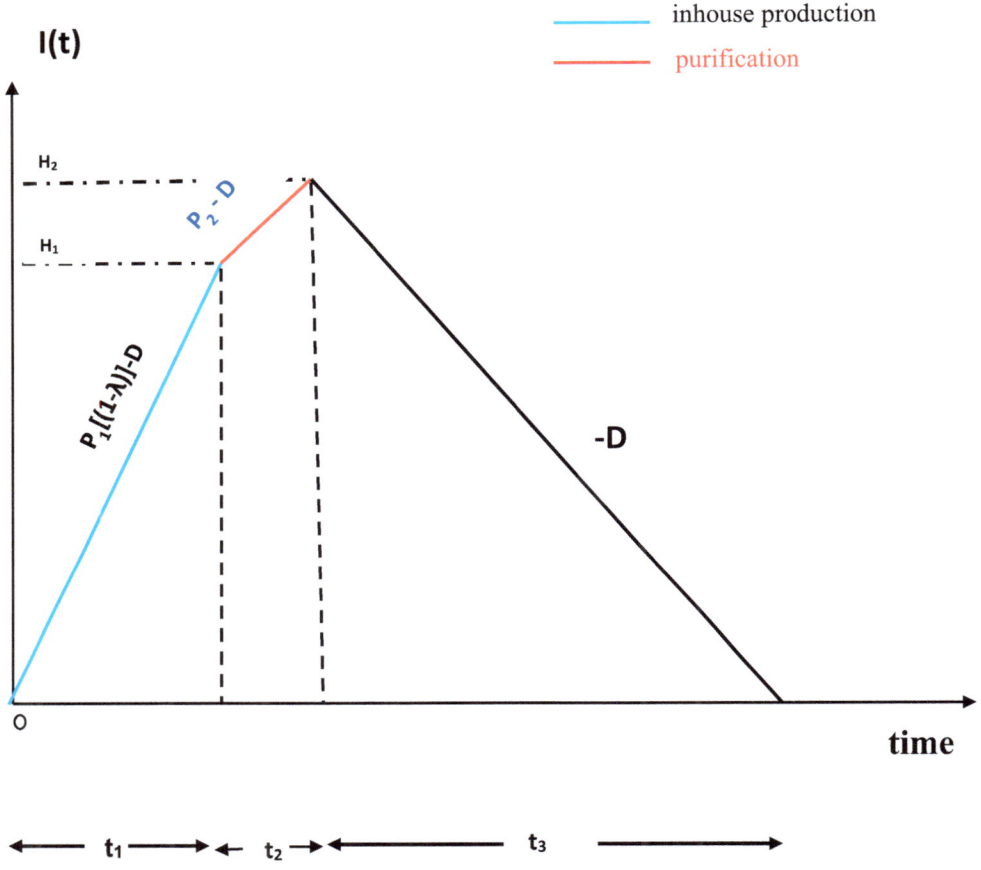

Fig. (2). Bioethanol inventory behaviour includes in-house production and purification at the same time.

$$COST_{insp} = (\Upsilon * D) + (\psi * E[\lambda] * D) + (INF) \tag{9}$$

Purification cost

To produce contaminant-free bioethanol, the purification process starts at a rate of P_2 in the biorefinery after the standard production process is completed. This purification step incurs additional costs to ensure the bioethanol meets the required purity standards (Fig. **2**).

$$COST_{purif} = CP * E[\lambda] * D \tag{10}$$

Inventory holding cost

At the biorefinery, the cost of holding inventory for impure bioethanol is captured in Equation (11), which reflects the cost associated with storage over time (Fig. **2**).

$$INV_{im} = h_2 * (E[\lambda] * \frac{Q}{2} * t_1) \tag{11}$$

The inventory holding cost at a biorefinery for pure bioethanol is given in Equation (12).

$$INV_{pu} = h_1 * [\frac{H_1 + E[\lambda] * Q}{2} * t_1 + \frac{H_1 + H_2}{2} * t_2] \tag{12}$$

The total inventory holding cost for bioethanol, accounting for both impure and purified forms during storage, is formulated as shown in Equation (13).

$$COST_{inv} = INV_{im} + INV_{pu} \tag{13}$$

Transportation Cost

Equation (14) estimates the total transportation cost, covering the movement of algal biomass from riverbanks to the biorefinery, as well as the distribution of bioethanol from the biorefinery to market demand zones.

$$COST_{trans} = \sum_{r=1}^{R}(CCT_r * Q_r) + \sum_{d=1}^{D}(CCT_d * Q_d) \tag{14}$$

Supply Constraint

Supply constraints (15) indicate that the quantity of algal biomass transported from a single riverbank to the laboratory must not exceed the available biomass. The yield parameter (a_r) and the area of the riverbank determine the availability of algal biomass. However, the yield parameter is considered uncertain and fuzzy due to the influence of weather conditions and insect activity. To address this uncertainty, the signed distance method is employed to defuzzify the yield parameter, providing a more reliable estimate for supply planning (Fig. **2**).

$$\sum_{r=1}^{R} Q_r \leq (A_r * \breve{a}_r) \qquad \forall r \tag{15}$$

Conversion Constraint

Eq. (16) represents the algal biomass to bioethanol conversion processes after liquid extraction.

$$\sum_{r=1}^{R}(Q_r \times \pi) = Q \tag{16}$$

Mass Balance Constraint

The quantity of bioethanol transferred to the market demand zones is less than the quantity produced.

$$\sum_{d=1}^{D} Q_d \leq Q \quad \forall\, d \tag{17}$$

Non-negative constraints

$$Q_r, Q_d, Q \geq 0 \quad \forall\, r, d \tag{18}$$

Solution Methodology

The solution methodology for the mathematical model requires an approach that can effectively manage a diverse range of variables. Therefore, classical methods are not appropriate. Numerous methodologies can be employed to achieve optimal results, as this model aims to minimize the total costs associated with the bioethanol smart production system. Obtaining a closed-form solution for this mathematical model is exceedingly time-consuming. Consequently, the fixed-point iteration technique, a fundamental aspect of Non-linear Programming (NLP), is utilized [14].

Numerical Example and Results

A mathematical illustration is provided to assess the algal-based bioethanol smart production model design using the proposed methodology. A uniform distribution is assumed for the impure bioethanol production rate P_1. Table **3** outlines the input parameters for the algal-based bioethanol smart production model. For this example, the production process involves 10 riverbanks (*R1, R2, R3, R4, R5, R6, R7, R8, R9, R10*), one biorefinery with a capacity of 20,000,000 gallons, and four market-demand zones for pure bioethanol (*D1, D2, D3, D4*).

Table 3. The associated input parameters for the algal-based bioethanol smart production model.

Riverbanks	A_r (acre)	a_r (ton/acre)	C_e ($/ton)	C_s ($/ton)	C_l ($/ton)
R1	88000	1.58	1.89	8.11	0.84
R2	87000	1.68	1.93	8.51	0.96
R3	85000	1.61	2.0	8.75	0.84

(Table 3) cont.....

Riverbanks	A_r (acre)	a_r (ton/acre)	C_c ($/ton)	C_s ($/ton)	C_l ($/ton)
R4	75000	1.55	1.93	8.51	0.87
R5	65000	1.70	1.83	8.11	0.92
R6	55000	1.51	1.89	8.52	0.97
R7	50000	1.65	2.0	8.51	0.90
R8	40000	1.68	1.93	8.72	0.96
R9	45000	1.58	1.89	8.11	0.85
R10	52359	1.58	1.89	8.75	0.84
Cost for liquid extraction					
C_{hr}	15.45 ($/ton)	C_{ex}	17 ($/ton)		
Cost for bioethanol production					
ρ	$155	δ	$105	α_1	0.83
α_2	0.03	Υ	0.02 ($/item)	ψ	0.02 ($/defective item)
INF	$156000	CP	30 ($/item)	h_1	10 ($/unit/unit time)
h_2	30 ($/unit)	π	91.7		
Cost for transportation					
CCT_r	0.3($/ton-km)	CCT_d	0.7 ($/gal-km)		

The distribution of the bioethanol production rate P_1 is uniform [15], then $E[\lambda]$ is

$$E[\lambda] = \int_{\mu_1}^{\mu_2} \mu \, f(\mu) \, d\mu$$

$$E[\lambda] = \int_{\mu_1}^{\mu_2} \frac{\mu}{\mu_2 - \mu_1} \, d\mu = \frac{\mu_1 + \mu_2}{2}$$

$$E[\lambda] = \frac{0.005 + 0.015}{2} = 0.01$$

And the probability distribution function:

$$f(\mu) = \left\{ \begin{array}{l} 0; otherwire \\ \dfrac{1}{\mu_2 - \mu_1}; \ \mu_1 \leq \mu \leq \mu_2 \end{array} \right\}$$

The primary objective of the mathematical model is to minimize the total cost of algal-based bioethanol smart production. This includes the costs related to algal biomass, variable bioethanol production costs, smart inspection costs, purification costs, inventory holding costs, and transportation costs. The optimal total cost derived from the model is $\$5.594795 \times 10^6$. Additionally, the optimal production quantity of bioethanol (Q) is determined to be 31,002,279 gallons, which is sufficient to meet the market demand for bioethanol.

This study presents valuable managerial insights for an algal-based bioethanol production model through total cost minimization. The incorporation of smart automation inspection methods is essential for achieving pure bioethanol by reducing impurities and enhancing overall process efficiency. For managers, this highlights the importance of adopting advanced automation technologies to enhance bioethanol quality and reduce purification costs (Fig. **2**).

Additionally, the model's use of a variable production rate offers significant flexibility, allowing managers to adjust production output in response to real-time fluctuations in demand or operational disruptions. This adaptability enhances resource allocation, reduces the risk of shortages, and optimizes the production schedule.

CONCLUSION

The findings of this study demonstrate the substantial potential of algal biomass for bioethanol production, offering both biofuel and value-added products. The smart bioethanol production system used in this study was highly effective in consuming the least amount of energy and minimizing the likelihood of unpredictable production results. The smart bioethanol production system was employed because the conventional production framework could not identify error-free contaminants in bioethanol, which could have led to the production process becoming uncontrollable. By utilizing the automation inspection framework to detect contaminants and improve bioethanol quality, the production system's uncontrollable situation was mitigated. The study presents a scalable model for the full conversion of algae into bioethanol, providing a foundation for future biofuel development. According to studies, there are opportunities for government action to support the bioethanol sector by providing incentives, localizing technology, and promoting cooperation between national and international stakeholders. The mathematical illustration's findings demonstrate that the high costs associated with harvesting and extracting algal biomass, particularly from riverbanks, remain the most significant economic barriers to scaling up production. The flexible mathematical model developed in this study can also be adapted for various biofuel production scenarios, providing a valuable tool for future design and planning in the renewable energy sector.

Future research can enhance the current model by incorporating more realistic environmental and operational variables, such as uncertainties in biomass supply and market demand. Additionally, future studies could address key limitations by considering feedstock scarcity in the presence of market competition and the involvement of multiple processing plant complexes. The model can also be extended to explore coordination mechanisms within the biofuel supply chain under various transportation modes and multi-disruption scenarios, thereby improving its applicability in real-world, dynamic contexts.

REFERENCES

[1]   V.S. Chang, and M.T. Holtzapple, "Fundamental factors affecting biomass enzymatic reactivity", In: *Twenty-First Symposium on Biotechnology for Fuels and Chemicals,* M. Finkelstein, B.H. Davison, Eds., Humana Press: Totowa, NJ, USA, 2000.
[http://dx.doi.org/10.1007/978-1-4612-1392-5_1]

[2]   M. Han, Y. Kim, Y. Kim, B. Chung, and G-W. Choi, "Bioethanol production from optimized pretreatment of cassava stem", *Korean J. Chem. Eng.,* vol. 28, no. 1, pp. 119-125, 2011.
[http://dx.doi.org/10.1007/s11814-010-0330-4]

[3]   D. P. Maurya, A. Singla, and S. Negi, "An overview of key pretreatment processes for biological conversion of lignocellulosic biomass to bioethanol", *3 Biotech,* vol. 5, no. 5, pp. 597-609, Feb 2015.
[http://dx.doi.org/10.1007/s13205-015-0279-4]

[4]   A. Rodrigues Gurgel da Silva, A. Giuliano, M. Errico, B.G. Rong, and D. Barletta, "Economic value and environmental impact analysis of lignocellulosic ethanol production: assessment of different pretreatment processes", *Clean Technol. Environ. Policy,* vol. 21, no. 3, pp. 637-654, 2019.
[http://dx.doi.org/10.1007/s10098-018-01663-z]

[5]   K. Gehlot, R. Sivakumar, and S. Ghosh, "*In Situ* distillation strategy to improve the sequential fermentation process using *Zymomonas mobilis* and *Pichia stipitis* for bioethanol production from kans grass biomass hydrolysate", *BioEnergy Res.,* vol. 15, no. 4, pp. 1958-1971, 2022.
[http://dx.doi.org/10.1007/s12155-021-10383-0]

[6]   A. Mishra, and S. Ghosh, "Saccharification of kans grass biomass by a novel fractional hydrolysis method followed by co-culture fermentation for bioethanol production", *Renew. Energy,* vol. 146, pp. 750-759, 2020.
[http://dx.doi.org/10.1016/j.renene.2019.07.016]

[7]   L.K. Singh, C.B. Majumder, and S. Ghosh, "Development of sequential-co-culture system (*Pichia stipitis* and *Zymomonas mobilis*) for bioethanol production from Kans grass biomass", *Biochem. Eng. J.,* vol. 82, pp. 150-157, 2014.
[http://dx.doi.org/10.1016/j.bej.2013.10.023]

[8]   R. Singh, M. Srivastava, and A. Shukla, "Environmental sustainability of bioethanol production from rice straw in India: A review", *Renew. Sustain. Energy Rev.,* vol. 54, pp. 202-216, 2016.
[http://dx.doi.org/10.1016/j.rser.2015.10.005]

[9]   N. Akhtar, D. Goyal, and A. Goyal, "Characterization of microwave-alkali-acid pre-treated rice straw for optimization of ethanol production *via* Simultaneous Saccharification And Fermentation (SSF)", *Energy Convers. Manage.,* vol. 141, pp. 133-144, 2017.
[http://dx.doi.org/10.1016/j.enconman.2016.06.081]

[10]   S.K. Singh, A. Chauhan, and B. Sarkar, "Strategy planning for sustainable biodiesel supply chain produced from waste animal fat", *Sustain. Prod. Consum.,* vol. 44, pp. 263-281, 2024.

[http://dx.doi.org/10.1016/j.spc.2023.10.012]

[11]   S.K. Singh, A. Chauhan, and B. Sarkar, "Sustainable biodiesel supply chain model based on waste animal fat with subsidy and advertisement", *J. Clean. Prod.*, vol. 382, p. 134806, 2023.
[http://dx.doi.org/10.1016/j.jclepro.2022.134806]

[12]   S.K. Singh, A. Chauhan, and B. Sarkar, "Resilience of sustainability for a smart production system to produce biodiesel from waste animal fat", *J. Clean. Prod.*, vol. 452, pp. 142047-142047, 2024.
[http://dx.doi.org/10.1016/j.jclepro.2024.142047]

[13]   B. Sarkar, B. Mridha, and S. Pareek, "A sustainable smart multi-type biofuel manufacturing with the optimum energy utilization under flexible production", *J. Clean. Prod.*, vol. 332, p. 129869, 2022.
[http://dx.doi.org/10.1016/j.jclepro.2021.129869]

[14]   B. Sarkar, B. Mridha, S. Pareek, M. Sarkar, and L. Thangavelu, "A flexible biofuel and bioenergy production system with transportation disruption under a sustainable supply chain network", *J. Clean. Prod.*, vol. 317, p. 128079, 2021.
[http://dx.doi.org/10.1016/j.jclepro.2021.128079]

[15]   B. Mridha, G.V. Ramana, S. Pareek, and B. Sarkar, "An efficient sustainable smart approach to biofuel production with emphasizing the environmental and energy aspects", *Fuel,* vol. 336, p. 126896, 2023.
[http://dx.doi.org/10.1016/j.fuel.2022.126896]

CHAPTER 12

Understanding Municipal Solid Waste Volatility in India

Priyanka Banerji[1,*] and **Mansi Yadav**[2]

[1] *The NorthCap University, Gurugram, Haryana, India*

[2] *K.R. Mangalam University, Sohna Road, Gurugram, Haryana, India*

Abstract: In addition to health and environmental concerns, the growing volume of Municipal Solid Waste (MSW) and the pursuit of Sustainable Development Goals (SDGs) highlight the need for sustainable MSW management. Improper waste disposal poses significant risks to public health and the environment. Due to the economic and ecological challenges associated with environmentally sound disposal methods, alternative waste management strategies are essential. Studies indicate that nearly 90% of MSW is improperly disposed of in landfills or open dumps, leading to environmental degradation, health hazards, and contamination of the food chain. In India, Urban Local Bodies (ULBs) face difficulties in managing large quantities of MSW due to high population density and inadequate infrastructure. Among the challenges are door-to-door garbage collection, MSW recycling techniques, and scientific waste treatment methods. Given these realities, the new Solid Waste Management Rules (SWM), 2016, were announced by the Union Ministry of Environment, Forests, and Climate Change (MoEF&CC) of India, aiming to modernize solid waste management nationwide. Several steps of waste management/treatments are being adopted, including incineration, pyrolysis, bio-refining, biogas plants, recycling, and composting. Composting is a sustainable, low-cost option for MSW management; however, a very small amount, 6–7% of MSW, was recycled through it. The present study emphasized a comprehensive review of the characteristics, production, collection, disposal, and effective treatment technologies of MSW practised in India. Some of the waste management/treatment processes used in India include incineration, pyrolysis, bio-refining and biogas facilities, recycling, and composting. Composting is a low-cost and sustainable method for managing Municipal Solid Waste (MSW); however, only 6–7% of MSW in India is currently recycled through this method. The present study focuses on a comprehensive examination of the characteristics, generation, collection, disposal, and treatment technologies associated with MSW in the Indian context.

Keywords: Development, Goals, Municipal, SDG, Solid waste, Sustainable, Volatility.

---

* **Corresponding author Priyanka Banerji:** The NorthCap University, Gurugram, Haryana, India; E-mail: priyankabanerji@ncuindia.edu

Harish Chandra Joshi, Anand Chauhan, Mikhail Vlaskin & Maulin P. Shah (Eds.)

INTRODUCTION

Rapid urbanization, population growth, and changing consumption patterns have led to an exponential increase in waste generation. India, with its burgeoning population and expanding cities, experiences significant fluctuations, or "volatility," in the production of MSW, which poses complex environmental, social, and economic challenges. The concept of waste volatility refers to the unpredictability and variability in waste generation, composition, and disposal. In India's dynamic society, socioeconomic shifts, including urbanization, migration, industrialization, and population pressure, have contributed to a notable rise in municipal solid waste. Only about 28% of the 80% of Municipal Solid Waste (MSW) that is collected is currently utilized. India generates approximately 70 million tonnes of MSW annually, and if current trends persist, this figure is projected to rise to 165 million tonnes by 2030 and 436 million tonnes by 2050 (Planning Commission Report, 2014). The ongoing and indiscriminate disposal of Municipal Solid Waste (MSW) is closely linked to unscientific practices, rapid urbanization, population growth, changing lifestyles, and a lack of ecological awareness. Open disposal of Municipal Solid Waste (MSW) has harmful effects on both human health and the environment [1]. Globally, the primary causes of MSW accumulation are unscientific waste collection practices and inadequate transportation infrastructure. This remains one of the primary environmental challenges facing Indian megacities and is crucial to the effective management of Municipal Solid Waste (MSW). It also includes tasks related to solid waste generation, storage, collection, transportation, processing, and disposal. Approximately 48–50% of MSW is organic, posing a significant problem for the nation [1]. Its inappropriate disposal is mainly caused by environmental contamination, hazards to human health, and a lack of disposal space. A tiny portion of MSW (6–7%) is intended for composting; however, the majority of MSW in Indian megacities is already disposed of in landfills [2]. Because of the increased creation of MSW per capita brought about by urbanisation, industrialisation, and economic growth, managing MSW is a severe problem for ULBs in Indian megacities. Thus, efficient solid waste management is a key concern for densely populated cities. Despite notable advancements in the social, economic, and environmental spheres, the state of MSW treatment in the nation stayed unchanged. Approximately 90% of residual waste was disposed of in the informal sector rather than in a proper landfill, contributing significantly to the extraction of value from MSW [2]. This review focuses on using MSW as an organic amendment for the reclamation of salt-affected soils. This article provides a comprehensive examination of Solid Waste Management (SWM) practices and regulations, with a focus on their application as organic amendments and fertilizers for sustainable crop production in India. There is an urgent need for scientifically grounded MSW management, supported by advanced systems and infrastructure [2]. Current technologies for managing solid waste are inadequate and have a negative impact on human health, environmental pollution, and the nation's economy [3].

Future research on these issues should focus on a global scale to minimize ecological disturbances, establish environmental standards, and assess the effectiveness of policies in different countries and their impact on the socioeconomic conditions of local populations.

Model Specification and Justification

To examine the volatility patterns in Municipal Solid Waste (MSW) generation across India, this study employs a GARCH (Generalized Autoregressive Conditional Heteroskedasticity) modeling approach. The GARCH model is well-suited for analyzing time-series data with volatility clustering, where periods of high and low variability alternate over time. Since MSW generation is influenced by seasonal trends, consumption patterns, and infrastructural changes, it demonstrates characteristics of time-dependent variability. The GARCH framework allows us to capture such heteroskedastic behavior, offering insights into both the persistence and magnitude of fluctuations in MSW data. This approach is particularly valuable for municipal planning and policy design, as it helps identify whether short-term shocks or long-term structural shifts drive fluctuations in waste generation.

Data Overview

The data represent Municipal Solid Waste (MSW) in India, measured in tons per day over 10 years from 2013-14 to 2023-24 (Table **1**). The data points reveal some significant fluctuations in waste generation trends:

Table 1. Municipal Solid Waste (MSW) data.

Municipal Solid Waste (Tons per day)	India (proxy Variable)
2013-14	142566
2014-15	141064
2015-16	135198.27
2016-17	119140.9
2017-18	43298.385
2018-19	43597.353
2019-20	43902.53447
2020-21	44209.85221

(Table 1) cont.....

Municipal Solid Waste (Tons per day)	India (proxy Variable)
2021-22	44908
2022-23	45509.5478
2023-24	46987.4567

Table **1**: Annual trends in Municipal Solid Waste generation in India (Tons per Day) from 2013-14 to 2023-24.

- 2013-14 to 2016-17: Waste generation starts relatively high, with a slight decrease over the years. From 2013-14 to 2016-17, MSW dropped from 142,566 tons/day to 119,140.9 tons/day.
- 2017-18: There is a dramatic drop in MSW generation to 43,298.385 tons/day, which is much lower than the previous years.
- Post-2017-18 (2018-19 to 2023-24): MSW generation gradually increased from 43,597.353 tons/day in 2018-19 to 46,987.4567 tons/day in 2023-24, which signifies a recovery or growth phase in waste generation over the last five years.
- The sharp decline observed in 2017–18 is particularly notable. It may be attributed to a combination of policy reforms, advancements in waste management practices, or external factors such as shifts in population growth, urbanization rates, or the implementation of waste reduction initiatives [4].

Initial Observations

- Sharp Decline in 2017-18: The substantial decline in MSW generation in 2017-18 requires careful investigation. This anomaly might be caused by:

Policy Changes: The introduction of stricter waste management policies or initiatives, such as waste-to-energy plants, recycling, and waste segregation.

Economic Factors: A downturn in certain economic sectors or a transition toward a more service-oriented economy may also influence waste generation, as such sectors typically produce less waste than industrial or manufacturing-based activities.

Behavioural Changes: Public awareness campaigns on waste reduction and environmental concerns could have had a significant impact.

- Post-2017-18 Recovery: From 2018-19 to 2023-24, MSW generation steadily increased. This recovery could be attributed to increased urbanization, growing populations, and the challenges of waste management in rapidly expanding urban areas.

Log-Likelihood Function

The log-likelihood function is a crucial statistic used in estimating model parameters [5 - 8]. In this case, the log-likelihood value is -70.8019. The log-likelihood value indicates how well the model fits the data:

- A more negative log-likelihood value typically indicates a poorer fit of the model to the data [9, 10].
- Contextual Analysis: In your case, this value could be used for model comparison. You can compare this log-likelihood value with that of other models or versions to assess whether a better fit is possible. If other models yield more negative log-likelihoods, that would suggest they are worse fits [11 - 14].

Key Variables in the Model

- ω (omega): 0.00000044
- α (alpha): 0
- β (beta): 1.02969440

These variables are estimated through the optimization process (in this case, through Solver in Excel). Here is a brief explanation of their possible interpretations:

- ω (omega): Often used in volatility models, this represents the baseline level or the long-term mean of a given variable.
- α (alpha): This variable is often used in autoregressive models (such as GARCH) to capture past shocks or changes.
- β (beta): Typically, this variable is used to represent the persistence of volatility or waste generation trends over time. The value close to 1 suggests that MSW generation patterns are strongly influenced by previous values (Table **2**).

Table 2. Final and reduced values for parameters ω, α, and β with corresponding gradients.

	Final	**Reduced**
Name	Value	Gradient
ω	4.38308E-07	0
α	0	-9.403876826
β	1.029694398	0

Sensitivity Report

The sensitivity report provides insights into the responsiveness of the solution to changes in the model parameters [15, 16]:

- ω (omega) has a value of 4.38308E-07, with a gradient of 0. This suggests that changes in this parameter have minimal impact on the solution, indicating it may be non-influential.
- α (alpha) has a significant negative gradient of -9.4, suggesting it plays an important role in the volatility of MSW generation. However, the exact value of $\alpha = 0$ suggests it might not be actively contributing to the model.
- β (beta) shows a gradient of 0, indicating that this parameter does not significantly influence the model's behavior for the given data.

These results are essential for fine-tuning the model. Understanding which parameters have a substantial effect (like α) will help guide future modifications to improve the model fit.

Data analysis using Residuals and Squared Residuals

Residuals represent the difference between the actual values of MSW and the predicted values from the model, while squared residuals help assess how much error is in the model's predictions:

- 2014-15: Residual = -1502, Squared Residual = 2,256,004—This indicates that the prediction for 2014-15 is off by 1502 tons/day, with a squared residual that shows the magnitude of this error.
- 2015-16: Residual = -5866, Squared Residual = 34,409,956—The squared residuals increase, which may indicate a growing error in the prediction as MSW generation continued to decline.
- 2016-17: Residual = -16058, Squared Residual = 257,859,364—The squared residuals grow even further, which could reflect a deviation from the expected values, possibly due to the aforementioned drop in MSW generation.
- 2017-18: Residual = -75842, Squared Residual = 5,752,008,964 — This is a huge deviation. Given the sharp drop in MSW, the model likely did not account for this large change, resulting in a very large squared residual.

These residuals and squared residuals will help you determine the overall accuracy of your model (Tables **3** & **4**). If the residuals are large for particular years, it suggests that the model might not be capturing all the underlying factors driving MSW generation [17, 18].

Table 3. Unconditional variance and parameter estimates for ω, α, and β

Unconditional Variance	570702204.95
ω	0.00000044
α	0.00000000
β	1.02969440

Table 4. Annual data on municipal solid waste generation, residuals, variance, and log-likelihood with conditional and unconditional standard deviations (2013-14 to 2023-24)

Date	Rate	Residual	Squared Residual	Lagged Squared Change	Conditional Variance	Log Likelihood Function	Conditional Standard Deviation time weighted	Unconditional Standard Deviation, no time weighted
2013-14	142566							
2014-15	141064	-1502	2256004.00		570702204.9			
2015-16	135198	-5866	34409956.00	2256004.00	587648863	-11.0440	24241.47	23889.37
2016-17	119140	-16058	257859364.00	34409956.00	605098742	-11.2425	24598.75	23889.37
2017-18	43298	-75842	5752008964.0	257859364.0	623066785	-15.6599	24961.31	23889.37
2018-19	43597	299	89401.00	5752008964	641568377.61	-11.0587	25329.20	23889.37
2019-20	43902	305	93025.00	89401.00	660619364.11	-11.0734	25702.52	23889.37
2020-21	44209	307	94249.00	93025.00	680236058.17	-11.0880	26081.34	23889.37
2021-22	44908	699	488601.00	94249.00	700435258.15	-11.1029	26465.74	23889.37
2022-23	45509.5478	601.5478	361859.76	488601.00	721234261.20	-11.1174	26855.80	23889.37
2023-24	46987.4567	1477.9089	2184214.72	361859.76	742650878.12	-11.1333	27251.62	23889.37
					LOG L	-71.1665		

Conditional and Unconditional Standard Deviation

Standard deviation measures the variation or spread of MSW generation over time [19 - 21]. The difference between conditional and unconditional standard deviations indicates the extent to which time-dependent factors influence waste generation [22 - 25].

- Conditional Standard Deviation: This value shows the time-weighted volatility of MSW generation. Over time, this trend increases, peaking in 2023-24, which may reflect the growing complexities in waste generation due to factors such as population growth, urbanization, and policy changes.
- Unconditional Standard Deviation: This measures the general spread of MSW generation, independent of time. It appears to be constant across the years, suggesting that the underlying variability in MSW is not heavily influenced by

time, but rather by consistent structural or systemic factors.

This is important for forecasting, as models that incorporate volatility can better predict future trends based on past variations (Fig. **1**).

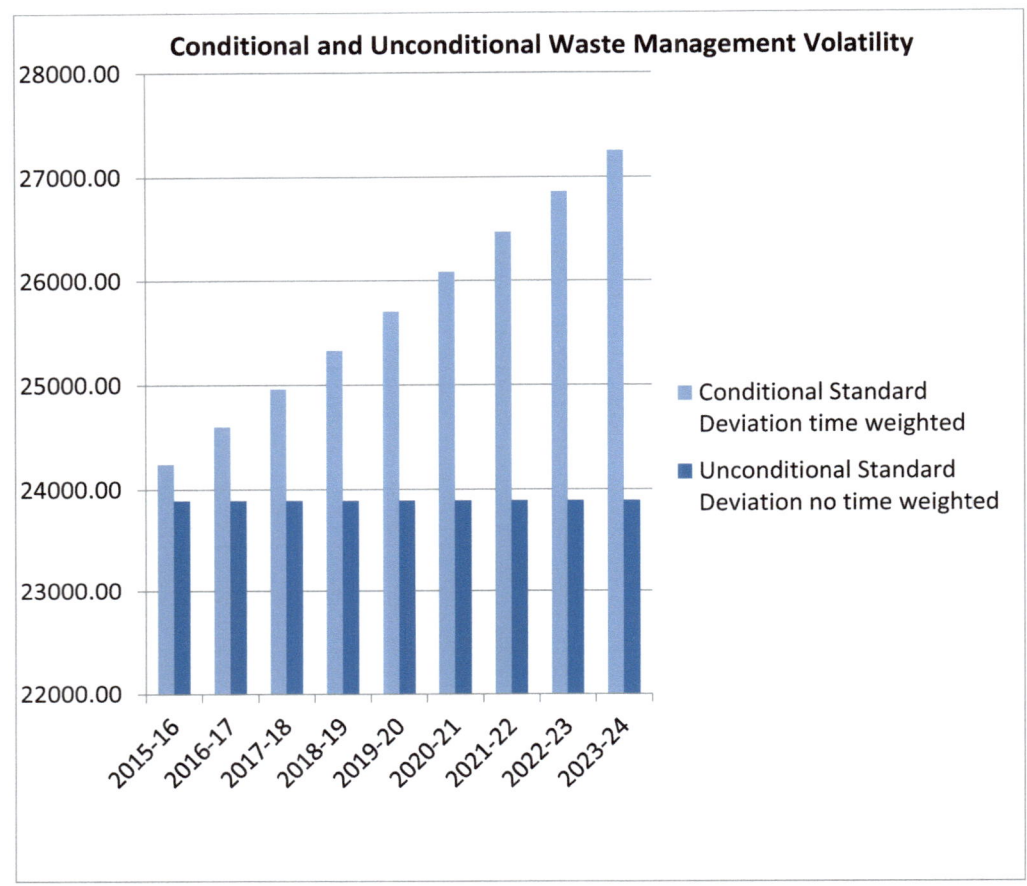

Fig. (1). Conditional and unconditional waste management volatility.

The graph illustrates the comparison between *Conditional Standard Deviation (time-weighted)* and *Unconditional Standard Deviation (no time-weighted)* for waste management volatility over the years from 2015-16 to 2023-24. Below are the key interpretations:

Consistent Growth in Volatility

Both the conditional and unconditional standard deviations show an upward trend, indicating that waste management volatility has increased over time.

Conditional vs. Unconditional Volatility

The conditional standard deviation (light blue bars) is consistently higher than the unconditional standard deviation (dark blue bars) across all years.

This suggests that when time-weighting factors are included (conditionally), the perceived volatility is greater, potentially highlighting the influence of temporal dependencies or trends in waste management dynamics.

Rate of Increase

The gap between conditional and unconditional standard deviations appears relatively consistent, indicating that the additional volatility captured by time-weighting remains proportional throughout the years.

Highest Volatility in Recent Years

The year 2023-24 shows the highest levels of volatility in both measures, reflecting increasing uncertainties or fluctuations in waste management processes or outcomes.

The increasing trend in both conditional and unconditional waste management volatility can stem from a variety of economic, environmental, and operational factors. Here is a deeper analysis of the possible contributors:

Over time, stricter environmental regulations and policies may have been introduced, leading to increased operational complexities and costs for waste management companies. These changes may lead to increased variability in operational efficiency and financial outcomes.

Economic uncertainties, such as inflation or fluctuating fuel prices, may have affected the cost structure of waste management operations. The conditional volatility captures these time-dependent shocks more prominently than the unconditional measure.

Changes in the composition of waste (e.g., increased e-waste or plastics) can affect the methods, costs, and effectiveness of waste processing. Conditional volatility may better capture these shifts as they evolve.

Rapid urbanization and population growth are likely to have increased waste generation, creating challenges in waste collection, segregation, and disposal. The surge in demand may have led to fluctuations in costs, resource allocation, and service quality.

The adoption of new technologies in waste management, such as automation or recycling innovations, may introduce short-term volatility. The implementation costs and varying success rates across time can lead to increased variability in the short term.

Events such as natural disasters, pandemics (e.g., COVID-19), or other large-scale disruptions could have caused spikes in volatility. Conditional measures are better equipped to detect these temporal anomalies, which could explain their consistently higher values compared to unconditional measures.

The upward trend in conditional volatility compared to unconditional volatility reflects that time-weighted measures capture dynamic changes and their impact over time. This suggests that fluctuations are not merely random but are influenced by temporal trends or cycles.

Recommendations for Waste Management

Based on your analysis, you can make several recommendations to improve waste management:

- Invest in Waste Reduction: The sharp decline in MSW generation in 2017-18 may be attributed to the success of waste reduction campaigns. Continue promoting waste segregation, recycling, and composting.
- Focus on Urbanisation: As the MSW generation rises post-2017-18, it is important to adapt waste management strategies for rapidly growing urban areas.
- Policy Improvements: Policymakers should consider scaling up successful programs from 2017-18 and implementing them across larger regions to stabilise and reduce the volatility in waste generation.
- Encourage investments in advanced waste management technologies (e.g., AI-driven waste sorting and bioconversion technologies) to reduce operational variability. Subsidies or incentives for waste management firms investing in sustainable innovations can help stabilise long-term operations.
- Develop contingency frameworks to address disruptions caused by natural disasters or pandemics. Preemptive strategies, such as stockpiling essential resources or creating flexible response systems, can reduce conditional shocks to the waste management system.
- Encourage investments in advanced waste management technologies (e.g., AI-

driven waste sorting and bioconversion technologies) to reduce operational variability. Subsidies or incentives for waste management firms investing in sustainable innovations can help stabilise long-term operations.

Practical Implications

- Develop subsidies or tax breaks for waste management firms during high-volatility periods identified through conditional measures. For example, during a natural disaster or economic shock, firms could be provided with financial relief to stabilize their operations.
- Equip waste management operators with training on financial literacy and digital tools to better navigate cost variabilities. Time-weighted measures suggest that operators may face increased financial uncertainties if they are not adequately prepared.
- Establish regional collaborations to share waste processing facilities, recycling centres, or landfill space. This can balance out disparities in local waste generation and management capacities, reducing systemic volatility.
- Utilize predictive modeling and time-series analysis to forecast periods of heightened volatility. Waste management firms should employ data-driven decision-making to prepare for operational fluctuations.

CONCLUSION

The findings from the GARCH-based modeling provide significant insights into the volatility dynamics of MSW in India. Our analysis reveals that the volatility in MSW generation is **time-persistent** and more strongly influenced by **structural and long-term trends**, such as urban population growth and consumption behavior, rather than short-term or seasonal shocks. This suggests that temporary interventions may have limited impact unless supported by sustained policy efforts and infrastructural development.

The analysis of MSW in India reveals two contrasting phases: a sharp decline in waste generation until 2017-18, followed by consistent growth from 2018-24. While policies like SBM likely contributed to early reductions, the subsequent rise underscores the challenges of urbanisation and increased waste generation. Effective waste management policies, combined with technological innovation and public awareness, will be essential to address the growing environmental burden of MSW.

By promoting integrated systems, leveraging predictive analytics, and fostering public-private partnerships, stakeholders can mitigate these volatilities and enhance the resilience of waste management operations. Additionally, empowering operators through training and raising community awareness will further contribute to reducing systemic uncertainties. The observed rise in both conditional and unconditional volatility in waste management highlights the increasing complexities and uncertainties within the sector. Conditional volatility, consistently higher than unconditional measures, indicates the significant impact of time-dependent factors such as regulatory changes, economic fluctuations, and technological advancements. These trends underscore the need for adaptive policies, operational flexibility, and strategic investments to stabilise waste management processes and ensure long-term sustainability.

Policy Takeaways

1. **Long-Term Investment:** Authorities must prioritize long-term investments in MSW infrastructure, especially in waste segregation, treatment, and scientific disposal mechanisms.
2. **Public-Private Partnerships:** Encouraging collaborations between municipal bodies and private enterprises can enhance operational efficiency, particularly in areas such as collection, recycling, and waste-to-energy projects.
3. **Behavioral Change and Awareness:** Implementing widespread educational campaigns and incentive programs can promote household-level behavioral change, which is crucial for adequate source segregation and composting practices.

Future Research Directions:

Further studies can investigate region-specific MSW volatility, examine rural-urban contrasts, and explore the integration of informal waste pickers into formal waste management systems, thereby enhancing both model precision and policy relevance.

REFERENCE

[1]   R. Goswami, S. Singh, P. Narasimhappa, P.C. Ramamurthy, A. Mishra, P.K. Mishra, H.C. Joshi, G. Pant, J. Singh, G. Kumar, N.A. Khan, and M. Yousefi, "Nanocellulose: A comprehensive review investigating its potential as an innovative material for water remediation", *Int. J. Biol. Macromol.,* vol. 254, no. Pt 3, p. 127465, 2024.
        [http://dx.doi.org/10.1016/j.ijbiomac.2023.127465] [PMID: 37866583]

[2]   S. Nanda, and F. Berruti, "A technical review of bioenergy and resource recovery from municipal solid waste", *J. Hazard. Mater.,* vol. 403, p. 123970, 2021.
        [http://dx.doi.org/10.1016/j.jhazmat.2020.123970] [PMID: 33265011]

[3]   M.E. Edjabou, J.A. Martín-Fernández, C. Scheutz, and T.F. Astrup, "Statistical analysis of solid waste composition data: Arithmetic mean, standard deviation and correlation coefficients", *Waste Manag.,* vol. 69, pp. 13-23, 2017.
        [http://dx.doi.org/10.1016/j.wasman.2017.08.036] [PMID: 28882426]

[4]   R.K. Annepu, "Sustainable solid waste management in India", In: *Columbia Univ*New York, 2012, 2012, pp. 1-89.

[5]   M. Alibeikloo, H. Khabbaz, and B. Fatahi, "Random field reliability analysis for time-dependent behaviour of soft soils considering spatial variability of elastic visco-plastic parameters", *Reliab. Eng. Syst. Saf.,* vol. 219, p. 108254, 2022.
[http://dx.doi.org/10.1016/j.ress.2021.108254]

[6]   J.A.B. Sinaga, M. Silalahi, E. Fatmawati, L. Judijanto, N. Saputra, and H. Herman, "Sensitivity analysis in operations research decisions: A case study on a mathematical model", *Library Progress International,* vol. 44, no. 2, pp. 443-448, 2024.

[7]   V.P. Singh, A. Kumar, C.S. Meena, S. Thangavel, and A. Ghosh, *Innovation in green building sector for sustainable future.* vol. Vol. 1. Sustainable Technologies for Energy Efficient Buildings, 2024.
[http://dx.doi.org/10.1201/9781003496656-1]

[8]   P. Gururani, P. Bhatnagar, P. Dogra, H. Chandra Joshi, P.K. Chauhan, M.S. Vlaskin, N. Chandra Joshi, A. Kurbatova, A. Irina, and V. Kumar, "Bio-based food packaging materials: A sustainable and Holistic approach for cleaner environment- a review", *Current Research in Green and Sustainable Chemistry,* vol. 7, no. 100384, p. 100384, 2023.
[http://dx.doi.org/10.1016/j.crgsc.2023.100384]

[9]   P. Chaudhary, S. Garg, T. George, M. Shabin, S. Saha, S. Subodh, and B. Sinha, "Underreporting and open burning – the two largest challenges for sustainable waste management in India", *Resour. Conserv. Recycling,* vol. 175, p. 105865, 2021.
[http://dx.doi.org/10.1016/j.resconrec.2021.105865]

[10]   A. Singhal, A.K. Gupta, B. Dubey, and M.M. Ghangrekar, "Seasonal characterization of municipal solid waste for selecting feasible waste treatment technology for Guwahati city, India", *J. Air Waste Manag. Assoc.,* vol. 72, no. 2, pp. 147-160, 2022.
[http://dx.doi.org/10.1080/10962247.2021.1980450] [PMID: 34554054]

[11]   M.D. Meena, M.L. Dotaniya, B.L. Meena, P.K. Rai, R.S. Antil, H.S. Meena, L.K. Meena, C.K. Dotaniya, V.S. Meena, A. Ghosh, K.N. Meena, A.K. Singh, V.D. Meena, P.C. Moharana, S.K. Meena, C. Srinivasarao, A.L. Meena, S. Chatterjee, D.K. Meena, M. Prajapat, and R.B. Meena, "Municipal solid waste: Opportunities, challenges and management policies in India: A review", *Waste Management Bulletin,* vol. 1, no. 1, pp. 4-18, 2023.
[http://dx.doi.org/10.1016/j.wmb.2023.04.001]

[12]   T.V. Ramachandra, H.A. Bharath, G. Kulkarni, and S.S. Han, "Municipal solid waste: Generation, composition and GHG emissions in Bangalore, India", *Renew. Sustain. Energy Rev.,* vol. 82, pp. 1122-1136, 2018.
[http://dx.doi.org/10.1016/j.rser.2017.09.085]

[13]   L. Chand Malav, K.K. Yadav, N. Gupta, S. Kumar, G.K. Sharma, S. Krishnan, S. Rezania, H. Kamyab, Q.B. Pham, S. Yadav, S. Bhattacharyya, V.K. Yadav, and Q-V. Bach, "A review on municipal solid waste as a renewable source for waste-to-energy project in India: Current practices, challenges, and future opportunities", *J. Clean. Prod.,* vol. 277, p. 123227, 2020.
[http://dx.doi.org/10.1016/j.jclepro.2020.123227]

[14]   R.M. Sebastian, D. Kumar, and B.J. Alappat, "A technique to quantify incinerability of municipal solid waste", *Resour. Conserv. Recycling,* vol. 140, pp. 286-296, 2019.
[http://dx.doi.org/10.1016/j.resconrec.2018.09.022]

[15]   S. Vyas, P. Prajapati, A.V. Shah, and S. Varjani, "Municipal solid waste management: Dynamics, risk assessment, ecological influence, advancements, constraints and perspectives", *Sci. Total Environ.,* vol. 814, p. 152802, 2022.
[http://dx.doi.org/10.1016/j.scitotenv.2021.152802] [PMID: 34982993]

[16]   A. Sharma, R. Ganguly, and A.K. Gupta, "Characterization and energy generation potential of municipal solid waste from nonengineered landfill sites in Himachal Pradesh, India", *J. Hazard. Toxic Radioact. Waste,* vol. 23, no. 4, p. 04019008, 2019.
[http://dx.doi.org/10.1061/(ASCE)HZ.2153-5515.0000442]

[17] S. Das, S.H. Lee, P. Kumar, K.H. Kim, S.S. Lee, and S.S. Bhattacharya, "Solid waste management: Scope and the challenge of sustainability", *J. Clean. Prod.,* vol. 228, pp. 658-678, 2019.
[http://dx.doi.org/10.1016/j.jclepro.2019.04.323]

[18] G. Sharma, B. Sinha, and H. Pallavi, Hakkim, B. P. Chandra, A. Kumar, and V. Sinha, "Gridded emissions of CO, NOx, SO_2, CO_2, NH_3, HCl, CH_4, $PM_{2.5}$, PM_{10}, BC, and NMVOC from open municipal waste burning in India", *Environ. Sci. Technol.,* vol. 53, no. 9, pp. 4765-4774, 2019.
[http://dx.doi.org/10.1021/acs.est.8b07076] [PMID: 31021611]

[19] H. Singh, V. Tripathi, Alka, H.C. Joshi, G. Kumar, G. Pant, K. Hossain, A. Ahmad, and M.B. Alshammari, "Water hyacinth (*Eichhornia crassipes* and *Epipremnum aureum*) - a potent tool for the removal of cadmium and chromium from industrial discharges", *Desalination Water Treat.,* vol. 315, pp. 432-445, 2023.
[http://dx.doi.org/10.5004/dwt.2023.30157]

[20] G.J. Stewart, W.J.F. Acton, B.S. Nelson, A.R. Vaughan, J.R. Hopkins, R. Arya, A. Mondal, R. Jangirh, S. Ahlawat, L. Yadav, S.K. Sharma, R.E. Dunmore, S.S.M. Yunus, C.N. Hewitt, E. Nemitz, N. Mullinger, R. Gadi, L.K. Sahu, N. Tripathi, A.R. Rickard, J.D. Lee, T.K. Mandal, and J.F. Hamilton, "Emissions of non-methane volatile organic compounds from combustion of domestic fuels in Delhi, India", *Atmos. Chem. Phys.,* vol. 21, no. 4, pp. 2383-2406, 2021.
[http://dx.doi.org/10.5194/acp-21-2383-2021]

[21] A. Jančauskas, N. Striūgas, K. Zakarauskas, R. Skvorčinskienė, J. Eimontas, and K. Buinevičius, "Experimental investigation of sorted municipal solid wastes producer gas composition in an updraft fixed bed gasifier", *Energy,* vol. 289, p. 130063, 2024.
[http://dx.doi.org/10.1016/j.energy.2023.130063]

[22] G.J. Stewart, B.S. Nelson, W.J.F. Acton, A.R. Vaughan, N.J. Farren, J.R. Hopkins, M.W. Ward, S.J. Swift, R. Arya, A. Mondal, R. Jangirh, S. Ahlawat, L. Yadav, S.K. Sharma, S.S.M. Yunus, C.N. Hewitt, E. Nemitz, N. Mullinger, R. Gadi, L.K. Sahu, N. Tripathi, A.R. Rickard, J.D. Lee, T.K. Mandal, and J.F. Hamilton, "Emissions of intermediate-volatility and semi-volatile organic compounds from domestic fuels used in Delhi, India", *Atmos. Chem. Phys.,* vol. 21, no. 4, pp. 2407-2426, 2021.
[http://dx.doi.org/10.5194/acp-21-2407-2021]

[23] Y. Pujara, P. Pathak, A. Sharma, and J. Govani, "Review on Indian Municipal Solid Waste Management practices for reduction of environmental impacts to achieve sustainable development goals", *J. Environ. Manage.,* vol. 248, p. 109238, 2019.
[http://dx.doi.org/10.1016/j.jenvman.2019.07.009] [PMID: 31319199]

[24] S.S. Hla, and D. Roberts, "Characterisation of chemical composition and energy content of green waste and municipal solid waste from Greater Brisbane, Australia", *Waste Manag.,* vol. 41, pp. 12-19, 2015.
[http://dx.doi.org/10.1016/j.wasman.2015.03.039] [PMID: 25882791]

[25] Y. Zhao, and H. Li, "Understanding municipal solid waste production and diversion factors utilizing deep-learning methods", *Util. Policy,* vol. 83, p. 101612, 2023.
[http://dx.doi.org/10.1016/j.jup.2023.101612]

SUBJECT INDEX

www.ingramcontent.com/pod-product-compliance
Lightning Source LLC
Chambersburg PA
CBHW041456280526
45792CB00004B/1028